Synthesis Lectures on Engineering, Science, and Technology

The focus of this series is general topics, and applications about, and for, engineers and scientists on a wide array of applications, methods and advances. Most titles cover subjects such as professional development, education, and study skills, as well as basic introductory undergraduate material and other topics appropriate for a broader and less technical audience.

Dora Musielak

Introduction to Rocket Propulsion for Astronautics

 Springer

Dora Musielak
The University of Texas at Arlington
Arlington, TX, USA

ISSN 2690-0300   ISSN 2690-0327 (electronic)
Synthesis Lectures on Engineering, Science, and Technology
ISBN 978-3-031-86140-6   ISBN 978-3-031-86141-3 (eBook)
https://doi.org/10.1007/978-3-031-86141-3

This Springer imprint is published by the registered company Springer Nature Switzerland AG
The registered company address is: Gewerbestrasse 11, 6330 Cham, Switzerland

If disposing of this product, please recycle the paper.

I dedicate this effort to the many unnamed engineers and technicians who work tirelessly to design, analyze, research, develop, and test rocket engines. To the smart people all over the world who are not afraid of challenges and continue pushing the limits of propulsion technologies to advance and ensure the rocket propulsion systems can take precious payloads to orbit and return safely to Earth the courageous astronauts, cosmonauts, taikonauts, and space navigators who entrust their lives to these engineering marvels we call rockets.

Preface

Rocket science is hard, but rocket propulsion engineering is orders of magnitude more challenging. The difficulty of rocket engineering cannot be overstated. Every rocket engine fully developed and extensively tested must operate flawlessly. Every system, every single subsystem, to the minutest seal and bolt, must work together as one for the rocket engine to propel a vehicle or a spacecraft.

A comprehensive coverage of propulsion engineering takes many specialized books to fully address the many designs, systems, and technologies that are incorporated to build a rocket engine. The study of rocket propulsion is intense, rigorous, and practicing propulsion engineers require discipline, effort, and time to gain the competences to become specialized. Whether working on analysis, design, ground or flight testing, the complexities of rocket propulsion mandate a multidisciplinary approach to address even the smallest aspect of the physical process involved and the demands on technologies to advance it.

For example, to better understand the complex processes involved in the conversion of the chemical energy of the propellants into kinetic energy to develop thrust, sophisticated analysis is performed involving numerical tools to model the many physical processes in the combustion chamber and supersonic nozzle, solving multi-dimensional nonlinear equations based on the conservation of mass, momentum, and energy. The turbulent reacting flows of propellants in the combustion chamber require realistic chemical kinetic models that include the interaction of the reacting propellant species (combustion), which must also consider the multi-phase nature of the flow (fuel and oxidizer injection, vaporization, and mixing). Advances in computational fluid dynamics (CFD) techniques make now possible to model and simulate these, and many other aspects of the complex flowfields found in the rocket chamber.

A high thrust rocket engine operates at high pressure and high combustion temperature, and thus, it experiences structural loads and very high heat transfer. Thermal management with passive and active cooling requires that we determine the heat flux and wall temperature resulting from the hot gas flow, and we carry out comprehensive heat transfer analysis. This mandates accurate computer simulations that couple thermal and structural load calculations of the solid structure with suitable models for the heat transfer to the cooling channels and chamber surfaces.

Because of those and many more technical challenges, most publications on rocket engineering are written for specialists. How did they arrive at this knowledge?

Introduction to Rocket Propulsion for Astronautics intends to provide a starting foundation. This book is for anyone who has only a basic understanding of engineering physics and wishes to begin a serious study of rocket propulsion. The emphasis of the coverage is on concepts, keeping the mathematical rigor to a minimum, assuming the reader has studied thermodynamics and Newtonian mechanics. This book will introduce the reader to modern propulsion technologies that are now developed to advance the capability of launch vehicles and spacecraft. My intent is to initiate readers to the field of study, provide the basic principles and a description of rocket engines, rocket motors, and thrusters, and highlight their key physical mechanisms or designs. My hope is that this effort enhances appreciation of rocket propulsion as the key field intended to enable astronautics and ensure direct exploration of space.

The reader will discover in this book descriptions of modern rocket propulsion and their propellants, such as those powering current launch vehicles and spacecraft because these systems exemplify the strong heritage on which new propulsion will be based and because these rocket engines characterize the multidisciplinary nature of the field.

I wrote this preface during an exciting time in the history of space exploration, as the United States and many other nations are pursuing new missions, starting with the return of humans to the Moon, followed by the first crewed voyages to Mars. To achieve these goals, NASA conceived the heavy-lift Space Launch System (SLS) and the Orion crew capsule as key elements in the architecture for more challenging interplanetary human missions, while private companies and foreign space agencies are developing reusable rocket propulsion capable of carrying crew and cargo to the Moon, Mars and beyond. These rocket engines and associated technologies are crucial for enabling the first human orbital mission around the Red Planet with attendant systems to allow for direct exploration of its moon Phobos, as well as conducting telerobotic exploration of the Martian surface. Future missions will culminate with permanent infrastructure to enable long-term exploration of the neighbor planet, fulfilling the ultimate goal for humanity to settle in other worlds.

The field and the emerging applications of rocket propulsion have grown and matured immensely since the original rocket systems were assembled to send the first satellite to orbit. It is thus impossible to cover all aspects of rocket engineering in an introductory book. However, I selected the most fundamental and important aspects related to rocket propulsion to provide a solid basis for appreciating its multidisciplinary nature. This coverage is treated in the following chapters:

Chapter 1 introduces the subject matter and discusses the overall scope and reach of astronautics. Chapter 2 provides the fundamental concepts of propulsion, starting with the definition of rocket thrust, followed by the development of the ideal rocket equation, and it establishes the performance measures we use to identify the capability of rocket engines.

In Chap. 3, I address basic concepts of orbital mechanics, just enough to give a perspective on the propulsion requirements for powering a launch vehicle to lift off and reach orbit, and for keeping a spacecraft in its orbital trajectory, maneuvering as needed to carry its deep space mission. I wanted to emphasize that the rockets we design and build will take a vehicle to regions of space far away from Earth, some still unexplored and all of them so different from home.

Chapter 4 is devoted to liquid propellant rocket engines. Here I discuss the main characteristics of liquid propellants (cryogenic, storable, hypergolic). The reader will find substantial performance data on rocket stages currently used to propel many launch vehicles and an overview of monopropellant hydrazine thrusters used for in-space applications.

Coverage of solid propellant propulsion is found in Chap. 5. Solid rocket motor (SRM) propulsion has a wide range of applications in astronautics, from lift-off boosters for launch vehicles to in-space maneuvering of spacecraft, and thus I provide fundamental concepts and a review of design requirements for the different applications.

Electric propulsion has become an integral part of the technologies for advancing astronautics and space exploration. Chapter 6 provides an overview of propulsion devices in which the acceleration of a propellant mass is accomplished by electrical heating and/or by electric and magnetic body forces, such as electrothermal thrusters and ion engines.

The book concludes with coverage of advanced nuclear rockets. I included Chap. 7 to discuss other forms of energy that can be used for propulsion, and then review proposed propulsion concepts that offer the potential to facilitate future travel within our Solar System.

I spent much effort into introducing and using a proper terminology. An example is the distinction I make between rocket, launch vehicle, and spacecraft, which are sometimes confused or incorrectly used. Although many refer to launch vehicles as "rockets," in this book I reserve the term exclusively for the propulsion system, e.g. rocket engines, solid rocket motors, and thrusters. To make a clear distinction between propulsion and vehicle, I define a "space launch vehicle" or SLV as the complete system designed to transport a payload to Earth orbit.

A note on units. As any engineer knows, it is absolutely crucial to understand measurements in standardized systems of units. The SI (Systems International) system of units (meter, kg, seconds), also known as metric system, is the measurement system in nearly all countries of the world. In the United States, most of the engineering and design and almost all the manufacturing are still being done in English engineering (EE) units (foot or inch, pounds, seconds). This poses a challenge for communicating data, drawings, and specifications, especially in rocket propulsion when, for example, one American company provides rocket thrust capability in "pounds," while engineers in the foreign counterpart group express that value in newtons. For instance, each of NASA's SLS solid rocket boosters is said to produce "3.3 million pounds of thrust at launch." ESA states that each of Ariane 6 main stage P120C solid rocket boosters can provide about "4500 kN of maximum thrust."

Hence, I avoid the use of the common term "pounds" that has become ingrained in some propulsion communities. In this book, I use pound-mass (lbm) and pound-force (lbf) to express mass and thrust, respectively. To facilitate exposure, I strive to express rocket performance in both sets of units, which of course requires conversion.

This book is intended for someone desiring to understand enough to undertake a formal study to become a professional rocket propulsion engineer. To this end, I attempt to explain the concepts as clearly as possible, adding the necessary equations but keeping derivations short or referenced elsewhere. I also included a few worked examples to illustrate important ideas.

The Recommended Reading and References at the end of each chapter provide a selection of recommended books, technical reports, and journal articles that will augment the presentation. These are books and articles which I have studied and consulted on different aspects of rocket propulsion. I include references that I find especially useful for students. Many titles are referenced in the different chapters, together with a few more works that were not mentioned but are especially important and that direct the reader to specific sources.

It is my sincere desire that reading this book will inspire readers to further study rocket propulsion.

Arlington, USA   Dora Musielak

Acknowledgments

First and foremost, I express my sincere thanks to NASA for its breathtaking discoveries, daring space missions, and technological developments that always inspire me. I am grateful for the knowledge I acquired from interacting with its distinguished researchers and engineers, and for the prestigious research fellowships NASA bestowed on me.

This book can only be completed by giving credit, especially in rocket propulsion and astronautics, as I have benefited from the enormous cumulative knowledge and technological developments that occurred in the last decades. First, I must recognize George P. Sutton, for I learned the fundamentals from his *Rocket Propulsion Elements*, a textbook still in my bookshelves. It is said that Sutton's seminal book guided generations of rocket scientists and engineers—indeed, I am one of them.

For teaching aerospace engineers, I also benefited from *Spacecraft Propulsion*, an excellent book by Charles D. Brown. Another book I consulted extensively while preparing lecture notes is *Rocket and Spacecraft Propulsion* by Martin J. L. Turner, who exposed me to the technologies developed by European propulsion engineers.

I thank many colleagues at Northrop Grumman, and at the former Alliant Techsystems (ATK) who shared technical data and practical insights regarding the engineering, analysis, and design of rocket engines. I also appreciate Northrop Grumman for providing me many of the pictures I use to illustrate Chap. 5. These images will give the novice a better understanding of the physical characteristics of actual solid rocket motors.

I acknowledge Stoke Space Technologies for permission to use the image of the reusable Nova vehicle (Chap. 1), and AIAA for permission to use data from Brown's book (Chap. 2). My sincere appreciation to Dr. Richard O. Ballard from NASA MSFC. Professor Oskar J. Haidn for allowing me to use the specific impulse data for several bipropellant combinations (Chap. 4). Professor Oscar Biblarz for giving me permission to use the performance data of solid rocket propellant combinations (Chap. 5).

I acknowledge propulsion pioneers and dedicated researchers, brilliant engineers, and scientists all over the world (too many to name individually) who have built the body of work synthesized in the following chapters. Hence, I recognize some of them in the citations to highlight some of their contributions. Of course, such inclusion may not do justice to their work, as the field of propulsion is extensive.

At Springer, I am deeply grateful to Editorial Director Charles Glaser (Chuck) for his unwavering support and wise counsel. And thanks also go to his editorial staff, Lokeshwaran Manickavasagam, Shalini Selvam, and Martina Wiese who helped me take this book to its final form.

To Dr. Alex Weiss, Chairman of the Physics Department at the University of Texas at Arlington (UTA), I am sincerely thankful for the opportunity to carry out my research in such intellectually stimulating environment. His kind encouragement means so much to me.

And to my family, words cannot express my gratitude for their steadfast love. To my intelligent and beautiful daughters Dasein and Lauren, thank you! and also to Dewey and Dan. Thanks especially to my brilliant husband Zdzislaw, for standing by me always, and at the same time accepting my independence to pursue my own scholarly work. In the end, we all converge on the pursuit of truth and universal enlightenment.

Contents

List of Figures

List of Tables

Introduction to Rocket Propulsion and Astronautics

> *We are born with a primordial desire to explore—to blaze new trails, map foreign lands, soar through the air, and travel through deep space so that we may answer profound questions about ourselves and find our place in this awesome universe.*
>
> —Dora Musielak

The development of powerful rockets during the twentieth century made spaceflight a reality and allowed humanity to conduct direct physical investigation of the cosmos. First and foremost, engineers designed rocket engines to produce the huge thrust forces to overcome Earth's gravity to lift a spacecraft above the ground, and then hurl it in a controlled manner into an orbital path above the surface of our planet. By the 1950s, the original integrated propulsion systems were ready to launch the first payloads into Earth's orbit.

The scientific history of humanity changed on 4 October 1957, when the former Soviet Union successfully launched Sputnik 1 satellite on the R-7 launch vehicle. This feat was matched by the United States on 31 January 1958 with the launch of Explorer 1 satellite, using a derivative of a Redstone vehicle. By the mid-1960s, even mightier rockets were assembled into propulsion systems to power the launch systems that took the first humans to the Moon. The Space Age was born, and the propulsion systems that achieved those magnificent achievements became the inspiration for the rockets we build today.

In this chapter I wish to briefly outline important aspects of astronautics and provide a basis for understanding the requirements of rocket propulsion based on the needs of space access and scientific exploration. The next sections introduce the principle of rocket

D. Musielak, *Introduction to Rocket Propulsion for Astronautics*, Synthesis Lectures on Engineering, Science, and Technology, https://doi.org/10.1007/978-3-031-86141-3_1

propulsion, and highlight major applications. The topics that follow are intended to answer basic questions such as

- What is the difference between a launch vehicle, a spacecraft, and a rocket engine?
- What is rocket propellant?
- What is the difference between rocket engine, rocket motor, and thruster?
- Why a single space mission requires several types of rocket propulsion?

1.1 Propulsion for Access to Earth Orbit and for Space Exploration

Rockets take spacecraft where they need to go, from the surface of the Earth to orbit, and to the destination in outer space that the mission of exploration requires. Rocket propulsion is a system designed to provide forces to a flight vehicle and cause it to accelerate (or decelerate), overcome drag and gravitational forces, or change flight direction. Rocket propulsion is needed to (a) boost a launch vehicle, (b) as upper stage to inject spacecraft into escaping trajectories; (c) to perform station-keeping maneuvers (making corrections to orbit position perturbations); (d) to propel planetary and lunar landers; (e) for in-space propulsion (attitude control, maneuvering); (f) for upper atmosphere sounding; (g) to boost suborbital vehicles for space tourism; and (h) for tactical/defense applications, including hypersonic vehicle testing.

A launch vehicle needs powerful rockets to provide the large velocity change to ascend from Earth's surface into orbit, overcoming the grip of our planet's gravity. A spacecraft, on the other hand, requires a propulsion system to adjust its orbital trajectory to reach its destination in deep space. The term spacecraft is used in this book to refer to any crewed or robotic vehicle designed for space missions. The term artificial satellites will be used to imply human-made systems intended for orbiting a central body. Spacecraft therefore include nano-satellites (0.1 m, 10 kg, 100 W), space stations (100 m, 10^5 kg, 100 kW), crewed vehicles (e.g. space shuttles, capsules), space probes, and orbiting astronomical telescopes. In some instances, I will use the term spaceships to refer to futuristic space transport concepts.

1.1.1 The Principle of Rocket Propulsion

A rocket engine is a propulsion device that expels a fraction of its mass at high speeds in a determined direction, exerting a propelling force called thrust. This thrust force impels a vehicle through the exchange of momentum, accelerating the vehicle to practical velocities. The time-integrated thrust—the impulse—is critically important for propulsion. The matter that is expulsed at high speed is known as *propellant*. The propellant can be (i) a cold gas expelled at high pressure, (ii) it can be a hot gas that results from chemical reactions (combustion), or (iii) a plasma resulting from a thermonuclear reaction, or by application of electromagnetic energy to a gas. The matter that is expulsed can also be the result of matter-antimatter annihilation.

When liquid or solid propellant substances undergo exothermal chemical reactions in a combustion chamber, we call the propulsion system "chemical rocket" and its exhaust consists of a hot stream of combusted gas. A cold, pressurized gas can also serve as propellant if it is expelled at high speed through a nozzle to provide thrust without any chemical reaction. Ions (charged atomic particles), nuclear particles, and even beams of light (photons) can also be rocket propellants. Hence, rockets that eject chemically reacting hot gases, high-pressure cold gases, or atomic particles all operate under the same physical principle—mass is accelerated and expelled, resulting in a reaction force to propel a spacecraft.

To facilitate this introduction to rocket propulsion, we assume it is a thermal rocket and its exhaust is a gas. Conceptually, a thermal rocket is a heat engine comprised of a pressure chamber where the propellant (working fluid) is heated, and a shaped converging-diverging duct at the rear of the rocket, called nozzle, through which the hot propellant gas is expanded and ejected at high velocity (Fig. 1.1). The nozzle converts the high pressure of the gases into kinetic energy. The result of the acceleration of the hot gas is a force exerted by the rocket in the direction opposite to that in which the gas is ejected.

Heating of the propellant can be accomplished by chemical reaction, by the application of electromagnetic energy, or by passing the propellant gas through a nuclear reactor. The change in momentum experienced by the hot exhaust gases in the rocket engine results in the generation of a forward thrust force.

The principle of rocket propulsion is relatively easy to understand, but a rocket engine is a very complex hardware system. A rocket is comprised of millions of interconnected pieces that are depending on each other to operate flawlessly (see Fig. 1.2 to get a sense

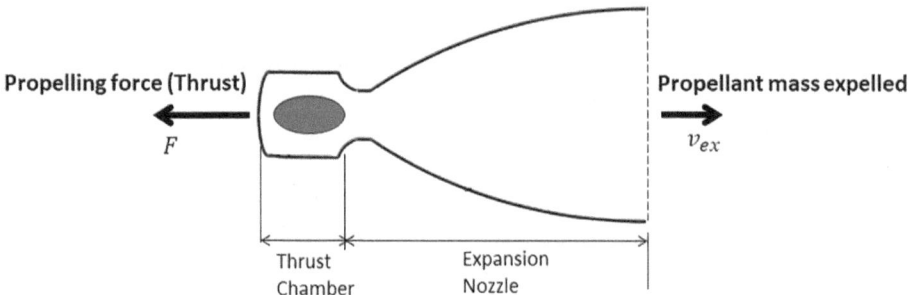

Propelling force (Thrust)

F

Propellant mass expelled

v_{ex}

Thrust Chamber

Expansion Nozzle

Fig. 1.1 Conceptual rocket engine. It can be viewed as a thrust control volume comprised of a pressure chamber with a converging-diverging nozzle

Fig. 1.2 Bipropellant (LOX/LH$_2$) engines: **a** RS-25 (Core stage: maximum thrust 418,000 lbf (1859.35 kN) at sea level); **b** RL10B-2 (Upper stage: thrust 24,750 lbf (110 kN) in vacuum)

of the actual hardware). This rocket then must be integrated into a launch vehicle or spacecraft equally complicated in design and operation.

Although it is customary to refer to launch vehicles as "rockets," in this book I reserve the term exclusively for the propulsion system, e.g. rocket engines, solid rocket motors, thrusters. To make a clear distinction between propulsion and vehicle, I define a "space launch vehicle" or SLV as the complete system designed to transport a payload to Earth orbit. The SLV includes propulsion systems, propellant tanks, structure, payload (crew spacecraft, cargo, satellite, lunar lander, space probe, space telescope, etc.), instrumentation and controls, a navigation system—every system needed to launch a payload. The SLV is propelled by powerful rockets capable of lifting the vehicle from the ground, overcoming the planet's gravitational force, and providing a velocity change sufficient to reach orbital velocity.

A space launch vehicle (SLV) is a multi-stage rocket propelled system designed to place payloads into orbit. All propulsion stages of a SLV are chemical rockets burning liquid or solid propellants. The first or booster stage yields the highest thrust and greatest total impulse. The thrust requirement becomes smaller with each subsequent stage; the payload is attached to the upper stage rocket, which is designed to position the spacecraft either in its final geocentric orbit or to inject it into an escape trajectory. The SLV's overall thrust depends on vehicle overall mass, which in turn depends on the payload mass and the mission.

1.1.2 Chemical Rocket Propulsion

Chemical rockets are propulsion systems that carry both fuel and oxidizer (the propellant) within the vehicle they propel. Rocket propulsion may be used anywhere in space vacuum as well as in a planetary atmosphere. Chemical rocket propulsion is used to solve the most important problem of astronautics—getting into space. Launch vehicles are comprised of chemical rocket engines burning liquid or solid propellants (or both) as these are the only type of propulsion with the thrust power to take payloads to orbit.

A liquid propellant rocket engine (LRE) uses liquid propellants that are fed under pressure from storage tanks into the combustion chamber where they undergo chemical reaction, and the hot gases are expanded through a converging-diverging nozzle, exhausting at very high speed. Bipropellant liquid rockets burn a mixture of fuel and oxidizer, such as liquid hydrogen (LH_2) and liquid oxygen (LOX). Other bipropellant combinations are LOX/RP-1, where RP-1 stands for Rocket Propellant-1 fuel, a highly refined form of kerosene, and LOX/LCH_4 where the fuel is methane, a bipropellant combination known as methalox. Figure 1.2 shows two important bipropellant rocket engines addressed in this book. The RS-25 used as the core stage engine for NASA's SLS, and the RL-10B-2 used as upper stage for SLS Block 1 version. The RS-25 evolved from the Space Shuttle Main Engine (SSME) that successfully provided liftoff thrust for the Orbiter in all 135 Space Shuttle flights.

A solid propellant rocket motor (SRM) burns a solid propellant within its combustion chamber or case. The propellant charge, called grain, contains all chemical elements for complete combustion. The resulting hot gases are then expanded through a nozzle. Figure 1.3 illustrates the main components of a solid propellant motor. This propulsion type can be as small as the STAR 3 TE-M-1082-1 developed as the transverse impulse rocket system (TIRS) for the Mars Exploration Rover (MER) program, yielding a thrust of 2051 N (461 lbf), or as large and powerful as the two strap-on solid rocket boosters

(SRBs) that provide 14,679 kN (3.3 million lbf) thrust to boost a launch vehicle from the ground. Chapter 5 is devoted to the study of solid rocket motors.

A hybrid propellant rocket uses both a liquid and a solid propellant. For example, a liquid oxidizing substance is injected into a combustion chamber filled with solid carbonaceous fuel grain; their chemical reaction produces hot combustion gases, which are then exhausted through a supersonic nozzle.

Important applications of chemical rockets are discussed in Chaps. 4 and 5. There we will focus on the main differences between rocket engines designed for launch vehicles and those for propelling spacecraft, also known as in-space propulsion systems. Both types are the focus of this book.

The energy and power requirements for launching a vehicle from the surface of the Earth are huge. A launch vehicle needs to accelerate to a very large velocity, to lift-off from the ground and achieve orbital velocity. The velocity change is at least $\Delta v = 7.6 \text{km/s}$ to achieve a low Earth orbit (LEO) at 400 km altitude. A powerful propulsion system with high thrust-to-weight ratio—must be greater than 1.0—is required for a launch vehicle to get off the ground, since to overcome the pull of Earth's gravity, the total thrust produced by the rockets must be greater than the vehicle's weight at launch. Only chemical rocket engines can meet such requirement as they can effectively convert the thermal energy of the propellants into kinetic energy and produce very high thrust. The NASA Space Launch System (SLS) achieves a thrust-to-weight value of 1.27 at launch using a sophisticated propulsion system that combines four RS-25 bi-propellant LOX/LH$_2$ rocket engines, delivering 1859 kN (418,000 lbf) thrust each, augmented with two strap-on solid rocket boosters that produce 16 MN (3.6 million lbf) thrust each.

The requirements for spacecraft propulsion are different from the propulsion for the launch vehicle. Once launched, a spacecraft requires its own propulsion system for attitude control, station-keeping, and orbit maneuvering. Station-keeping refers to the maneuver using onboard rockets to make periodic corrections to orbit perturbations, making position adjustments to maintain its required orbit. Every 21 days the James Webb Space Telescope (JWST) fires onboard thrusters to maintain its orbit around its equilibrium position. Orbit maneuvering includes plane changes, Hohmann transfers, circularization, orbit trim, and other orbit corrections, each with a different requirement for propulsion thrust and capabilities. In-space maneuvers are discussed in detail in Chap. 3.

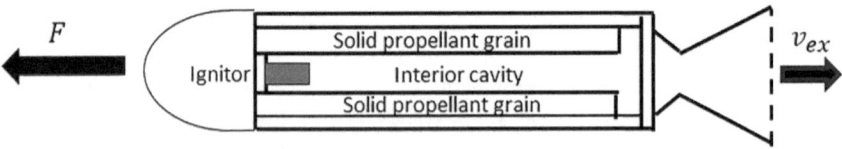

Fig. 1.3 Conceptual representation of a solid rocket motor (SRM). There are of many different types and sizes, varying in thrust from ~2 N to over 8 MN (0.4 lbf to over 2 million lbf)

The requirements for in-space or spacecraft propulsion can be met by low thrust rockets, also known as thrusters. There are different types of thrusters: (1) cold-gas propulsion, mainly used for roll and attitude control; (2) Hydrazine-based thrusters, for attitude control, station-keeping, and orbit corrections; (3) solid propellant rocket motors, for orbit maneuvering; or (4) electric thrusters for station-keeping (orbit maintenance), and as primary propulsion for interplanetary missions.

1.1.3 Cold Gas Rocket Propulsion

A cold-gas rocket is the simplest propulsion device. It uses a pressurized cold gas as its sole thermodynamic energy source. To produce thrust, the cold propellant gas (e.g. nitrogen, helium, hydrogen, argon) is stored at very high pressure and then expanded through a converging-diverging nozzle. A typical cold gas system includes the storage tank, a gas-loading valve, a filter, and a pressure regulator (pressure relief and controlling valves). The specific impulse of a cold gas rocket is typically 50 to 120 s. Cold gas thrusters are considered for in-space maneuvers where the total impulse requirement is 1200 N-s (5000 lbf-sec), making them adequate for reaction control systems (RCS) (spacecraft roll control and attitude control). Cold gas thrusters were incorporated in the reusable first stage of the SpaceX Falcon 9 vehicle to help in the attitude control while returning to land.

1.1.4 Electric Propulsion

Electric propulsion (EP) is a form of low-thrust propulsion wherein charged particles or plasmas are accelerated using electrostatic or electromagnetic forces and ejected at very high exhaust velocities. This results in high specific impulse, as high as 10 times greater than the specific impulse of chemical propulsion systems. There are many and varied electric propulsion concepts, and are typically classified in three fundamental types: electrothermal, electromagnetic, and electrostatic systems. The most advanced EP systems to date are the ion thrusters and Hall thrusters, both electrostatic systems used in space exploration missions. Dawn was the second ion-propelled spacecraft developed by NASA sent to study Vesta and Ceres, two of the largest asteroids in the Solar System. Chapter 6 is devoted to the study of electric propulsion.

1.1.5 Advanced Rocket Propulsion

The term "advanced rockets" is used in this book to describe futuristic concepts conceived to increase propulsion performance beyond that of current chemical rockets or electric thrusters. Advanced rockets may include chemical air-breathing rockets now in development to propel single stage space spaceplanes, or concepts that utilize nuclear energy and matter-antimatter annihilation energy and having the potential to revolutionize or exceed our capability for spaceflight, including applications to interstellar missions.

Nuclear Propulsion. For this concept nuclear energy is the sole energy source for a rocket. In principle, a propellant is introduced into a nuclear reactor, heated to a very high temperature, and then forced through the nozzle, expanding to very high velocity. This is known as nuclear thermal propulsion, as it employs the nuclear reactor as the agent for heating an inert low-molecular-weight propellant such as hydrogen. A nuclear thermal rocket is comprised of a nuclear reactor, and a thermal management system utilizing the propellant as the coolant for the core. The heat generated by fission is absorbed by the propellant, and the resulting hot gas is expanded through a nozzle. Conceptually, nuclear rockets are extensions of liquid propellant rocket engines. The propellant gas is heated by energy derived from transformations within the nuclei of atoms. While in chemical rockets the energy is obtained from within the propellants, in nuclear rockets, the power source is separate from the propellant. Other aspects of nuclear energy utilization for rocket propulsion are discussed in Chap. 7, along with the description of advanced or revolutionary concepts conceived for interplanetary missions.

1.2   Definition of Outer Space

Outer space is the region that extends from the upper boundary of Earth's atmosphere and throughout the vast universe. This means that outer space begins where the last layer of air around our planet vanishes. But, where exactly? To study spaceflight we must reach a consensus and define formally where outer space begins.

The Earth's atmosphere has no definite end; it slowly thins out into the near imperfect vacuum of interplanetary space. For every 4.8 km (3 miles) of altitude over sea level, the air density is approximately halved. Humans can live without artificial aids at heights of 4.8 to 6.4 km (3 to 4 miles) given time to adapt to those environments. However, 8 km (5 miles) establishes the limit of human endurance for sustained periods. A human being can in fact climb the tallest mountain in the world without breathing gear, but would be unable to live normally at the highest elevations. Hence, as far as unprotected humans are concerned, even 16 km (10 miles) above sea level is already "space." On the other extreme, 161 km (100 miles) altitude is not high enough for an artificial satellite to be positioned in a permanent stable Earth orbit. Below this altitude no spacecraft is capable of prolonged free flight, as it would encounter increasing atmospheric drag and cause it to fall.

1.2.1 The Kármán Line

Outer space is considered to begin at the Kármán line, an invisible boundary at an altitude of 100 km (62 miles) above the Earth's surface. Named after the Hungarian-American engineer Theodore von Kármán, the Kármán line was adopted by the International Astronautic Federation (IAF) to separate aeronautics (aerodynamic flight, or flight through atmospheric air) and astronautics (navigation into and through deep space). In this region, where the gas density is very thin ($\rho \approx 0.6 \times 10^{-6}$kg/m^3; $p = 0.03$Pa; $T = 195$K), the flow transitions from a continuum to a rarified, free molecular regime, i.e., when the molecular mean free path λ is greater than a characteristic spacecraft dimension L.

The dimensionless ratio $K = \lambda/L$ is called the Knudsen number. The Knudsen number serves as a criterion for the division into various flow regimes. Thus, for $K < 10^{-2}$ the flow is classified as continuum flow; for $10^{-2} < K < 10^{-1}$, slip flow; for $10^{-1} < K < 10$, transition flow; and for $K > 10$, free-molecular flow. In some cases, primarily for blunt bodies (representative of some re-entry vehicles), free-molecular flow may exist down to $K = 3$. At altitudes near the Kármán line (100 km), the Knudsen number typically begins to exceed 1 indicating that the atmosphere more accurately corresponds to a rarefied, free molecular flow regime than a continuum flow regime.

Aircraft require the air in the atmosphere for aerodynamic flight—the greater the gas density the higher its lift—and for propulsion—the oxygen in the air is ingested by jet engines and mixed with the onboard fuel to provide the thermal energy to produce thrust, the force necessary to accelerate the vehicle. Spacecraft, on the other hand, if surrounded by a dense atmosphere cannot maintain orbital speed around Earth without significant forward thrust (thus making the free fall, or orbiting, concept meaningless). In the 1950s, Kármán deduced that a vehicle would have to fly faster than orbital velocity to have sufficient aerodynamic lift from the air to stay aloft at that altitude. At an altitude of 160 km artificial satellites are subjected to sufficient atmospheric drag that begin to fall. Atmospheric drag is the resistance offered by the air to any object moving through it. If a spacecraft comes within 120 to 160 km over the Earth's surface, atmospheric drag will cause it to fall, spiraling down for some time, and its final disintegration will occur at an altitude of about 80 km. The deterioration of a spacecraft's orbit due to atmospheric drag is called *orbit decay*.

The gravitational acceleration of gravity decreases with altitude according to the inverse square law:

$$g(h) = g_0 \frac{R_E^2}{r^2} = g_0 \frac{R_E^2}{(R_E + h)^2}$$

where R_E is the Earth's equatorial radius, g_0 is the acceleration of gravity on the Earth's surface, and h is the altitude relative to sea level.

On the surface of the Earth, the acceleration of gravity is $g_0 = 9.80665$ m/s^2. Hence, at the Kármán line, the gravitational acceleration is $g = 9.516$ m/s^2 (97% of the gravitational acceleration on the ground).

Suborbital flights can reach an altitude of 80 km (50 miles) or even cross the Kármán line, but a suborbital vehicle will not orbit the planet, rather it will quickly return to Earth in a parabolic descent trajectory. In general, a suborbital flight is defined as that in which its trajectory intersects the surface of the gravitating body from which it was launched (planet or Moon). Hence, the suborbital vehicle will not become an artificial satellite nor will it reach escape velocity. In this book, we will not study propulsion specifically designed for suborbital spaceflight.

For Earth re-entering spacecraft, *atmospheric entry-interface* is located at an altitude of 121.92 km (76 miles or 400,000 ft) when the vehicle encounters the highest reaches of the sensible atmosphere as it moves at hypersonic speed, greater than 11 km/s. Sensible atmosphere is defined as that part of an atmosphere that offers resistance or drag to a body passing through it. Re-entering Earth's atmosphere, a spacecraft experiences temperatures of about 1900 °C (3500°F), a re-entry heating caused by the intense pressure and friction of moving through the air at speeds of about 11,265 km/h (7000 mph).

1.2.2   Deep Space: Interplanetary and Interstellar Space

Deep space is usually reserved to characterize spaceflight beyond Earth orbits. Our immediate region of deep space is known as cislunar space. This is a huge region of space subject to the gravitational influence of the Earth-Moon system. Extending at least to~2 million km, cislunar space is 1728 times larger than the volume of space within 1 GEO radius (35,786 km above the Earth's equator). Cislunar space is of great interest to astronautics, serving as a bridge between present crewed missions and future interplanetary missions, by enabling and reducing risk for future human missions to the Moon, Near-Earth Asteroids (NEAs), Mars, and other deep space destinations.

The region of space between planets or moons is called interplanetary space. Interplanetary space extends to the outer edge of the Solar System where it strikes interstellar space and forms the heliosphere. The heliosphere is a bubble of magnetism from the Sun, inflated to gigantic proportions by the solar wind, the stream of energetic charged particles emanating from the Sun. The bubble, considered the outermost atmospheric layer of the Sun, extends to about 100 astronomical units (AU), which is beyond the orbit of Pluto, marking the boundary that separates our Solar System from interstellar space. One AU is the distance from the Earth to the Sun, which is about 1.496×10^8 km (93 million miles). The boundary between interplanetary space and interstellar space is known as the *heliopause*. This is the theoretical boundary where the Sun's solar wind is stopped by the interstellar medium. The heliopause is found at about 90 to 120 AU from the Sun.

After crossing the orbit of Pluto, the farthest planetary body gravitationally bound to our Sun, an object finally escapes the control of the Sun and enters interstellar space. There is a very large region of space before we find the closest star to our Solar System, Proxima Centauri, located at approximately 4.0×10^{13} km (268,000 AU) from the Sun.

1.3   Space Launch Vehicles (SLVs)

Direct exploration of the Universe depends on our ability to access space reliably, safely, and economically. This requires a unique space launch capability. The launch vehicle chosen for a particular mission whether scientific or commercial, with a crew or robotic, initially depends on the size, weight, payload, destination, and mission objectives. The payload may be an astronaut crew, satellites, an interplanetary or lunar spacecraft, instruments and equipment for exploration, or a combination of all.

A space launch vehicle (SLV) can be expendable (used only once), partially reusable, or fully reusable. The term "reusable" refers to several different levels of reuse. Some launch vehicle concepts incorporate one reusable stage working in combination with other expendable stages that are discarded after completing their task.

The SLV can also be classified according to its capacity to lift a payload to designated low Earth orbits, or to place it in trans-lunar injection (TLI) condition, an escape trajectory that leads a spacecraft to intersect the orbit of the Moon, or to place it in escape trajectories for interplanetary missions. Table 1.1 summarizes the different SLV categories and the payload capacities to LEO.

Electron, a two-stage, partially reusable launch vehicle developed by Rocket Lab, is a representative of the small-lift category. Powered by Rutherford engines burning LOX (liquid oxygen) and RP-1 (refined kerosene), Electron can launch payloads of up to 300 kg (661 lbm), servicing the commercial small satellite launch market. Electron is often flown with a Kick Stage (upper stage), which serves as in-space propulsion to deploy payloads to orbit. Rocket Lab is an American aerospace company with a wholly owned New Zealand subsidiary.

Table 1.1 Low Earth Orbit (LEO) launch space vehicle capability

SLV performance class	Payload capacity to LEO	
	kg	ton
Small	Less than 2000	< 2
Medium	2000–10,000	2–10
Mid heavy	10,000–20,000	10–20
Heavy	20,000–50,000	20–50
Super heavy	Greater than 50,000	> 50

Pegasus XL, an air-launched multistage launch system developed by Orbital Sciences and later built and launched by Northrop Grumman, is also a small class vehicle. Comprised of three solid propellant stages and an optional monopropellant fourth stage, Pegasus can take small payloads of up to 443 kg (977 lbm) to LEO. Pegasus is released from its carrier aircraft at approximately 12 km (39,000 ft) using a first stage wing and a tail to provide lift and altitude control while in the atmosphere. A Pegasus XL carried NASA's Interstellar Boundary Explorer (IBEX) to its high Earth orbit (HEO) in 2008 to image the interaction region between the Solar System and interstellar space.

Representative heavy and super heavy launch systems are depicted in Fig. 1.4 to illustrate their relative size and payload capacities. The former Space Shuttle, which had a capacity to carry payloads up to 27,500 kg to LEO, was a heavy SLV. Its successor, NASA's Space Launch System (SLS) is much more powerful, designed to launch more than 27,000 kg to translunar orbit (TLO). The SLS massive 130-metric-ton-configuration is the most capable launch vehicle in NASA's history, with an initial payload capability of 70 ton and subsequent versions up to 130 ton. Towering a staggering 98 m or 384 ft tall and weighing 2.6 million kg or at liftoff, the SLS Block I can provide 39.1 MN (8.8 million lbf) of thrust.

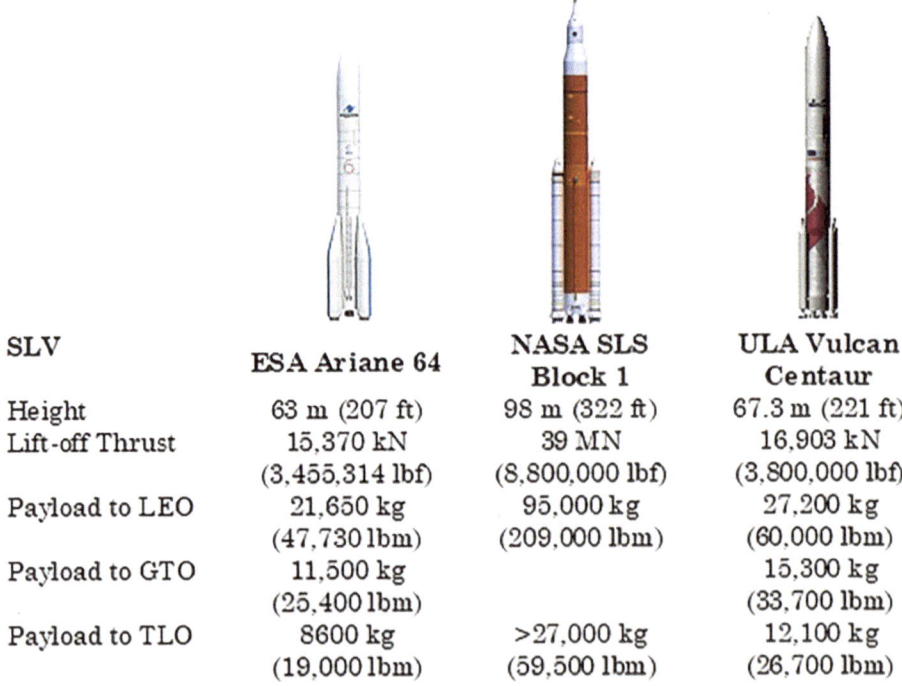

SLV	ESA Ariane 64	NASA SLS Block 1	ULA Vulcan Centaur
Height	63 m (207 ft)	98 m (322 ft)	67.3 m (221 ft)
Lift-off Thrust	15,370 kN	39 MN	16,903 kN
	(3,455,314 lbf)	(8,800,000 lbf)	(3,800,000 lbf)
Payload to LEO	21,650 kg	95,000 kg	27,200 kg
	(47,730 lbm)	(209,000 lbm)	(60,000 lbm)
Payload to GTO	11,500 kg		15,300 kg
	(25,400 lbm)		(33,700 lbm)
Payload to TLO	8600 kg	>27,000 kg	12,100 kg
	(19,000 lbm)	(59,500 lbm)	(26,700 lbm)

Fig. 1.4 Representative heavy and super heavy space launch systems

As more challenging space missions are conceived, launch vehicle suppliers will reconfigure their propulsion system to maximize the payload mass delivered to a desired orbit.

1.3.1   Expendable Launch Vehicles

An expendable launch vehicle (ELV) is a single-use transportation system comprised of several rocket stages. The entire vehicle is discarded, piece by piece, during its ascent until the last rocket stage remains, which is used to deploy the payload (crew, spacecraft, or cargo) to its assigned orbit. Throwing away stages is done in order not to carry and accelerate parts of the vehicle that are no longer needed. This is not the only method to access space but it has proven to be reliable and practical.

In the United States, the National Aeronautics and Space Administration (NASA) and the Department of Defense (DoD) have created government-industry partnerships to develop space transportation launch technology and the supporting infrastructure. The activity in the commercial and military segments benefit the scientific goals of our space exploration program, as human spaceflight, space observatories, and robotic missions beyond LEO are activities that are only possible with a viable space launch infrastructure.

Demand for launching commercial communications satellites has grown significantly through the early twenty-first century and sparked plans to develop new launch vehicles of all types. United Launch Alliance (ULA), a company funded by a public–private partnership of Boeing and Lockheed Martin with the U.S. government, has provided launch services using the expendable vehicles Delta IV Heavy and Atlas V. The Atlas V can deliver 20,050 kg to LEO and 8200 kg to GTO. Although the Delta IV Heavy could carry 25,800 kg to LEO and 12,400 kg to GTO, its role in the American space launch capability ended in 2024. After six decades, the last ULA's Delta IV Heavy launched the NROL-70, a classified payload for the U.S. National Reconnaissance Office (NRO). Retiring the Delta IV was decided by ULA in favor of its newly Vulcan Centaur, which flew a near-perfect first mission in January 2024. ULA developed the Vulcan SLV to replace both Deltas.

Capable of launching payloads to all high energy geocentric orbits and to TLI (Translunar Injection), the Vulcan Centaur is a new expendable SLV available in four standard offering configurations to satisfy the different launch requirements. These include variants with zero, two, four and six solid rocket boosters (SRBs). For NASA's Peregrine cargo mission to the Moon, the vehicle is identified as VC2S, which stands for Vulcan Centaur, powered with 2 SRBs, and configured with a Standard payload fairing. The main booster stage is powered by two Blue Origin's BE-4 rockets (nominal sea level thrust 2.45 MN or 550,000 lbf each) burning liquefied natural gas (LNG) (liquid methane) and liquid oxygen (LOX). To power its second upper stage, the VC2S uses two Aerojet Rocketdyne RL10C engines, burning liquid oxygen (LOX) and liquid hydrogen (LH$_2$), providing 106.76 kN or 24,000 lbf nominal thrust. The Vulcan Centaur draws most of its heritage from the Atlas SLV.

Expendable launch vehicles designed to carry people into space and return them to Earth are Russia's Soyuz, and China's Long March 2. The Soyuz is part of the Soyuz/Molniya/Vostok family of launch vehicles developed and manufactured by Progress Rocket Space Centre in Samara, Russia. With over 1900 flights since its debut in 1966, the Soyuz has the most launches in history. The three-stage Soyuz-2 is the most modern variant today for placing payloads into LEO, incorporating uprated LOX/RP-1 engines for the first-stage boosters. RP-1 or Rocket Propellant 1 is hydrocarbon, a highly refined form of kerosene with chemical formula CH1.953. The Soyuz carries medium-class payloads and Soyuz capsules to the International Space Station (ISS). Russia's human missions with Soyuz TMA capsules are launched in the Soyuz FG configuration. Currently, Russia is developing the Angara (Russian: Ангара), multi-stage launch vehicles intended to lift payloads into LEO, with masses between 3800 kg (8400 lbm) and 24,500 kg (54,000 lbm).

The European Space Agency (ESA) developed the Ariane family of launch vehicles operating since 1982. The Ariane 5 could deliver 18,000 kg to LEO and 6800 kg to GTO. The new Ariane 6 is a heavy-lift launch system, available in two versions, depending on the required performance: Ariane 62 with two strap-on solid boosters, and Ariane 64 with four. ESA also developed a smaller launch vehicle named Vega designed to place 1500 kg payloads into LEO and into polar orbits. The single body Vega is designed with three solid propulsion stages and a liquid propulsion upper module used for attitude and orbit control, and satellite release. Unlike most small launchers, Vega can place multiple payloads into orbit.

The People's Republic of China (PRC) has developed several versions of its Long March (LM) launch vehicle, including several configurations capable of launching satellites to LEO, GTO, and Sun-synchronous orbits. The largest launching capacity of the LM rockets reached 9200 kg for LEO, and 5100 kg for GTO.

In 2003, China launched the first crewed spacecraft named Shenzhou. A second successful piloted mission was carried out in 2005, and a third in 2008 with a crew of three. The Shenzhou is launched atop the Long March 2F vehicle, which incorporates systems like those on Soyuz and Apollo, including a crew emergency escape tower. The Long March CZ-2E booster called the 2F carries the crew vehicle into orbit. Tiangong ("heavenly place") is China's space station, which hosts three astronauts for periods of six months at a time. Today (2024) China is expanding its space exploration capability with lunar sample return missions, human trips to Mars, and a projected Moon base initiative. China, together with Russia, is developing an International Lunar Research Station (ILRS), planning to make it available to other counties.

India conducted its first successful orbital space launch in 1980. Its expendable Polar Satellite Launch Vehicle (PSLV) and the Small Satellite Launch Vehicle (SSLV) can place small satellites in LEO. Through its space agency, the Indian Space Research Organisation (ISRO), India also developed a larger, three-stage vehicle capable of reaching GTO known as Geosynchronous Satellite Launch Vehicle (GSLV). The GSLV uses Russian rocket engines with cryogenic propellant. The three-stage Launch Vehicle Mark-III (LVM3) will launch crewed missions under the Indian Human Spaceflight Programme. India is also developing a new genre of launch systems, including a re-usable two-stage concept that incorporates a winged-aircraft like vehicle boosted by a conventional vertical rocket system.

Japan developed several launch vehicles to answer diversified launch needs, including the H-IIA and H-IIB, and the Kounotori2 HTV2, a cargo transporter to the International Space Station (ISS). Through its independent administrative institution, the Japan Aerospace Exploration Agency (JAXA), Japan successfully conducted the first launch of its two-stage H-2 launch vehicle in 1994, the first all-Japanese rocket capable of putting satellites in geostationary orbit. The H-IIA 212 can deliver 17,280 kg payloads to LEO and 7500 kg to GTO. The H-IIA can be reconfigured to enhance its capability by adding SRB-A solid rocket booster and Castor 4AXL solid strap-on booster (SSB) to its basic configuration.

Israel has satellite programs both for reconnaissance and commercial purposes. The Israel Space Agency (ISA) launched its first satellite, the Ofeq-1, on 19 September 1988, from Palmachim Airbase in Israel. Since then, ISA has continued advancing commercial space activities, using the three-stage Shavit (Hebrew: "comet"), a small SLV capable of sending payloads into LEO. Since the mid-90 s ISA and NASA work under a cooperation agreement, which has resulted in one Israeli astronaut participating on a NASA mission.

In 2014, the United Arab Emirates (UAE) established the UAE Space Agency and has successfully completed numerous space projects. In July 2020, UAE became the fifth country in the world to launch a probe to Mars, and in December 2022, the Rashid Rover was launched on a mission to the Moon. Although the rover crashed into the lunar surface and was destroyed, UAE's technology effort is remarkable, providing lessons for future missions.

NASA Space Launch System (SLS)

The newest expendable SLV developed in the U.S. is NASA's Space Launch System (SLS), a super heavy-lift launch vehicle. Known as the *Moon Rocket*, the SLS is intended to expand human presence to celestial destinations beyond LEO. In 2017, NASA established the Artemis program intended to reestablish a human presence on the Moon for the first time since the Apollo 17 mission in 1972. Artemis includes robotic and human Moon exploration utilizing the SLS, the Orion spacecraft, a Lunar Gateway space station, and the commercial Human Landing Systems. Partnering with European Space Agency (ESA), Japan Aerospace Exploration Agency (JAXA), and Canadian Space Agency (CSA), NASA's long-term goal is to establish a permanent base camp on the Moon and ultimately to culminate with crewed missions to Mars.

The launch capability (transporting astronauts or lift cargo) for deep space missions will rely on the flexibility of the SLS, a vehicle conceived to evolve into increasingly more powerful configurations. Each succeeding SLS block variant will be more capable through upgrades to the engines, boosters, and the upper stage, an approach that will provide an adaptable launch vehicle for a variety of human and robotic deep space missions, rather than requiring the development of entirely new rocket propulsion to increase performance.

The first four versions of the SLS are Blocks 1, 1A and 2, each utilizing some components derived directly from Space Shuttle hardware and others being developed specifically for the new vehicle. The SLS crewed version will be serve as a backup launch system for supplying and supporting the ISS crew requirements not met by other launch vehicles.

NASA SLS Block I Propulsion System

Engine/Motor	Propellant	SLS stage	Thrust	Number	Remarks
RS-25	Cryogenic liquid LOX/LH$_2$	Core stage	1.86 MN 418 klbf (sl) 512 klbf (vac)	4	Engine manufactured by Aerojet Rocketdyne; stage manufactured by Boeing
SRB	Solid PolyButadiene AcryloNitrile (PBAN)	Strap-on boosters	16.01 MN each 3.6 Mlbf each (Total 7.2 Mlbf)	2	Five-segment SRBs manufactured by Northrop Grumman
RL 10-2B	Cryogenic liquid LOX/LH$_2$	Upper stage ICPS	110 kN 24,750 lbf	1	Engine manufactured by Aerojet Rocketdyne; Stage manufactured by Boeing/ULA

The first SLS version, identified as Block 1 (Fig. 1.5), includes the Orion Crew Capsule. The core stage is powered by twin five-segment solid rocket boosters, and four RS-25 liquid propellant rocket engines. Designed and manufactured by Aerojet Rocketdyne, the RS-25, evolved from the Space Shuttle Main Engine (SSME), is now the SLS Core Stage Engine. With a cluster of four RS-25 engines, the core stage delivers 1859 kN (418,000 lbf) thrust at sea level, burning a mixture of propellants consisting of hydrogen and oxygen, which are stored and delivered in a cryogenic liquid state.

The SLS is designed to carry payloads to orbit, using the same core stage. The upper stage uses the RL10 engine to position the Orion spacecraft on TLI, and eventually to

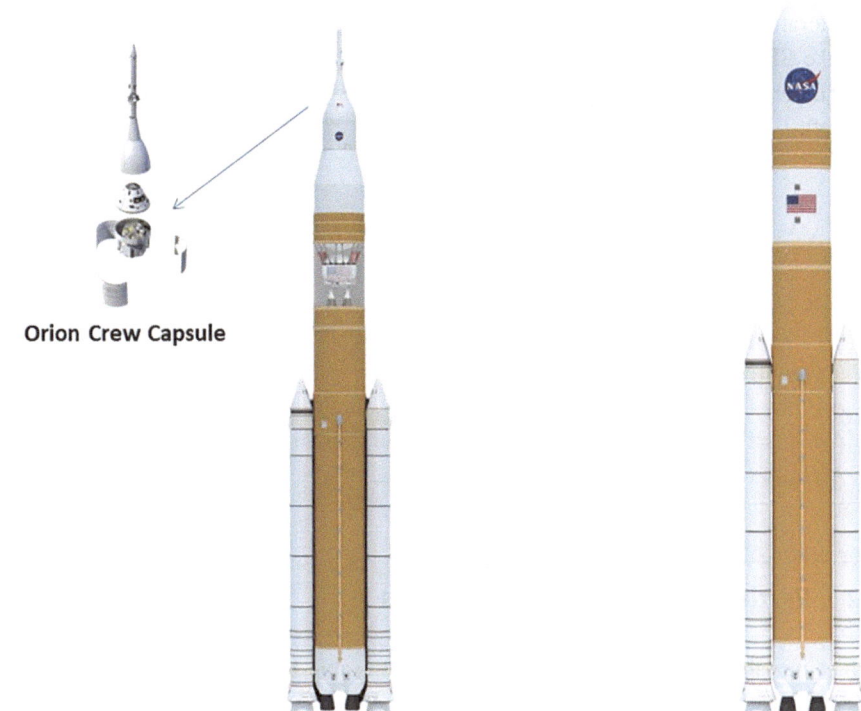

Orion Crew Capsule

Fig. 1.5 NASA Space Launch System (SLS) designed to carry different payload mass. The left version is designed to carry the Orion crew vehicle (Block 1 Crew Payload to TLI: > 27 t); the right version will carry only cargo (Block 2 Cargo Payload to TLI: 46 t). *Credit* NASA

send astronauts to deep-space destinations. Manufactured by Aerojet Rocketdyne, the RL10 engine produces vacuum thrust of 110 kN (24,750 lbf). The performance of the SLS individual rocket engines will be examined in Chaps. 4 and 5.

In November 2022, NASA launched Artemis I, the first mission of the Artemis Program. The SLS carried the uncrewed Orion spacecraft to an Earth orbit from where it performed the translunar injection (TLI) maneuver to begin its 25-day flight around the Moon. Orion performed a powered lunar flyby 169.78579 km (105.5 miles) from the Moon surface to move on a distant retrograde orbit (DRO). The Orion spacecraft splashed down in the Pacific Ocean off the coast of Baja California on 11 December 2022, completing a 2.3 million-kilometer (1.4 million-mile) spaceflight. The successful Artemis I mission tested the Orion module, SLS propulsion as well as ground systems at the Kennedy Space Center, laying the foundations for subsequent missions.

Future SLS versions will use five RS-25E engines with upgraded boosters, and an 8.4 m diameter upper stage with three J-2X rocket engines. NASA may add a 5 m class fairing with a length of 10 m or greater to allow heavy payloads for deep space missions.

The SLS Block 1B version will be used beginning with Artemis IV. This version includes the Exploration Upper Stage (EUS) comprised of four RL10 engines. This stage upgrade will make SLS capable of sending 42,000 kg (92,594 lbm) of payload into TLI for the Block 1B cargo configuration, and 3800 kg (83,766 lbm) when it carries the crewed Orion spacecraft.

NASA Artemis Missions

Artemis was the twin sister of Apollo and the goddess of the Moon in Greek mythology; she was also referred to as the "torch bringer." Artemis now personifies NASA's path to the Moon.
Artemis I—First uncrewed flight around the Moon in 2022.
Artemis II—First crewed (4 astronauts) flight around the Moon in 2025.
Artemis III—First lunar landing planned for 2026.

1.3.2 Partially and Fully Reusable SLVs

The first reusable launch vehicle powered by a reusable rocket propulsion system ever developed was NASA's Space Transportation System (STS), known simply as the Space Shuttle. The STS had the capability of launching a crew and of performing other missions such as satellite deployment and recovery, research in pressurized modules, repair, and space station support.

Following the Apollo pioneering lunar landings, NASA conceived the idea of a reusable launch system, a spaceplane that would allow routine trips to a station orbiting in LEO, and occupied by 12 to 24 people. The station was intended to assure a permanent human presence in space. NASA believed that reusable space shuttles could also serve as multi-purpose satellite delivery vehicles with the potential to completely replace Atlas-Centaur, Delta, and Titan launch systems.

After numerous iterations for a fully reusable space launcher, the final version of the STS reduced to three major components: the Orbiter, an external tank (ET), and two solid rocket boosters (SRBs). The 37-m-long Orbiter (the actual space shuttle) was designed to carry up to seven crewmembers and the payload into LEO and return. The SRBs and the ET were part of the propulsion system required to boost the Orbiter into space.

The Orbiter was launched in a vertical position, with a huge thrust force provided by two solid rocket boosters (SRBs) (the first stage), and the Orbiter's three Space Shuttle Main Engines (SSMEs) (the second stage). The total thrust for liftoff was about 30.16 MN; each of the three liquid propellant Rocketdyne SSMEs provided almost 1.752 MN at 104% power, for a total of 5.255 MN of thrust for lift off.

The two (SRBs) each generated 12.5 MN of thrust at liftoff, which was 83% of the total thrust needed. To achieve orbit, the Shuttle had to accelerate from zero to a speed almost 8 km/s (28,968 km/h; 18,000 mph). After burning their solid propellant (about 123 s after ignition), the SRBs separated from the ET and Orbiter and deployed a system of self- contained parachutes. When the SRBs hit the water, the parachutes were jettisoned and the rocket casings were towed back to port by recovery ships.

The external tank (ET) contained the propellants for the Orbiter SSMEs from liftoff until main engine cutoff (MECO). The ET was 47 m (153.8 ft) tall and 8.4 m (27.6 ft) in diameter, and contained separate containment for liquid oxygen and liquid hydrogen, (390,000 gallons liquid hydrogen and 145,000 gallons liquid oxygen). The two SRBs were attached to the ET at launch; they provided additional initial ascent thrust. The ET was the only element of the SST that was expendable. After the SSMEs shut down (~8.5 min after liftoff), the ET was jettisoned, entered the Earth's atmosphere, broke up, and the surviving pieces fell in a remote ocean area.

The Shuttle Orbiter was designed as a spaceplane, with aerodynamic surfaces so it could reenter Earth's atmosphere and fly unpowered to assigned landing strips. The wings helped the vehicle to glide and bank like an airplane during a great part of the return flight phase.

Between 1981 and 2011, the STS was the sole U.S. means for launching humans into orbit. The Space Shuttle fleet flew 135 missions, helped construct the International Space Station, placed in orbit the Hubble Space Telescope, and inspired generations. The NASA SST program ended with the final landing on 21 July 2011. Despite the challenges it experienced, the Space Shuttle provided a great leap in capability for human space access.

Did you know? The American Space Transportation System (STS) was propelled by the first reusable rocket propulsion system in the world, and the first to utilize solid propellant rockets to boost the propulsion capability for lift-off of a crewed reusable Shuttle Orbiter vehicle on missions to LEO.

At the turn of twenty-first century, the U.S. private sector companies, including Space Exploration Technologies Corporation (SpaceX), began developing their own launch capability, setting technological goals centered on launch vehicle reusability and affordability. Founded in 2002, SpaceX conceived multi-stage systems intended to launch payloads to geocentric orbits. The vehicles incorporate reusable first stage rockets that, after burnout, separate from the ascending SLV and are guided back to land upright on the launch pad. Such recoverable, reusable rocket stages can reduce considerably the cost of a typical launch. Following stage separation, the first stage flies back and lands nearly 1000 km downrange on a moving ship. The second stage engines ignite while the fairing separates and the payload is delivered safely to orbit. SpaceX's reusability was demonstrated in

Fig. 1.6 Nova, a reusable two-stage medium-lift launch vehicle. *Credit* Stoke Space Technologies

2014. A historic vertical landing was achieved on 21 December 2015, when the first-stage booster of Falcon 9 Flight 20 successfully landed at Cape Canaveral.

Stoke Space Technologies, a private American company, is developing a fully-reusable two-stage medium-lift launch vehicle called Nova (see Fig. 1.6). For the reusable first stage, Stoke is developing 7 full-flow-staged combustion rocket engines (the same cycle as SpaceX's Raptor engine), burning LOX/LCH$_4$ (methalox), each engine able to produce more than 100,000 lbf of thrust. The reusable second-stage uses a single LOX/LH$_2$ engine with several thrust chambers distributed around the circumference of the vehicle, integrated into an actively cooled base heat shield. This actively cooled engine and integral re-entry heat shield eliminates the need for passive insulators like brittle ceramic tiles that have required detailed inspections and lengthy refurbishments on other space vehicles designed for reuse. In 2023, Stoke successfully completed a vertical takeoff and vertical landing (VTVL) developmental test flight of its reusable second stage. The test demonstrated Stoke's novel LOX/LH$_2$ engine, regeneratively cooled heat shield, and differential throttle thrust vector control system, as well as its avionics, software, and ground systems.

For space tourism, Blue Origin developed New Shepard, a fully reusable suborbital single stage vertical take-off/vertical landing (VTVL) vehicle. Named in honor of Alan Shepard who, in 1961, became the first American to travel into space, New Shepard consists of a booster rocket, and a crew capsule large enough to accommodate up to six passengers. On July 2021, New Shepard successfully completed its first crewed suborbital mission reaching the Kármán line and returning to Earth intact; the capsule returned to Earth via parachute, while the booster returned and landed vertically on the same launchpad it took off from. The pilotless New Shepard is controlled entirely by on-board computers.

Blue Origin is also developing a massive reusable SLV rocket family called New Glenn. This vehicle is intended to launch satellites and people into space. Named for the first American to orbit Earth on the Mercury capsule, John Glenn, the two-stage heavy lift New Glenn SLV is intended for LEO launches and beyond. The first stage is fully reusable, comprised of a cluster of seven BE-4 rocket engines, with a combined thrust of 17.1 MN (3.85 million lbf), designed for a minimum of 25 flights.

In Russia, the partially reusable medium lift Soyuz-7 is a methane–fueled, orbital launch vehicle currently in the design concept stage of development by the Roscosmos State Corporation. This is a proposed family of new, more efficient launch vehicles intended to replace the legacy Soyuz.

To increase the payload capability of its Falcon 9, SpaceX conceived the Falcon Heavy, a partially reusable, 2.5 stage derivative with a structurally strengthened Falcon 9 as the "core" component, a second stage on top of the center core, and two Falcon 9 with aerodynamic nose–cones mounted outboard, which serve as strap-on boosters for added payload capability. Falcon Heavy can lift payloads of 63,800 kg (140,700 lbm) to LEO, and 26,700 kg (58,900 lbm) to geosynchronous transfer orbit (GTO). On 14 October 2024, Falcon Heavy launched NASA's Europa Clipper, a unique spacecraft to search for alien life on Jupiter's moon Europa.

Today, the most futuristic fully reusable SLV is SpaceX's Starship concept. Comprised of the Starship spacecraft and the super heavy rocket booster—collectively referred to as Starship—this two-stage super heavy SLV is designed to transport both crew and cargo to LEO, the Moon, Mars and beyond. Currently under development and flight testing, Starship will be the world's most powerful SLV ever conceived. Some call it "Starship megarocket."

The launcher first stage or booster will be powered by 33 Raptor engines to deliver 74.5 MN maximum thrust fueled by cryogenic liquid methane reacted with LOX. The second stage, the Starship spacecraft, will be powered by six engines: three Raptor engines, and three Raptor Vacuum (Rvac) engines, a variant of Raptor with an extended, regeneratively-cooled nozzle for higher specific impulse in the vacuum of space. At launch, the vehicle could achieve 74.5 MN (16.7 million lbf) of maximum thrust. It is projected to lift at least 100 tonnes of payload to LEO. To validate the Starship system SpaceX plans to take humans to the Moon and return them safely to Earth.

The vision for Starship is to be the most powerful launch system ever developed, capable of transporting up to 100 people on long-duration, interplanetary flights. SpaceX engineers consider using propellant tanker vehicles (the Starship spacecraft without the windows) to refill the Starship spacecraft in LEO prior to departing for Mars. Their studies indicate that refilling on-orbit enables the transport of up to 100 tons all the way to Mars.

1.4   Crew Transportation Spacecraft

The first humans to reach space were transported in capsule-type spacecraft designed to be launched atop expendable multi-stage rocket launch vehicles. A space capsule has a simple conical shape for the main section, without any wings or other features to create lift during atmospheric re-entry. The re-entry spacecraft is designed to accommodate a crew and adequate life support system, and is equipped with a propulsion system to allow the spacecraft to separate and maneuver in space. The crew capsule is designed with a thermal protection system for shielding it during re-entry, and incorporates a parachute to facilitate descent. Capsules have been used in most of the human space programs to date. Table 1.2 provides the basic characteristics of current crew capsules, including the

Table 1.2 Reentry spacecraft for human spaceflight

Spacecraft	Crew	Crewed flights	Launch Vehicle	Length (m)	Diameter (m)	Dry mass (ton)	Habitable volume (m^3)
China SHENZHOU	1–3	2003–present	CZ-2F	8.65	2.8	7.84	8
Russia SOYUZ TMA	3	1967–present	Soyuz MS-25	4.4	2.2	7.10	8.5
NASA ORION	4	2024–present	SLS		5.0	7.80	9
SPACEX DRAGON	7	2020–present	Falcon 9	8.10	3.7	9.50	10
BOEING STARLINER	7	2024–present	Atlas V	5.03	4.56	13.0	11

Russian Soyuz TMA capsule, China's Shenzhou, and spacecraft designed by U.S. private companies.

In 2015, NASA selected SpaceX and Boeing as providers of the new crew transportation systems to provide human access to the ISS via the commercial (non-government) sector. In 2020, SpaceX began providing crew transportation service to NASA using the Dragon spacecraft launched atop the two-stage Falcon 9 (block 5).

SpaceX designed Dragon, a reusable capsule spacecraft that can deliver cargo and is also configured to carry a crew of seven, or a combination of cargo and a small crew. Positioned atop a Falcon 9 launch system, Dragon is a conventional blunt-cone ballistic capsule with a hinged nose-cone cap which opens to reveal a standard ISS Common Berthing Mechanism. The heat shield is designed to withstand re-entry velocities from potential lunar and Martian space flights. Dragon is equipped with eighteen Draco thrusters for orbital maneuvering and attitude control, and two solar array wings for power.

For NASA, SpaceX's Dragon has flown crews of astronauts to and from the ISS, and it has flown private missions. The world's first all-tourist spaceflight, called Inspiration4, carried four passengers through Earth's low orbit for three days. In September 2024, Dragon took a crew of four non-professional astronauts to perform the first spacewalk, part of Polaris Dawn Mission. The Dragon crew achieved an orbit up to 1400 km (870 miles) from Earth's surface, passing through the Van Allen radiation belts, to assess how that environment affects their bodies. This is an important study since future missions to Mars will expose astronauts to immense amounts of space radiation. When the hatch was opened for the spacewalk, Dragon was on an orbit of 736 km (457 miles) altitude, moving at a speed of about 25,000 km/h (at least 15,500 mph).

Boeing designed the CST-100 Starliner capsule to transport up to seven astronauts to the ISS. It is of similar shape and capacity as the Dragon capsule. The Starliner spacecraft consists of a reusable capsule (up to 10 missions), and an expendable service module (used only once). The capsule accommodates seven passengers, or a mix of crew and cargo but for NASA missions to the ISS it will carry four passengers and a small amount of cargo. Boeing's Starliner includes a pusher abort system to provide safe crew escape throughout the launch phase of the mission.

The NASA Artemis program was conceived to take humans back to the Moon—and eventually Mars—and return them to Earth safely onboard the Orion spacecraft. The Orion spacecraft, which is attached to the Interim Cryogenic Propulsion Stage (ICPS), the upper stage of SLS, forms part of NASA's advanced spaceflight vehicles with emergency abort capability, systems to sustain the crew during long space travel, and provide safe re-entry from deep space.

Shaped like the Apollo capsule, Orion has an improved, larger, blunt-body module with a diameter of 5 m (16.5 ft), and a habitable volume of 9 m^3 (316 ft^3) to accommodate a four-member crew. The Orion spacecraft consists of four main elements: (1) the Crew Module (CM), a capsule sized to provide a habitable pressurized volume to support four crew members and cargo, (2) the Service Module (SM) consisting of the European Service Module (ESM) and the Crew Module Adapter (CMA) which is required for propulsion, heat rejection and power generation, (3) the Spacecraft Adapter (SA) which is a structural connection to the launch vehicle, and (4) the Launch Abort System (LAS) to provide abort capability to transport the CM away from the an out of control launch vehicle while still on the launch pad and move the crew capsule far enough out to achieve a safe landing under parachutes.

Orion has a requirement for full abort coverage from T-5 min to lift off through main engine cutoff. The launch abort system (LAS) is attached to the top of the Orion CM during launch, to be jettisoned once the spacecraft safely reaches the edge of space. The LAS consists of a tower where three rocket motors are located, and a fairing assembly which protects the crew capsule. Together, the CM and the LAS constitute the SLS launch abort vehicle (LAV).

The Orion design uses technology and knowledge gained from previous programs, but incorporates modern materials, manufacturing techniques, and avionics. The Orion CM has a liftoff mass of about 10,387 kg (22,900 lbm) and a nominal landed mass of approximately 9299 kg (20,500 lbm). The CM is a stand-alone module for re-entry and landing. It was designed to maintain stable hypersonic, transonic, and subsonic atmospheric flight, providing thermal protection from the heat of re-entry, store and distribute power during re-entry and descent, provide guidance, navigation, and attitude control, and to provide thermal control from SM separation through landing. Orion's Thermal Protection System (TPS) was optimized to limit the peak temperature of the spacecraft structure and systems to acceptable levels during atmospheric entry. These features were successfully tested during the uncrewed Artemis I mission in 2022, ensuring that the crew in Artemis II mission will be returned safely to Earth.

Artemis II will be the first crewed mission of NASA's Orion spacecraft on its mission to the Moon launched by the SLS into a lunar orbit. It is intended to test the Orion spacecraft, its habitability and adequacy of life support systems. Taking four astronauts, Orion will perform a lunar flyby test and return the crew to Earth. The astronauts will inhabit the Orion Crew Module (CM) for the duration of the voyage, up to 21 days. The European Service Module (ESM) designed by ESA will provide electricity, water, oxygen, and nitrogen to the CM and will keep it at the right temperature and on course. Figure 1.7 depicts the Orion spacecraft attached to the ESM. This 4 m long cylindrical service module is unpressurised and is powered by a main rocket engine (26.7 kN or 6000 lbf thrust OMS-E) and 32 thrusters for in-space maneuvers (see Chap. 4). The service module includes 8.6 tonnes of propellant to power one main rocket engine and the smaller thrusters in the RCS system. During launch, the ESM will be attached to the Spacecraft Adapter and will be connected to the CM by the CM Adapter.

At the time of this writing (2024), NASA had selected the crew for the Artemis II mission, scheduled the launch in 2026. We should note that after returning from the Moon, Orion will enter Earth's atmosphere with a speed greater than 11 km/s, and then it will carry out a precision-guided skip entry. Orion will target a water splashdown off the west coast of California for expeditious recovery and return to port. After that, Artemis III will be the first human mission to land on the Moon in the vicinity of the South Pole. This is an area still largely unexplored, critical for future lunar crewed missions because of the presence of abundant water, ice, and minerals. The Moon explorers will gather the working experience to eventually land on and colonize Mars.

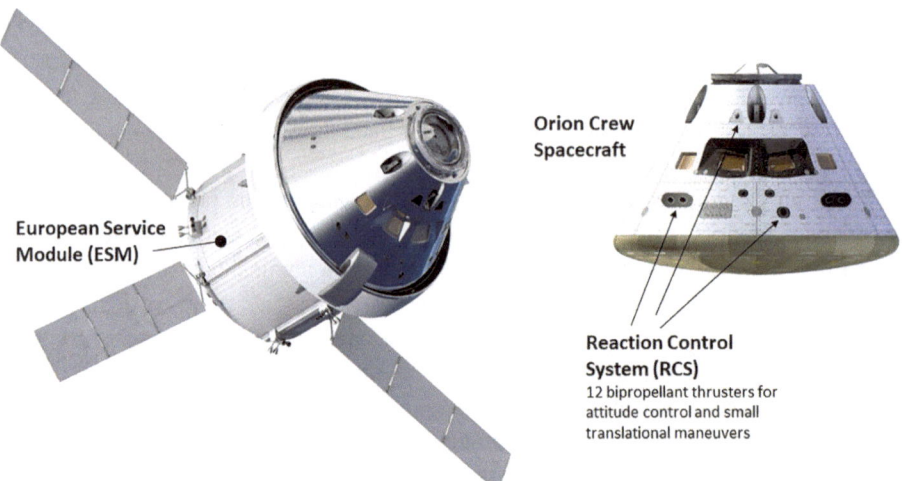

Fig. 1.7 Artistic depicting of the Orion crew spacecraft. Attached directly below the crew module is the ESM to provide propulsion, power, thermal control, and water and air for four astronauts. The solar array spans 19 m, designed to provide enough energy to power the mission. *Credit* NASA/ESA

Did you know? **Artemis I was the first NASA mission to use Wi-Fi in lunar orbit.** During the SLS launch countdown, the flight control team at NASA's Johnson Space Center in Houston activated a set of cameras looking back at Orion from the wingtips of the solar arrays. The cameras sent photos over Wi-Fi®, even as Orion lost communication with Earth as it voyaged behind the Moon.

1.5   Spaceflight

We define spaceflight as motion into or through outer space; that is, when a vehicle travels beyond our planet's atmosphere. It can be as short as access to low Earth orbit, a three-day trip to the Moon, or as long as a voyage across the planets and beyond. Spaceflight involves a journey either by automated or robotic spacecraft or voyages by people. For the latter, I will adopt the term *human spaceflight*.

Traveling in outer space is a rather complex endeavor. The gravitational fields of the Sun and the planets dominate interplanetary space within our Solar System. The Earth and all other objects in space move in unseen yet well-defined orbital paths and are subject to gravitational forces.

The first phase of spaceflight is the launch of the spacecraft. To overcome the grip of Earth's gravity, we need a powerful propulsion system producing the required force to ensure the vehicle lifts off the ground and reaches orbit. If the propulsion is not powerful enough, then the vehicle cannot escape the gravitational field and falls back to the ground. On the other hand, if the rocket thrust allows the launcher to depart with sufficiently high speed, the vehicle can escape the gravity of the planet and continue its travel forever, unless it is slowed down by its own propulsion, or by the effect of gravity of a large body (such as a planet or moon) on its path.

The vehicle must be launched at the right velocity, which we call the "escape velocity," and intersect a determined orbit. With the equations given in Ch. 4, you will find the escape velocity, denoted v_{esc}, for a mission to low Earth orbit is 9.7 km/s or 34,920 km/h. That is almost forty times the cruise speed of a commercial jet airplane flying within the atmosphere.

In outer space, once a spacecraft is set in motion it will continue in this state of motion unless it fires its propulsion system to change it. This requires that the spacecraft carry sufficient propellant (source of energy) on board to accomplish any maneuver to accelerate, decelerate or stop.

Another characteristic of spaceflight is that the energy required to move depends on the direction of motion. For a spacecraft to escape the Solar System, moving in the opposite direction in which the Earth orbits around the Sun, it would require a velocity more than four times as high as if moved in the same direction as our planet moves. The Earth rotates and moves at an average speed of 29.8 km/s (18.5 miles per second) around the Sun. Moreover, changes of direction in space require enormous amounts of energy. Hence, the direction of a space voyage must be chosen carefully. A launch to LEO from the U.S.A. can be made from the eastern launch site to take advantage of the Earth's rotational velocity, and to provide the minimum orbit inclination. Orbit inclination is the angle between a reference plane and another plane or axis of direction.

In a typical launch to geosynchronous transfer orbit (GTO), the acceleration is done to place the spacecraft on a 280-km circular parking orbit with speed of 7.737 km/s. A parking orbit is a temporary, intermediate orbit, not a final destination, where a spacecraft stays until ready to transfer to another orbit or prepare for an escape trajectory. For the eastern launch, the vehicle contributes about 7.329 km/s, and the rotation of the Earth contributes 408 m/s.

For interplanetary spaceflight, both our point of departure and our destination are large bodies moving at different and very high speeds around the Sun. This makes navigation rather challenging, as we must know where the bodies are with respect to each other when we begin the voyage and where our destination will be so that we know the type of transfer orbit to get there. This requires we know the motion of the planet or moon we are trying to reach. The spacecraft must adapt its velocity to the velocity of the target planet or moon. And it does not matter whether the spacecraft is moving slower or faster than the destination because in either case we must accelerate or decelerate and use the same amount of energy. Hence, all interplanetary spaceflight will require twice as much energy: the energy to put the spacecraft in motion, and the energy required to stop it.

A spacecraft going on a mission from our planet to the Moon needs an escape velocity of about 11.2 km/s (40,320 km/h) to leave Earth's gravitational field. The return flight requires less rocket power, as the escape velocity from the Moon is just 2.4 km/s (8640 km/h). This is because the Moon is smaller (about a quarter the diameter) and less massive (about one percent the mass) than the Earth. Objects that move at speeds below 0.71 times the speed of escape ($v < 0.71v_{esc}$) cannot reach a stable orbit. At a speed equal to 0.71 times the speed of escape ($v = 0.71v_{esc}$), the orbit is circular, and at a greater speed, the orbit becomes an ellipse until it reaches the speed of escape and then the orbit becomes a parabola. If a vehicle reaches Sun escape velocity, it will leave the Solar System and follow a trajectory in interstellar space.

1.6   Spacecraft Orbits

Reaching space from Earth means getting to an altitude above the atmosphere and intersecting a stable planetary or lunar orbit. These orbits can be classified in many ways. They can be classified according to (1) the central body, e.g. geocentric, heliocentric, selenocentric (circling the Moon); (2) altitude (for geocentric orbits); (3) inclination; (4) eccentricity; and (5) synchronicity.

Since most of the routine space operations concentrate on Earth orbits, we will place especial emphasis on propulsion designed to reach geocentric orbits. A heliocentric orbit (also called circumsolar orbit) is an orbit around the barycenter of the Solar System, located within or very near the surface of the Sun. In addition to all planets, comets, and asteroids in the Solar System, spacecraft can also follow heliocentric orbits. The Kepler telescope that searches for planets outside the Solar System (exoplanets) was placed on a heliocentric Earth-trailing orbit (HETO) with a period of 372.5 days, as this orbit provides the optimum approach to meeting the scientific objectives of the planet hunter mission.

1.6.1 Low, High Earth Orbits, and Geosynchronous Transfer Orbits

Earth-centered or geocentric orbits can be classified according to their altitude: (1) Low Earth orbit (LEO) that extends from about 100 km (below which satellites cannot remain in orbit) to about 2000 km (1240 miles) (above which the intensity of the trapped radiation belts make it more difficult for satellites to operate); (2) Medium Earth orbit (MEO) ranges in altitude from 2000 km (1200 mi) to 35,786 km (22,236 mi); (3) High Earth orbit (HEO) any orbit higher than 35,786 km (22,236 mi). A Very Low Earth Orbit (VLEO) is one at an altitude of approximately 300–350 km, an orbit of interest for geophysics, as it allows a satellite for precise imaging of the Earth. Given the rapid orbital decay of objects below 200 km, the commonly accepted definition for LEO is between 160 and 2000 km (100–1240 miles) above the Earth's mean sea level. At 160 km, one revolution takes approximately 90 min, and the circular orbital speed is 7.808 km/s (17,466 miles per hour or 25,617 ft/s).

The lowest LEO working orbit for long term satellites is at approximately 260 km, as below that altitude spacecraft experience so much atmospheric drag (force that opposes motion through a fluid) that they require continuous propulsion to overcome the retarding force.

The ISS maintains an orbit with an altitude ranging between 330 and 435 km (205 and 270 mi) above Earth's surface, and the Chinese Tiangong-2 Space Laboratory orbits at a nominal altitude of 370 km. Starlink satellites orbit at a nominal altitude of 550 km (342 mi), while the Hubble Space Telescope orbits at a nominal altitude of 596 km. A low Earth orbit can also be used as a first stop for spacecraft on missions to other destinations, in which case LEO is known as a parking or transfer orbit.

Geosynchronous or geostationary orbit (GEO) is a prograde orbit around Earth's equator (low inclination) with a period of 23 h 56 min 4 s. Communications satellites covering large, specific regions are often placed into GEO because the speed required to keep an object in this orbit matches the speed of Earth's rotation, so in effect satellites appear to remain fixed above a single location. To achieve GEO, a spacecraft is first launched into a geosynchronous transfer orbit (GTO). The standard GTO is 185 km (100 mi) perigee

by 35,786 km (19,323 mi) apogee, with inclination of 27 degrees. At GTO apogee altitude, the spacecraft uses its own propulsion system to circularize its orbit to attain GEO.
Perigee is the point in the orbit nearest to the Earth, while apogee is the farthest point in
the orbit.

A high Earth orbit (HEO) has an altitude entirely above that of a geosynchronous orbit,
with a radius greater than 35,786 km. This high-altitude orbit is farther away from Earth's
radiation belts, with small gravity gradient effects, and it requires modest launch vehicle
and spacecraft propulsion requirements (using lunar gravity assist). A highly elliptical
orbit is a HEO of high eccentricity with a low-altitude perigee, often under 1000 km (540
mi) and a high-altitude apogee, over 35,786 km (19,323 mi).

There is an intermediate orbit with a 12-h period also known as *half-geosynchronous*.
The Global Positioning System (GPS), an array of satellites specifically designed for the
GPS navigation system, is placed in this orbital regime. Intermediate orbits avoid the
dangerous inner radiation belt; however, being deeper in the outer belt than geostationary
satellites, spacecraft in those intermediate orbits experience a substantial higher electron
flux.

A Molniya orbit, used by the Russian Molniya ("lightning") communications satellites,
is a highly elliptical orbit with an inclination of 63.4 degrees and an orbital period of
precisely one half of a sidereal day. A satellite placed in a Molniya orbit spends most of
its time over a designated area of the Earth due to the apogee dwell.

A polar orbit is an orbit that passes over both poles of a planet. Geocentric polar orbits
are 90-degree inclination orbits, meaning that the orbit is at 90-degrees to the plane of
the equator. Since the orbital plane is nominally fixed in space, the Earth rotates below
a polar orbit, allowing the spacecraft low-altitude access to virtually every point on the
surface below. Thus, during a 12-h day a satellite in a polar orbit can observe all points
on the planet, making it useful for mapping or surveillance operations. To achieve a polar
orbit requires more energy (more propellant) than is needed for a low inclination orbit.
This is because a launch vehicle cannot take advantage of the "free ride" provided by
the planet's rotation, and therefore the vehicle must provide all the energy for attaining
orbital speed.

Keep in mind—For geocentric orbits, the radial coordinate r equal to the sum
of the Earth's radius R_E and the altitude above the Earth's mean surface h; thus,
$r = R_E + h$, where the mean equatorial radius is $R_E = 6378$km. The equation that
relates the distance of a satellite of mass m from the Earth's center, r, to its speed,
v, in a circular orbit is given by the equality of Newton's second law and the law
of universal gravitation written as:

$$G\frac{mm_E}{r^2} = \frac{mv^2}{r}$$

where m_E is the mass of the Earth, and G is the universal gravitational constant with value $G = 6.674 \times 10^{-11} \text{N} \cdot \text{m}^2 \cdot \text{kg}^{-2}$ in the SI System.

1.7 Astronautics and Space Exploration

Having the rocket power to place satellites in orbit made it possible to develop space telescopes that revolutionized astronomy. With orbiting telescopes, scientists can peer deeper into the cosmos without the distortion caused by the Earth's atmosphere that, until 1990, significantly blurred the images of distant objects—this is why stars appear to twinkle to the human eye. NASA's Great Observatories together with automated spacecraft and space probes changed our perspective of the Universe.

The Great Observatories are four space-borne telescopes designed to conduct astronomical studies over many different wavelengths (visible, gamma rays, X-rays, and infrared). They overlap the operation phases of the missions to enable astronomers to make contemporaneous observations of an object at different spectral wavelengths. The four observatories are the Hubble Space Telescope, the Compton Gamma Ray Observatory, the Chandra X-ray Observatory, and the Spitzer Space Telescope. Except for the Compton (safely deorbited on 4 June 2000), all other observatories are still in operation today.

The Hubble Space Telescope (HST), perhaps the better-known observatory in orbit, was deployed by a NASA Space Shuttle in 1990. From a nominal altitude of 560 km over the surface of the Earth, the Hubble has made celestial observations and detailed measurements of unprecedented scientific value, revealing the breathtaking beauty hidden in the depths of the cosmos. Moving in orbit around the Earth at 28,163.52 km/h (17,500 mph), the Hubble has made more than a million observations and snapped over 570,000 astonishing gorgeous images of thousands of celestial objects in its twenty-five years of operation.

Launched on Christmas Day 2021, the James Webb Space Telescope (JWST) orbits around the SEL_2, the second Euler-Lagrange equilibrium point of the Sun-Earth system situated at 1.5 million km away from Earth, a distance beyond the Moon itself. This far away location allows the Webb to stay in line with Earth as it moves around Sun. Developed by NASA and launched on Ariane 5, ESA's launch vehicle, the JWST can peer through clouds of gas and dust, using infrared imaging to detect and identify the first galaxies to form in the Universe, to trace the dynamics of galaxies, and to study stellar and planetary system formation. Figure 1.8 illustrates the relative positions of the Hubble's LEO orbit, and the Webb's heliocentric orbit. Orbiting at SEL_2 allows the JWST to stay in line with the Earth as it moves around the Sun, ensuring its sunshield protects the telescope from the light and heat of the Sun and Earth (and Moon).

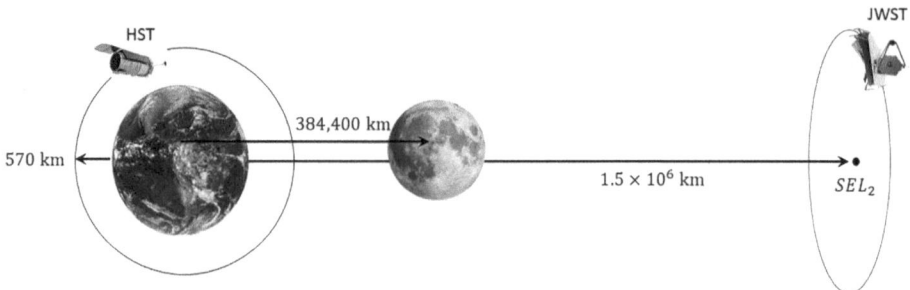

Fig. 1.8 Cislunar space. HST on its LEO orbit 570 km over Earth's surface, and JWST orbits SEL$_2$, the second equilibrium point on the Sun-Earth system at 1.5 million kilometers from Earth (not to scale)

> ***Did you know?*** On 25 December 2021, the James Webb Space Telescope was launched on an Ariane 5 two-stage SLV. After orbit injection, Webb travelled about a month to reach the second Sun-Earth equilibrium point SEL$_2$, 1.5×10^6km (940,000 miles) from Earth. In this orbit, Webb stays in line with Earth as it moves around Sun. A large sunshield protects telescope from light and heat emanating from Sun, Earth, and Moon.

Many spacecraft have been sent on missions to explore the Solar System. Robots have journeyed across interplanetary space, going to survey the planets and their moons, to comets and asteroids, and have probed our mother star. Space robots have gone where no human being has gone before and revealed many secrets of the Universe.

A probe is a robotic spacecraft that travels through space to examine phenomena or explore alien environments and collect science information. These probes are designed to send data back to Earth so that we learn about their findings. Spacecraft designed for missions to planets or moons in the Solar System can be orbiters, landers, or rovers, depending on whether the probes are required to observe or explore directly the surface of a body. A flyby spacecraft passes near a planet or moon, it images the landscape and makes observations of temperature, radiation, and other physical features of the planet's environment. An orbiter is a probe that moves around a planet or moon and performs similar observations as a flyby probe but its mission is much longer by remaining in the same orbit. NASA's Lunar Crater Observation and Sensing Satellite (LCROSS) is an example of an orbiter space probe that took data of the Moon and confirmed that there is water in the southern lunar crater Cabeus.

Space probes have been sent to study characteristics of planets, moons, asteroids and even to collect samples of comets. Since 1970, a variety of robotic spacecraft have been launched from Earth on missions to land on planets, moons, comets, and asteroids. These include several highly successful probes have been sent to the farthest regions of our Solar System, including the twin Pioneer, the twin Voyager, and New Horizons.

Many missions have launched spacecraft to orbit other bodies in the Solar System, including the Moon and Mars. At least 18 spacecraft have operated in Mars' orbit, including NASA's Odyssey (the first to orbit another planet), ESA's Mars Express (acting as relay for NASA rovers), and more recently the Emirates Mars Mission (Hope), China's Tianwen 1, and India's Mars Orbiter Mission (MOM), also called Mangalyaan, launched in 2013 and orbiting Mars since 24 September 2014. In 2022, contact was lost with MOM, and so 7 orbiters remain active. On 29 July 2015, the Mars Reconnaissance Orbiter was placed into a new orbit to provide communications support during the arrival of the InSight Mars lander in 2016.

NASA's Perseverance rover is part of the recent exploration of Mars. On 18 February 2021, the rover made its harrowing landing at Jezero Crater. Almost immediately, it began an expedition to collect a geologically diverse set of rock samples that could help answer the question if Mars once had ancient microbial life. Since February 2021, Perseverance has identified those rocks, cored, sealed them, and set several sample tubes down on the crater floor for a future mission to pick up and return to Earth for further study.

In 2020, NASA issued a call for information with the Lunar Terrain Vehicle Services (LTVS) to support the Artemis program. In response, General Motors in partnership with Lockheed Martin designed a prototype autonomous rover called Lunar Mobility Vehicle (LMV). The requirements for the next-generation vehicle include longer range and have more powerful batteries than previous rovers, as it must carry astronauts to the Moon's south pole, which has more difficult terrain and is colder and darker. For 14 days, the temperature is about 121 °C (250°F) and is very bright. Then it is dark for 14 days and the temperature drops to about -183 °C (-298°F). Thus, the lunar rover must survive those extreme conditions, hibernating or going dormant, and then awakening on command. Another challenge is the lunar dust. The abrasive lunar regolith is electromagnetic and it adheres to surfaces, wheels, and can damage the equipment.

Other private companies are also pursuing design of autonomous vehicles to facilitate more comprehensive, long-term research on the Moon. Several countries have already attempted to land their robotic spacecraft. In 2019, China landed a vehicle on the far side of the Moon. In April 2023, the Japanese company Ispace attempted landing of its Hakuto-R spacecraft but failed.

Did you know? China landed uncrewed probes on the far side of the Moon, first in 2019, and in 2024. With Chang'e-6 mission, a robot landed on a crater close to the Moon's south pole. Chang'e-6 automatic capsule returned rare lunar rocks, specimens that will help scientists answer key questions about how planets are formed.

A flyby spacecraft is one that enters a planet's gravitational sphere of influence and does not impact the planet or go into orbit around it. The sphere of influence (SOI) is the region around every celestial body that controls its gravitational influence on smaller bodies near them, a concept very useful to define spacecraft interplanetary trajectories. When there is no impact and no drop into a capture orbit around the planet, then the spacecraft will simply continue past periapse (the closest point in the orbit to the orbited body) on a flyby trajectory, exiting the sphere of influence with the same relative speed it entered, but with the velocity vector rotated through the turn angle. Flyby missions can be conceived to make crucial observations when a probe goes past a planet or moon while exploring large regions of the Solar System, or simply to use the gravitational energy of celestial bodies to change the trajectory of a spacecraft without using onboard propellant. The gravity assist flyby technique can add or subtract momentum to increase or decrease the energy of a spacecraft's orbit without using rocket propellant. This technique was first demonstrated with NASA's Mariner 10 Venus/Mercury mission in 1973–74, and then successfully adapted for the missions of discovery that revealed unknown characteristics of the outer giant planets.

Launched in 1972, Pioneer 10 was the first flyby probe to study Jupiter and its environment. Jupiter is the fifth planet from the Sun, 715 million km (444 million miles) away from the Earth (on average), and it is the largest in the Solar System. After launch, Pioneer 10 reached a speed of 14.48 km/s (52,140 km/h; 32,400 mph) needed for the flight to Jupiter, making it the first fastest human-made probe to leave the Earth's sphere of influence; it was fast enough to cross Mars' orbit in just 12 weeks. On 15 July 1972, Pioneer 10 entered the asteroid belt. This is a wide region located between the orbits of Mars and Jupiter that contains millions of asteroids ranging widely in size from about 940 km in diameter (one-quarter the diameter of our Moon) to bodies that are less than 1 km across. Pioneer 10 reached its destination on 3 December 1973 and took the first images of Jupiter.

In April 1973, NASA launched the twin Pioneer 11, the first spacecraft to fly by Saturn. After observing its rings, and moons, Pioneer 11 acquired additional gravitational energy from Jupiter, and then followed an escape trajectory from the Solar System. The gravity-assist maneuver gave the spacecraft a small fraction of the planet's orbital energy to help it change direction and speed. When the last signal was received in 1995, the robot was 44.7 AU from the Sun. Pioneer 11 is now headed toward the constellation Aquila (The Eagle). If it does not disintegrate by age, cosmic dust, or other interstellar effects, it would take about 4 million years for Pioneer 11 to reach the nearest stars in the Sagittarius constellation.

On 22 January 2003, Pioneer 10 sent its last signal, a final farewell before leaving us forever after a 30-year travel through the Solar System. When the last, very weak signal was received, the robot was about 82 times the nominal distance between the Sun and the Earth. At that distance, it takes approximately 11 h and 20 min for a radio signal to reach the Earth. The spaceship now drifts away in interstellar space, heading in the direction of the red star Aldebaran in the constellation Taurus (the Bull). Aldebaran is about 68 light-years away. To reach that star, Pioneer 10 would have to maintain its current speed for more than two million years!

Two other twin spacecraft named Voyager 1 and 2 were launched in 1977 on a mission to explore the outer Solar System. Voyager 2 was launched first, on 20 August 1977, and 16 days later (5 September), Voyager 1 followed, launched on a faster, shorter trajectory to reach Jupiter and Saturn, ahead of Voyager 2. The Voyager mission was conceived to explore four planets, benefiting from the unique orbital arrangement of the outer giant planets, which would occur in the late 1970s and the 1980s (and would not repeat for another 175 years). With Jupiter, Saturn, Uranus, and Neptune positioned in such manner would allow a spacecraft to swing from one planet to the next, using gravity assist to increase its velocity with a minimum of onboard propellant and visit all four planets in much less time. For example, using flyby maneuvers the flight time to Neptune was reduced from 30 years to 12.

Each Voyager spacecraft at launch consisted of a mission module (planetary vehicle) and a propulsion module, which provided the final energy increment to inject the mission module onto the Jupiter trajectory. The propulsion module, with its large solid-propellant rocket motor, weighed 1207 kg (2660 lbm); it was jettisoned after the required velocity was attained. For attitude stabilization and for trajectory correction maneuvers, the Voyager included 16 hydrazine (N_2H_4) mono-propellant thrusters. The 16 thrusters on the mission module each delivered 0.889 N (0.2 lbf) thrust. Four were used to execute trajectory correction maneuvers; the others, in two redundant six-thruster branches, were used to stabilize the spacecraft on three axes. Only one branch of attitude control thrusters was needed at any time.

Did you know? **Voyager 1 is the most distant spacecraft**, about 17.5 billion kilometers away from the Sun at a northward angle. Pioneer 10, the next most distant, is about 15.4 billion kilometers away from the Sun on the opposite side of the solar system. Voyager 2 is about 14.2 billion kilometers away from the Sun on a southward trajectory, on the same side of the Solar System as Voyager 1. Pioneer 11 is about 12.4 billion kilometers away from the Sun. New Horizons is about 3 billion kilometers away from the Sun, on its way to Pluto.

The Pioneer 10, Pioneer 11, Voyager 1, and Voyager 2 are the first spacecraft that went past the edge of the Solar System, moving into interstellar space. It took 33 years for Voyager 1 to reach the end of our Solar System. Both Voyager have enough energy to continue operating until the available electrical power will no longer support science instrument operation. At this time science data return and spacecraft operations will end.

Although no longer communicating with Earth, both Pioneer 10 and 11 move on inter-stellar trajectories, each carrying a return map should any intelligent life ever intercept and send them back.

New Horizons is the first flyby space probe designed to study Pluto, its moon Charon, and objects in the Kuiper Belt. On 19 January 2006, New Horizons was launched directly into an Earth-and-solar escape trajectory with a speed of about 16.26 km/s (58,536 km/h; 36,373 mph). After a brief encounter with asteroid 132,524 APL, New Horizons proceeded to Jupiter, making its closest approach, 2.3×10^6 km (1.4 million miles), on 28 February 2007. This Jupiter flyby provided a gravity assist that increased the spacecraft's speed by 4 km/s (14,000 km/h; 9000 mph). On 15 January 2015, New Horizons began its approach phase to Pluto, and on 14 July it was 12,500 km (7800 mi) above the surface, making it the first spacecraft to explore the dwarf planet and its moon Charon. It came as close as 28,800 km (17,900 mi) to Charon. New Horizons is the first spacecraft to explore a second Kuiper Belt Object up close.

All space robots accomplish their missions of exploration in the Solar System, thanks to the rocket propulsion systems that were engineered so beautifully and accurately by humans. More recently, one of the most anticipated missions is NASA's Europa Clipper, launched in October 2024 to search for alien life on Jupiter's moon Europa. The spacecraft will travel 2.9 billion km (1.8 billion miles) to reach Jupiter in April 2030. What Europa Clipper will find on Jupiter's moon could change what we know about life in our Solar System.

1.8   The Birth of Astronautics: A Brief Historical Perspective

Astronautics was born in 1958 with the launch of the first artificial satellite, but the ideas of spaceflight may be as ancient as humanity. From the beginning of our history, human beings have raised their eyes skywards, awestruck by the exquisite and mysterious beauty of the stars, trying to make sense of what they could see but could not touch. Through the ages, spaceflight must have been a dream of many, igniting ideas to make it a reality. In the twentieth century, at last, the first humans soared high, strapped to thundering rocket vehicles that left the confines of Earth's atmosphere and reached outer space; a few privileged astronauts made it to the Moon.

Reading the fable of Daedalus and Icarus in Greek mythology one contemplates an inherent yearning for human spaceflight. Through the centuries, other thrilling tales took people's imagination much farther. The first science fiction stories written by Lucian of Samosata in the second century A.D. considered the notion of humans traveling to the Moon. In the nineteenth century, Jules Verne wrote his famous space sagas, making readers wonder and dreaming of going to the Moon. By then, the scientific theories that led to spaceflight had been laid by Copernicus, Kepler, and Newton.

The scientific foundation of spaceflight began to take shape in the sixteenth century when modern astronomy was born. First, Polish astronomer Nicholas Copernicus determined that the Sun is the center of our planetary system. Copernicus heliocentric model, in which the Earth and the planets move around the Sun, began an unprecedented scientific revolution. Scholars continued asking more probing questions, including German astronomer Johannes Kepler who wondered if the Sun exerted a force on the planets and caused them to move. That was before the telescope was invented, and so Kepler was only aware of five planets besides the Earth—Mercury, Venus, Mars, Jupiter, and Saturn—which were known since ancient times.

Kepler discovered that the planetary orbits were not circles, as astronomers had believed since the time of Ptolemy. Kepler determined that the paths followed by the planets were ellipses and that the Sun was at a focus of the elliptical orbits. Kepler also ascertained that the velocity of a planet in its orbit increased or decreased while moving closer or farther from the Sun. Finally, Kepler found how the velocities of the planets are related. These observations became Kepler's laws of planetary motion, which described the motion in the entire Solar System and paved the way for Newton's work.

Isaac Newton discovered that all motion in the Universe obeys three principal laws. He also found that the force the Sun exerts on the planets is the same force of gravity that keeps us firmly on the ground. In 1687, Newton published, *Philosophiae Naturalis Principia Mathematica*, Latin for "Mathematical Principles of Natural Philosophy," known simply as the *Principia*. In this monumental book, Newton presented a mathematical model of the world, stating the physical laws that govern the motion of all bodies on Earth, and the motion of the distant planets.

Newton described gravity with the law of universal gravitation, asserting that gravity is a force that should behave in similar ways regardless of where we are. Newton realized that the gravitational force accounts for falling bodies on Earth as well as the motion of the Moon and the planets in orbit. This was a revolutionary conclusion, as it extended the influence of earthly behavior to the realm of the heavens. With the three laws of motion and universal gravitational law, Newton laid the ground for classical mechanics.

Others after Newton, especially Euler, Lagrange, and Laplace, developed the mathematics and scientific principles of Newtonian mechanics to grow the branches of physics that were combined to establish the foundation of astronautics, including the operating principle of rocket propulsion.

In the twentieth century, many rocket scientists, propulsion engineers, mathematicians, astronomers, scientists, and physicists combined their work and vision to transform our world. Today, we attribute the progress and spectacular achievements in space exploration to their efforts, so many to name here, bright individuals who applied scientific principles, developed the engineering tools and the technologies to make spaceflight possible.

1.9   Pioneering Rocket Science and Aerospace/Astronautics Engineering

The turn of the twentieth century ushered a new revolutionary era for humankind, one that would take people to the Moon and would give everyone on Earth a closer look of the stars.

Practical *aeronautics* and theoretical *astronautics* were born in the same year. In 1903, while the American Wright brothers achieved the first powered flight through the air in the first aircraft of their own design, Konstantin E. Tsiolkovsky, a Russian school teacher, conceived clear ideas for spaceflight.

When Tsiolkovsky published "The Exploration of Cosmic Space by Means of Reaction Devices" he considered rockets to propel a spacecraft. Tsiolkovsky calculated the speed required for a minimal orbit around the Earth and predicted that it could be achieved with a multistage rocket fueled by liquid propellants. However, Tsiolkovsky never built or experimented with rockets. The first rockets for space exploration were built years later by America's foremost rocket engineer, Dr. Robert Goddard. In 1912, Goddard began to study solid propellant rockets, and in 1919, he published "A Method of Reaching Extreme Altitude" to explain how scientific instruments could be sent into the stratosphere using a rocket-powered vehicle. In 1926 Goddard built the world's first liquid propellant rocket.

In Germany, Hermann Oberth was thinking along the same lines. In 1923, when his doctoral dissertation on spaceflight was rejected at the University of Heidelberg, Oberth published a 92-page article entitled "The Rocket into Interplanetary Space." Six years later, he wrote a book on the same topic. Oberth's work undoubtedly inspired many of his compatriots, some of which worked for the German army at Kummersdorf near Berlin. Among those rocket scientists was Werner von Braun, one of the developers of the first American space launchers. In 1937, engineers led by von Braun began research to perfect the liquid-propellant rocket. Eventually they built the V-2, the world's first long range ballistic missile, launched in 1946 for the first time at White Sands, and finally was used with devastating consequences during the end of World War II.

France also had its spaceflight pioneer. Robert Esnault-Pelterie was an engineer, aviator, aircraft designer, and spaceflight theorist. He became interested in space exploration and studied rockets. On 8 June 1927, Esnault-Pelterie gave a talk before members of the French Astronautics Society on "The Exploration of the Very High Atmosphere by Rockets and the Possibility of Interplanetary Travel." In this talk, published a year later, Esnault-Pelterie considered the exploration of outer space using chemical rocket propulsion. He also discussed technical issues related to the thrust, guidance and control of a vehicle that could be built for interplanetary travel. In 1927, Esnault-Pelterie coined the term "astronautics" (in French *astronautique*), a lovely word that evokes the allure of spaceflight which means—literally—navigating among the stars.

In all parts of the world people were fascinated by the possibilities of outer space. Thus, it is not surprising to discover a rocket pioneer in Latin America. Between 1946 and 1970,

Ricardo Dyrgalla, a developer of liquid and solid propellant rockets, helped Argentina and Brazil to establish their own rocket programs. Ricardo Dyrgalla, whose birth name was Ryszard Dyrgalla, was born in Poland in 1910, but immigrated to Argentina in 1946 upon accepting a job offer from the Argentinean Army. He proposed the development of an aerial-launched vehicle powered with a liquid rocket engine. The Tábano vehicle—named in honor of Teófilo Tabanera, the founder of the Argentine Interplanetary Society (AIS) and member of the International Astronautical Federation (IAF)—was built and flight-tested in 1950 with a glider prototype, and later with the rocket engine installed. Dyrgalla also developed the Prosón, a solid-propellant rocket intended for meteorological research.

The flight of that first rocket built by Goddard became pivotal in the development of liquid propellant rocket propulsion, as it established the basic elements of engineering propulsion integration that we find in today's rocket engines. As pointed out by historians, Goddard's 1926 rocket design was based on in-depth analytical techniques, which he and his team verified thorough methodical testing before flight. It would take another three decades for a powerful launch system to be ready.

In the mid 1940s and 1950s, Earth-to-orbit rocket propulsion was under intense development in the Soviet Union and in the United States. In 1950, the first high-thrust American engine was developed by North American Aviation (NAA), originally known as the XLR43-NA-1 and then evolved into the Redstone engine, which generated a thrust of 75,000 lbf. In 1955, a division of NAA became Rocketdyne to focus on developing rocket engines fueled with kerosene. Meanwhile, Aerojet began developing engines fueled with hypergolic propellants.

By the late 1950's, the former Soviet Union had developed many different engines, including the LOX-Kerosene RD-107 and RD-108 engines, under the leadership of Valentin Glushko, a brilliant engineer who became program manager of the Soviet space program (1974 until 1989). The RD-107 engines were used as boosters and the RD-108 in the central core of the R-7, the first launch vehicle to reach orbit designed by Sergei Korolev. The R-7 vehicle was 34 m (112 ft) long, had a diameter of 10.3 m (34 ft), and weighed 280 metric tons. Korolev was the lead Soviet rocket engineer during the competition to reach space that developed between the United States and the Soviet Union.

The National Aeronautics and Space Administration (NASA) was established in 1958, with a distinctly civilian (rather than military) orientation encouraging peaceful applications in space science. That same year, Pratt & Whitney began to develop the RL10, the world's first hydrogen fueled rocket engine (see Fig. 1.2). It was first ground tested in 1959, and its first successful flight was in 1963, when a pair of RL10s boosted a vehicle into orbit around the Earth. The RL10 became the rocket engine for NASA's upper-stage Centaur space launch vehicle, and through the last decades it has played a vital role in placing hundreds of government and commercial spacecraft launched into LEO and beyond. As an upper stage, the RL10 has helped send spacecraft to explore every planet in our Solar System, including Voyager 1 and Voyager 2, the first two spacecraft to reach interstellar space.

In 1959, Rocketdyne began work on the F-1 engine fueled with RP-1. Producing a thrust of 7770 kN (1,746,000 lbf), the F-1 was selected by NASA to power the first booster stage of the three-stage Saturn V launch vehicle. In 1962, Rocketdyne began development of the J-2 engine, burning cryogenic liquid hydrogen (LH$_2$) and liquid oxygen (LOX) propellants. The J-2 engine produced 1033.1 kN of thrust in vacuum and was used as second stage on NASA's Saturn IB launch vehicle.

A Heavy Lift Vehicle, Saturn V had a propulsion system comprised of five F-1 LOX/ RP-1 engines clustered in its first stage, five J-2 LOX/LH$_2$ engines powering the second stage, and one J-2 engine for the third stage, as depicted in Fig. 1.9. Saturn V was the most powerful launch vehicle ever developed by NASA in the 1960s, conceived under the Apollo program for taking humans to the Moon. As part of the Apollo program, the first flight of Saturn V was on 9 November 1967 without a crew to test its capability. The first Saturn V carrying a crew was Apollo 8 on a mission to only orbit the Moon. The first mission to land astronauts on the lunar surface was Apollo 11, which took place on 16 July 1969. The Saturn V also launched astronauts to land on the Moon as part of the Apollo 12, 14, 15, 16 and 17 missions.

The Space Age was born with the launch of the first Sputnik in 1957. The Sputnik 1 was the first artificial satellite to orbit Earth, launched by the Soviet Union using a derivative from the R-7 vehicle powered by the RD-108/107 rockets. The R-7 was a single stage vehicle comprised of a central sustainer engine (RD-108) as the central core, surrounded by four strap-on liquid rocket engines (RD-107) to boost the thrust at lift-off. This launch system was a modified Soviet missile, and in its adapted form it became the basis for the R-7 family of space launchers. The sustainer stage and the four boost

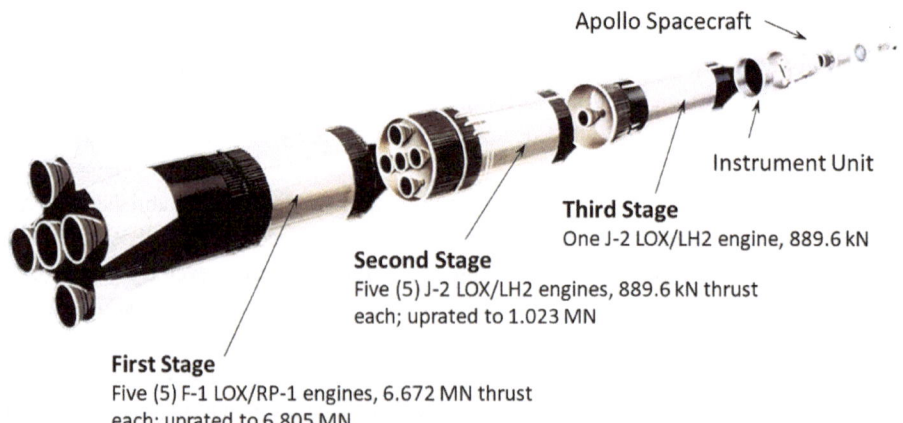

Apollo Spacecraft

Instrument Unit

Third Stage
One J-2 LOX/LH2 engine, 889.6 kN

Second Stage
Five (5) J-2 LOX/LH2 engines, 889.6 kN thrust
each; uprated to 1.023 MN

First Stage
Five (5) F-1 LOX/RP-1 engines, 6.672 MN thrust
each; uprated to 6.805 MN

Fig. 1.9 Saturn V launch vehicle developed by NASA in the 1960's to take humans to the Moon

engines used liquid oxygen (LOX) and kerosene, delivering a total lift-off thrust of more than 4.5 MN, to take its payload to an elliptical orbit with an apogee of 939 km and perigee of 215 km over Earth's surface. With a mass of 83.6 kg (183.9 lbm), Sputnik 1 was a 58.0 cm (22.8 in)-diameter aluminum sphere, having a primary function to place a radio transmitter into orbit.

Meanwhile in the U.S., von Braun turned the Atlas and Titan ICBMs into launch vehicles for human spaceflight and launching satellites. The Jupiter-C vehicle—a special modification of the U.S. Army's Redstone ballistic missile—was used to launch Explorer-1, America's first satellite. The same Redstone then launched the Project Mercury, the first program in the U.S. intended for sending astronauts to LEO, starting in 1958. Mercury included several uncrewed test flights. Of the six crewed missions, two were suborbital and four were orbital missions. The orbital missions were launched using Atlas vehicles.

The first men ever to reach space were Russian cosmonaut Yuri Gagarin, and American astronaut Alan Shepard. On 12 April 1961, Gagarin launched into orbit aboard Vostok 1, the first spacecraft to carry a human into space, and circled the Earth for 108 min. Shortly after, on 5 May 1961, Shepard performed a 15-min sub-orbital flight. Mercury MA-6 (Friendship 7) capsule was launched on 20 February 1962 carrying astronaut John Glenn. A year later, Russian Valentina Tereshkova became the first woman to reach orbit.

The first Moon visitors were American astronauts Neil Armstrong and Edwin (Buzz) Aldrin. They landed on the Moon in 1969 as part of the historical American mission Apollo 11. Shepard became the fifth man to walk on the Moon. The last lunar landing occurred in 1972 during the Apollo 17 mission. It was the eleventh crewed space mission in the NASA Apollo program, and the sixth and final lunar landing. During the American Apollo program, twenty-four astronauts left Earth's orbit and flew around the Moon (Apollo 7 and Apollo 9 just made it to LEO). Some of those courageous explorers landed on the powdery surface of the Moon, some drove a lunar rover, and each one collected soil and rock samples that were brought home for study.

The first human-rated SLVs were expendable and incorporated capsule-type spacecraft to transport the crew. Then in 1969, after the successful Apollo program, a paradigm change led to new proposals to transport astronauts to LEO. Following the successful Apollo Moon mission, Rocketdyne won a contract to design and build the main liquid propulsion engines for the Space Shuttle orbiter. The RS-25 became known as Space Shuttle Main Engine (SSME), the same now propelling the core stage of the SLS. When designed in the early 1970's, the reusable was the most advanced rocket engine in the world. At the time of development, the SSME had the highest specific impulse, and could be throttled between 60 and 109% of rated thrust.

The successful first flight of the NASA Space Transportation System (STS)—the Space Shuttle—on 12 April 1981 marked the beginning of a new era in human space activities for the United States, an era which lasted three decades. The STS program built a fleet of four vehicles named Columbia, Challenger, Discovery, and Atlantis. A fifth orbiter, Endeavour, was built to replace Challenger which was destroyed in an accident during launch which killed seven astronauts on 28 January 1986. Challenger exploded 73 s after launch because of the failure of a seal between two segments of an SRB. In 2003, Columbia disintegrated as it returned to Earth from a 16-day science mission. NASA resumed access to space missions with the remaining spaceflight-worthy Orbiters.

Used principally to transport astronauts to LEO, the Space Shuttle was crucial for launching the structures of the International Space Station (ISS) and helping its construction. The STS serviced the Mir (Russian space station in LEO from 1986 to 2001), and carried out some of the most important scientific missions for space exploration in history: on April 1990, the Space Shuttle Discovery (mission STS-31) deployed the Hubble Space Telescope (HST) to its LEO. Nine years later, the Space Shuttle Columbia carried the Chandra X-Ray Observatory to its orbit. It also performed numerous servicing missions including the repair the HST, and it deployed many satellites. The Space Shuttle was the first system to routinely reuse expensive spaceflight hardware. It was also the first system to retrieve hardware from space. The Space Shuttle achieved 133 successful missions to LEO; its last flight was on 21 July 2011.

Twenty-two years after the first man reached space, Sally Ride became the first American woman to fly into LEO. After Ride's historic flight aboard the Space Shuttle Challenger in 1983, many other women joined the ranks of space explorers.

During the Apollo days, the United States and the Soviet Union led the space race, developing the technologies for space exploration. In 1975, the establishment of the European Space Agency added its scientific and technological resources to the space superpowers. The Agenzia Spaziale Italiana (ASI) was founded in 1988 to coordinate all of Italy's efforts and has contributed significantly to space exploration by building scientific instruments that are aboard NASA and ESA probes bound for exploration of Mars, Jupiter, and Saturn.

The China National Space Administration (CNSA) was founded in 1993 to manage the People's Republic of China's space activities. In 2003, the CNSA launched their first crewed spacecraft named Shenzhou. A second successful piloted mission was carried out in 2005, and a third in 2008 with a crew of three. Chinese astronauts are known as yuhangyuans or taikonauts. The word *yuhangyuan* means space navigator, while the word taikonaut is derived from *taikong*, the Chinese word for space. During the three-day mission, the three taikonauts launched a small satellite and conducted their country's first spacewalk. China's space exploration program includes human missions to the Moon and the asteroids, and a robot mission to Mars will follow. China operated a space station called Tiangong 1 from 2011 to 2018.

In 2014, India's Mars Orbiter Mission successfully entered the orbit of the Red Planet. Launched by the Indian Space Research Organisation (ISRO), the Mars orbiter's primary objective is to develop the technologies required for designing, planning, management, and operations of an interplanetary mission. India became the first nation to arrive on its first attempt and the first Asian country to reach Mars.

Other countries, including Japan, Iran, Malaysia, and Turkey, have spaceflight programs and are developing their own launch capabilities. In Latin America several countries
have also established space exploration plans. México, Costa Rica, and Argentina have
national space agencies to carry out activities of space exploration. In the next decades
the number of nations capable of developing a space industry will increase, and the
exploration of space will accelerate. I believe that these capabilities will increase more
international cooperation and ease political conflicts among countries.

Astronautics is the science of spaceflight or space travel, concerned with the design,
building, and the operation of vehicles that travel through cislunar, interplanetary
or interstellar space. Astronautics literally means "navigating among the stars,"
whereas cosmonautics means navigation through the cosmos.

Both branches of engineering (aeronautics and astronautics) have developed at different
pace due in part to the required distinct propulsion systems to travel past the edge of
Earth's atmosphere and to escape our planet's gravitational field. Astronautical Engineering has evolved in the last seven decades as the interdisciplinary branch of engineering
that deals with spacecraft designed to move or work entirely beyond the Earth's atmosphere. In general, Aerospace Engineering addresses the design, construction of aircraft
and spacecraft, thus dealing with both aeronautics (aerodynamic flight, or flight through
atmospheric air) and astronautics (spaceflight). Astronautics requires specialized study.
The field developed in part to technological advances not foreseen in the early 1950s,
making space exploration part of academic education, and it accelerated progress in
observational astronomy to a remarkable extent. Areas of specialization include orbital
mechanics, spacecraft and launch vehicle design, satellite control and dynamics, space
physics, and propulsion.

Propulsion: Field of engineering focusing on propulsion systems required to propel
a vehicle to and in space. Propulsion includes chemical rockets, nuclear propulsion,
electric propulsion, solid propellant rockets, ion propulsion and others. Propulsion
also deals with the operations to launch spacecraft into space, rocket propellants,
orbit transfers and maneuvering spacecraft using reaction control jets and other
means of thrusting.

Today, whether considering human spaceflight or robotic exploration, the motivation,
and goals for designing new launch systems, rocket propulsion, and spacecraft remain
the same: explore the Moon, Mars, the outer planets, and survey the entire Universe to
expand our understanding of that vast mysterious cosmic world and define our place in

it. Advances in propulsion, controls, and structural lightweight composite materials, and many other technologies are changing the paradigm in launch system and rocket engine design. In the next chapters, we will study the fundamental concepts governing rocket propulsion and will be exposed to the designs and technologies that make possible the exploration of the Universe and human space travel.

Glossary

Earth-to-Orbit (ETO) Propulsion Any rocket engine that can power a launch vehicle intended to carry payloads from the surface of the Earth to a minimum stable parking orbit and for any upper-stage for transferring the payload to its operational mission location or inject it into a lunar or interplanetary trajectory. To date, only chemical rockets (liquid propellant rocket engines or combination of LREs with solid propellant rocket motors) have the high thrust per weight ratio to lift a launch vehicle from ETO.

Cold Gas Thruster The simplest type of rocket that relies on gas under pressure as its only source of thermodynamic energy.

LEO Low Earth Orbit. Any orbit found at altitudes above 100 km (to ensure satellites remain in orbit) and below 2000 km (since the intensity of the trapped radiation belts make it difficult for satellites to operate). A low LEO is a position reached by launch systems prior to orbital adjustments that are typically made using perigee kick motor (PKM) and apogee kick motor (AKM) propulsion.

Propellant The source of energy of a rocket engine, the stored matter that after being processed in the thrust chamber is then expulsed at high velocity. The propellant can be: (i) a cold gas expelled at high pressure, (ii) a hot gas that results from chemical reactions (combustion), or (iii) hot gas or plasma resulting from a thermonuclear reaction or by application of electromagnetic energy to matter.

Rocket Engine A self-contained device which ejects a small fraction of its propellant mass at high velocity in a determined direction, and in so doing produces a thrust reaction force intended to accelerate a vehicle to practical velocities. A rocket engine generates thrust by accelerating a high-pressure gas to supersonic velocities in a converging-diverging nozzle. In chemical rockets, the high-pressure gas is generated by high-temperature or chemical decomposition of propellants.

Solid Rocket Boosters (SRBs) Solid rocket motors that operate in parallel with the first or main stage liquid propellant engines for the initial segment of ascent flight to provide additional thrust needed for an SLV to escape Earth's gravitational pull.

Spacecraft A vehicle designed to move in space beyond the outer layers of Earth atmosphere. It can be a space probe, an orbiter, a space telescope, a space station, a crew or cargo capsule, or an interplanetary or interstellar vehicle.

Space Launch Vehicle (SLV) A vehicle powered by a carefully integrated multi-stage rocket system and associated propellant tanks designed to place payloads into orbit.

The propulsion system consists of powerful chemical rockets that include a combination of bipropellant main stage rockets and solid propellant rocket boosters. The thrust magnitude of a SLV at launch depends on overall vehicle mass, which in turn depends on structure mass, payload mass and mission.

Starlink Satellite network developed by SpaceX to provide low-cost internet to remote locations. Having a lifespan of 5 years, the current V2 Starlink satellite version weighs approximately 800 kg (1760 lbm) at launch. As of August 2024, there are 6350 Starlink satellites in orbit, of which 6290 are working. SpaceX eventually hopes to have as many as 42,000 satellites in this megaconstellation.

Thruster A low thrust (< 10 kN) propulsive device used by spacecraft for station-keeping maneuvers, attitude control, in the reaction control system (RCS) or long duration, low-thrust acceleration maneuvers. Thrusters may be fueled by monopropellants or by bipropellants.

Recommended Reading

1. Artemis III Science Definition Team Report, NASA, Dec. 7, 2020.
2. Griffin, M.D., French, J.R. (2004). *Space Vehicle Design*, Second Edition (AIAA Education) AIAA (American Institute of Aeronautics & Astronautics; 2 edition (January 1, 2004).
3. Creech, S., Guidi, J., and Elburn, D. (2022). "Artemis: An Overview of NASA's Activities to Return Humans to the Moon." IEEE Aerospace Conference, Big Sky, MT, 2022.
4. Gruntman, M. (2004). *Blazing the trail: the early history of spacecraft and rocketry*. Reston, Va.: American Institute of Aeronautics and Astronautics. ISBN 978-1-60086-872-6. OCLC 774285730.
5. Halchak, J.A., Cannon, J.L., and Brown, C. (2018). *Materials for Liquid Propulsion Systems*. Chapter 12 in Aerospace Materials and Applications, Ed. Bhat, B.N., AIAA Progress in Astronautics and Aeronautics, August 31, 2018. https://doi.org/10.2514/4.104893.
6. Hale, F.J. (1994). *Introduction into Space Flight*. Prentice-Hall. ISBN 0-13-481912-8.
7. Hammond, W.E. (2001). *Design Methodologies for Space Transportation Systems* (AIAA Education Series) (September 1, 2001).
8. Handbook of Space Technology (2011). Eds. W. Ley, K. Wittmann, W. Hallmann. John Wiley & Sons, Ltd.
9. Hunley, J. D. (2008). *US Space launch vehicle technology: Viking to space shuttle*, University press of Florida, 2008, ISBN 978-0-8130-3178-1.
10. Jenkins, D.R. (2001). *Space Shuttle: The History of the National Space Transportation System. The First 100 Missions*, 3rd Edition (May 11, 2001).
11. Johnson, N.L. (2012). "A new look at the GEO and near-GEO regimes: operations, disposals, and debris." Acta Astronaut. 80, 82–88 (2012).
12. NASA Space Launch System (SLS) Fact Sheets. https://www.nasa.gov/humans-in-space/space-launch-system/sls-fact-sheets/.
13. NASA Lunar Exploration Program Overview, Artemis Plan, September 2020 https://www.nasa.gov/sites/default/files/atoms/files/artemis_plan-20200921.pdf.
14. Nova Space, https://www.stokespace.com/.

15. Sackeim, R. L. (2003). "Spacecraft Chemical Propulsion." In Encyclopedia of Physical Science and Technology (Third Edition), 2003.
16. Schartz, W. T., Cannova, R. D., Cowley, R. T., and Evans, D. D. (1979). "Development and Flight Experience of the Voyager Propulsion System." AIAA Paper 79-1334.
17. Smith, M., et al. (2020). "The Artemis Program: An Overview of NASA's Activities to Return Humans to the Moon." IEEE Aerospace Conference, 2020, pp. 1-10, https://doi.org/10.1109/AERO47225.2020.9172323.
18. Space Fostering Latin American Societies (2021). Annette Froehlich, Editor. Springer Cham. Southern Space Studies, https://doi.org/10.1007/978-3-030-73287-5.
19. Sutton, G.P. (2006). *History of Liquid Propellant Rocket Engines.* American Institute of Aeronautics and Astronautics, Inc. Illustrated edition (November 1, 2005). https://doi.org/10.2514/4.868870.
20. ULA: https://www.ulalaunch.com/rockets/vulcan-centaur.
21. Voyager Mission Summary. Frequently Asked Questions, JPL: https://voyager.jpl.nasa.gov/frequently-asked-questions/fact-sheet/.

Rocket Propulsion Fundamentals

2

Rockets produce thrust by expelling propellant mass at high velocity relative to the vehicle they propel. The thrust produced may be used to accelerate a launch vehicle, upper stage, or spacecraft to high velocity, countering the effects of drag (while moving within an atmosphere) and orbital perturbations (while moving in orbit about Earth, or any celestial body), or to change or control the attitude of spacecraft by producing torque about the system's center of mass.

—Dora Musielak

In this chapter we review the basic concept of rocket propulsion and derive the rocket equation, outlining the main performance parameters that characterize a rocket engine. Then we introduce the model of thermal rocket, which is based on gas dynamics and thermodynamics principles to describe processes inside a rocket chamber and its nozzle. The equations found in this chapter will help the reader gain a basic understanding of the high-temperature and pressure gas expansion flow processes in the rocket engine. These flow relationships apply to chemical rocket propulsion (burning liquid and solid propellants), cold gas thrusters, nuclear rockets, electric propulsion (solar-heated and resistance or arc-heated electrical rocket systems), and any propulsion system that utilize gas expansion as the mechanism for ejecting matter at high velocity. It is expected that the reader will have a basic acquaintance with both elementary thermodynamics and fluid mechanics.

The material covered is intended to answer questions such as

© The Author(s), under exclusive license to Springer Nature Switzerland AG 2025   45
D. Musielak, *Introduction to Rocket Propulsion for Astronautics*, Synthesis Lectures on Engineering, Science, and Technology, https://doi.org/10.1007/978-3-031-86141-3_2

- What are the main performance parameters that characterize a rocket engine?
- Why space launch vehicles must carry so much propellant?
- What design and operational parameters affect rocket thrust?
- Why can't we reach low Earth orbit (LEO) with one-stage space launch vehicle (SLV)?
- How does the engine thrust depend on the rocket nozzle design?

2.1   The Principle of Rocket Propulsion

A rocket engine is a propulsion device that expels matter at high speeds, exerting a propelling force, pushing it forward through the exchange of momentum. Consisting of a thrust chamber, and a converging-diverging nozzle, the rocket generates thrust by accelerating a high-pressure gas to supersonic velocities in the nozzle. The gaseous matter that is expulsed at high speed is known as *propellant*. The propellant can be a cold gas expelled at high pressure, or it can be a hot gas that results from chemical reactions (combustion), or from a thermonuclear reaction, or it can be plasma produced by application of electromagnetic energy to the propellant gas (see Chap. 6).

When liquid or solid propellant substances undergo exothermal chemical reactions in a combustion chamber, we call the engine a chemical rocket; its exhaust consists of a hot stream of combustion gas. A cold, pressurized gas can also serve as propellant if it is expelled at high speed through a nozzle to provide thrust without any chemical reaction. Ions (charged atomic particles) and nuclear particles can also be rocket propellants. Hence, rockets that eject chemically reacting hot gases, high-pressure cold gases, or atomic particles all operate under the same physical principle—mass is accelerated and thrown out, resulting in a reaction force to propel a vehicle moving within an atmosphere or in space.

It is common for us to idealize the flow in a rocket engine and use basic thermodynamic principles to predict rocket performance. Such idealization is referred to as thermal propulsion. The trends obtained for the main performance parameters are indicative of actual performance, and their values are found to be within a few percent of measured values.

2.2   Conservation of Momentum for One-Dimensional Steady-State Motion

Rocket propulsion is best explained with the principle of conservation of linear momentum. The momentum of an object is the product of its mass m and its velocity $\vec{v}$. The momentum vector is expressed as $\vec{p} = m\vec{v}$.

Let us visualize a rocket loaded with propellant as an isolated system of mass m, at rest in some reference frame. The conservation principle requires that the total momentum of the rocket-propellant system be zero, i.e. $\vec{p} = 0$. When the rocket fires, expelling gases at high speed, the vector momentum remains zero. That is, when the propellant is heated in the propulsion chamber, it releases a stream of hot gases, and the momentum of these hot gases at the rear of the rocket is balanced by the forward momentum gained by the rocket itself as it accelerates, moving forward. This simple application of momentum conservation explains how a rocket can accelerate in space. The total momentum of the rocket in a vacuum is given by $\vec{p} = \vec{p}_{rocket} + \vec{p}_{gas} = 0$. All methods devised to produce a thrust force for propulsion purposes are based on the principle involving the time rate of change of momentum.

Consider the rocket at some point during its one-directional flight, with a small amount of propellant mass dm leaving the nozzle with a velocity $-v_{ex}$ (relative to the rocket), as depicted in Fig. 2.1. The negative sign is there because this amount of propellant mass is moving in the opposite direction to the rocket. As a result of ejecting the small amount of mass, the rocket increases its velocity by a small amount $+dv$. The motion of the rocket under the influence of all external forces $\mathbf{F}_{ext}$ is given by applying Newton's second law,

$$\sum \vec{\mathbf{F}}_{\text{ext}} = \frac{d\,(m\vec{v})}{dt} = \frac{d\vec{\mathbf{p}}}{dt} \tag{2.1}$$

where $\vec{\mathbf{p}}$ is the total momentum of the rocket system, and $\sum \vec{\mathbf{F}}_{\text{ext}}$ is the net external force exerted on it.

If the net external force is known, Eq. (2.1) can be integrated in time to yield an expression for the velocity of the rocket as a function of time. This expression emphasizes that a rocket-propelled vehicle derives its acceleration from the discharge of propellant mass: the rocket throws propellant mass away in one direction, imparting momentum to the vehicle, which moves in the opposite direction.

Let us consider a spacecraft moving in deep space, where gravity effects and other external forces can be neglected. We assume that thrust is the only unbalanced force

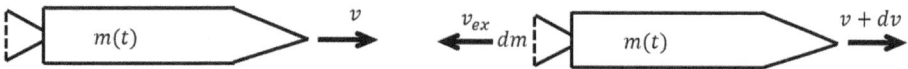

Fig. 2.1   The motion of a rocket spacecraft in the absence of resisting forces

acting on the spacecraftdrag is zero, and weight is balanced by centrifugal forces. These assumptions describe a condition in space flight. The vehicle has a total mass m (including rocket engine, structure, propellant, and payload) and is moving with velocity v relative to a non-accelerating frame of reference. The motion is steady and one-dimensional (see Fig. 2.1). Further assume that as the vehicle is accelerated its mass decreases at the propellant mass flow rate of dm/dt.

Initially, at some time t before firing its rocket, the vehicle moves with velocity v with respect to some reference frame, and the momentum is mv. The rocket engine is then fired for a short time Δt, and during this interval the vehicle mass changes by a small amount Δm and its speed by Δv. At the end of the interval, the vehicle is moving with a velocity $v + \Delta v$ so that at the end of rocket firing the momentum is $(m + \Delta m)(v + \Delta v)$. As noted earlier, Δm is a negative quantity, since the rocket is losing propellant mass, exhausted from the rear end nozzle, moving with a velocity v_{ex} relative to the vehicle.

The propellant gas is expelled by the rocket in the opposite direction of the motion of the vehicle, hence its velocity is $(v - v_{ex})$. This is approximately true, since the speed of the vehicle changes during the period of the engine firing. However, taking the limit as $\Delta t \to 0$, our approximation becomes exact. At the same time, the momentum of the exhaust gas is given by $-\Delta m(v - v_{ex})$. Therefore, the total momentum of the vehicle after engine firing is equal to the momentum of the vehicle plus the momentum of the exhaust gas:

$$p = (m + \Delta m)(v + \Delta v) - \Delta m(v - v_{ex}). \tag{2.2}$$

Since we assume that there are no other external forces acting on the vehicle, the final momentum is equal to the initial momentum; that is,

$$(m + \Delta m)(v + \Delta v) - \Delta m(v - v_{ex}) = mv,$$

which we reduce to

$$m\Delta v + \Delta m \Delta v + \Delta m v_{ex} = 0.$$

Dividing by the time interval Δt, we obtain:

$$m\frac{\Delta v}{\Delta t} + (v_{ex} - \Delta v)\frac{\Delta m}{\Delta t} = 0$$

In the limit $\Delta t = 0$, the ratios $\Delta v/\Delta t$ and $\Delta m/\Delta t$ become derivatives, while $\Delta v = 0$; therefore, we obtain the differential equation:

$$m\frac{dv}{dt} + v_{ex}\frac{dm}{dt} = 0 \tag{2.3}$$

where m is the time varying (instantaneous) mass of the vehicle; dv/dt is the vehicle acceleration; v_{ex} is the velocity of the exhaust stream relative to the rocket; dm/dt is the rate of change of spacecraft mass due to propellant expulsion.

Equation (2.3) is the time rate of momentum change for the right side of Eq. (2.1). It requires that an increase in vehicle speed $+dv/dt$ be accompanied by a decrease in the mass of the rocket $-dm/dt$. In general, the equation of motion for the rocket is given by the projection along the direction of $\vec{v}$ (tangent to the path of motion) of the vector Eq. (2.1). Therefore, we write the **general equation** of a rocket system in vector form as

$$\sum \vec{F}_{ext} = m\frac{d\vec{v}}{dt} + \vec{v}_{ex}\frac{dm}{dt} \tag{2.4}$$

where the magnitudes of $\vec{v}$ and $\vec{v}_{ex}$ are v and v_{ex}, respectively, which are parallel to each other and have opposite direction; the left side of this equality contains all external forces acting on the rocket system, including the gravitational force $\vec{F}_g$ (obtained from Newton's law of universal gravitation), and aerodynamic drag force $\vec{F}_D$, which is very important for a launch vehicle during ascent through Earth's atmosphere. See Chap. 3.

Other forces such as aerodynamic lift (important for winged vehicles), and those due to a rotating Earth, which lead to Coriolis and centripetal acceleration, have an important effect in high-speed flight in the vicinity of our planet. For an accurate and realistic analysis of rocket vehicle motion, all external forces acting on the vehicle must be included.

2.3   Rocket Thrust

Conceptually, a thermal rocket is a heat engine comprised of a pressure or thrust chamber where the propellant (working fluid) is heated, and a converging-diverging shaped duct at the rear of the rocket, called nozzle, through which the hot propellant gas is expanded and ejected at high velocity v_{ex} (Fig. 2.2). The term *thrust chamber* is used here for the section of the rocket where the propellant is processed at high pressure before exhausting through the supersonic nozzle. The result of the acceleration of the hot gas is a force exerted by the rocket in the direction opposite to that in which the gas is ejected. In other words, the change in momentum experienced by the hot exhaust gases results in the generation of a forward thrust force.

Since dm (in the second term of Eq. 2.4) denotes the mass change of the rocket, which accounts for the mass of spent propellant ejected through the nozzle, then $\dot{m} = dm/dt$ denotes the propellant mass flow rate ejected from the rocket. And since dm/dt is changing by rocket expelling mass the thrust force F is exactly in opposite direction to the exhaust velocity.

For constant exhaust velocity v_{ex} and mass flow rate $dm/dt = \dot{m}$, the magnitude of the thrust force (for a steady-state rocket engine moving in a homogeneous atmosphere, neglecting localized boundary-layer effects) is simply

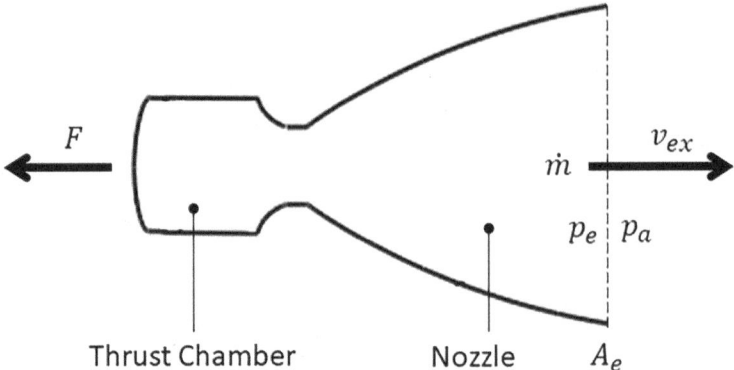

Fig. 2.2 Idealized rocket engine

$$F = \dot{m}v_{ex}. \tag{2.5}$$

This expression, known as the *momentum thrust*, assumes that the velocity of the exhaust gas acts along the nozzle centerline. We measure the performance of a rocket engine in terms of its spent propellant exhaust. The momentum thrust Eq. (2.5) is true for any rocket engine, and it shows thrust can be higher by expulsing the propellant gas with a higher velocity v_{ex} or by ejecting the gas at a greater flow rate ($\dot{m}$), or both. The units of thrust are newton (N) in the international system of units (SI), or in pounds-force (lbf) in the English Engineering system.

Equation (2.5) states that the thrust of the rocket engine is independent of the flight velocity; however, thrust is greatly affected by the pressure of the external environment and the design of the exhaust nozzle, which contributes an additional thrust term as the propellant mass is accelerated and exhausted, having a pressure p_e at the nozzle exit plane with cross sectional A_e (where the exhaust gases are ejected from the rocket). This is illustrated in Fig. 2.2. The external ambient pressure p_a is assumed to be uniform around the rocket engine.

According to the momentum theorem, the net force on the gas must be equal to the momentum flux out. At the nozzle exit area A_e there is an unbalance of the local pressure p_e of the hot gas exhausting stream and the external ambient pressure p_a. The net axial thrust force is the integral of the pressure forces acting on the chamber and nozzle in the axial direction, $F = \oint p dA$. There is also a retarding force acting on the gas flow, yielding an additional term consisting of the product of the nozzle exit area A_e and the difference between the exhaust pressure p_e and the ambient pressure. The total thrust force becomes

$$F = \dot{m}v_{ex} + (p_e - p_a)A_e \tag{2.6}$$

where the subscript e represents the nozzle conditions at the exit plane.

Hence, the thrust force in a rocket engine is comprised of two terms: the first term is the *momentum thrust* (product of the propellant mass flow rate and the exhaust velocity relative to the vehicle), which was derived by applying the conservation of momentum to the rocket-propellant system; the second term is called *nozzle pressure-area thrust*, as it depends on the nozzle design and its ability to exhaust the propellant gas at pressure p_e that may or may not match the ambient or external pressure p_a.

Since $v_{ex} > 0$, the momentum term always provides thrust, whereas the pressure-area term can increase or decrease the thrust. For optimum expansion the nozzle exit pressure should be equal to the atmospheric pressure ($p_e = p_a$). In such condition, the second term vanishes and Eq. (2.6) reduces to the momentum thrust, $F = \dot{m}v_{ex}$.

Since ambient pressure is not constant across Earth's atmosphere, Eq. (2.6) states that rocket thrust varies with altitude, an outcome of critical importance when we consider a core or first stage engine of a launch vehicle that must operate over a wide range of conditions during its ascent from sea-level to orbital altitude. In other words, the pressure term of Eq. (2.6) will affect the magnitude of the rocket thrust.

Operating in a vacuum, the rocket thrust is simply

$$F = \dot{m}v_{ex} + p_e A_e \tag{2.7}$$

It is clear from this expression that for any given rocket engine, thrust is greater in a vacuum than at sea level by an amount equal to $p_e A_e$, a pressure-area term that is fixed by nozzle design. High-area-ratio rocket engines have a low exit pressure and operate at nearly optimum expansion in a vacuum. For example, the main engines of the Space Shuttle Orbiter produced about 25% more thrust in vacuum than at sea level, due to the pressure thrust term difference between sea level atmospheric pressure and near zero vacuum pressure at LEO.

In a real rocket nozzle, the actual exhaust velocity v_{ex} is not uniform over the entire exit cross-section, and therefore it does not represent the entire thrust magnitude. Thus, we assume a uniform axial velocity known as *effective exhaust velocity* to define the average velocity at which propellant mass leaves the nozzle:

$$C = v_{ex} + (p_e - p_a)A_e/\dot{m} \tag{2.8}$$

The term $(p_e - p_a)A_e/\dot{m}$ indicates that the rocket exhaust velocity is not optimum for operation at all altitudes, as the ambient pressure varies accordingly. Let us consider two possible conditions for nozzle expansion:

- Over-expansion: $p_e < p_a$. This is the case for a rocket nozzle at lift-off. Because many launch pads are near sea level, the atmospheric pressure has the maximum value and it can cause shock waves to form at the nozzle's lip. These shocks represent losses, reducing kinetic energy from the flow, lowering the exhaust velocity, and thus decreasing the overall thrust.

- Under-expansion: $p_e > p_a$. In such situation, the nozzle has not fully expanded the exhaust gas, and has not converted all the enthalpy of the flow into kinetic energy (velocity). This happens when a rocket operates in a vacuum, because p_e is always higher than the vacuum pressure—a spacecraft requires an infinitely long nozzle to expand the flow to zero pressure. In Chap. 4 we discuss nozzle design measures to deal with such situation.

Only when the exiting gas static pressure matches the ambient pressure, $p_e = p_a$ the effective exhaust velocity is equal to the actual gas velocity exhausting the nozzle, $C = v_{ex}$. We can easily determine the effective exhaust velocity on a test stand by measuring thrust and propellant flow rate. This common practice is more accurate than attempting to determine the actual exhaust velocity profile. Nevertheless, it is important to note that the second term of the right-hand side of Eq. (2.8) is small compared to v_{ex}.

The effective exhaust velocity C applies to all rocket propulsion that thermodynamically expand hot gas in a nozzle. In a spacecraft rocket with a high-area ratio nozzle, the exit pressure p_e is very low and the nozzle operates very nearly at optimum expansion in a vacuum. Thus, it is common practice to use the effective velocity to adjust the nozzle exhaust velocity to compensate for the small pressure-area thrust produced by the real vacuum engine. Therefore, Eq. (2.6) may be expressed in terms of the nozzle *effective exhaust velocity*:

$$F = C \cdot \dot{m} \tag{2.9}$$

2.4   Rocket Propulsion Performance

Rocket performance is based on two parameters: *total impulse*, which depends on the thrust, and *specific impulse*, which depends on the exhaust velocity of the propellant mass it expels.

2.4.1 Total Impulse

A rocket engine exerts a thrust F during a time interval t_b (thrusting time or burn time). It also delivers a total impulse I defined as

$$I = \int_0^{t_b} F(t)dt \tag{2.10}$$

For constant thrust, $I = F \cdot t_b$. If thrust is not constant, as in solid propellant rockets, impulse is determined with the area under the thrust-time curve (see Chap. 5). Impulse is typically given in N-s or lbf-sec, depending on engine manufacturer. For example, a rocket engine delivering 45 kN thrust in 10 s exerts a total impulse of $I = 450{,}000$ N $\cdot$ s.

The total impulse is proportional to the energy released by all the propellant in the rocket. In general, the longer the distance or the faster the trip, the higher the total impulse must be. In general, first stage rockets require to impart the highest total impulse. As we will find in Chap. 5, the solid rocket boosters of launch vehicles exert total impulses as high as 298,000,000 lbf-sec (1325.57 N-s).

Since propellant flow rate multiplied by thrusting time is the propellant mass m_p, we can also write the total impulse delivered by the rocket as

$$I = m_p v_{ex} \tag{2.11}$$

where v_{ex} represents the effective exhaust velocity (see Eq. 2.8).

2.4.2   Specific Impulse

For a given space mission, a certain amount of propellant must be prescribed to carry out all thrusting requirements. To establish the performance of the rocket, we define the *specific impulse* I_{sp}, a parameter representing the total impulse delivered per unit weight of propellant. Since impulse is the effect of a force applied for a very short time, we define the specific impulse simply as the ratio of the total impulse and the propellant weight consumed during the thrusting time interval:

$$I_{sp} = \frac{I}{\dot{w}_p} = \frac{\int_0^{t_b} F(t)dt}{\dot{m}_p g_0} \tag{2.12}$$

where g_0 is the standard acceleration of gravity ($g_0 = 9.80665$ m/s^2 at the surface of the Earth).

In this form, we refer to I_{sp} as the *specific thrust* because it represents the thrust per propellant weight flow rate consumed. Using Eq. (2.5) with constant v_{ex} thrust, and omitting the subscript for the propellant mass flow rate, the specific impulse can also be written as

$$I_{sp} = \frac{F}{\dot{m}g_0} = \frac{\dot{m}v_{ex}}{\dot{m}g_0} = \frac{v_{ex}}{g_0} \tag{2.13}$$

Both definitions are identical if F and $\dot{m}$ are constant. In the SI system, the units of I_{sp} are newton-seconds/newton, or kg-m/s per kg-m/s/s; cancellation of units produces seconds in both. Specific impulse indicates how much impulse is possible with a unit weight of propellant. That is why we use I_{sp} as a parameter to compare the performance of rocket engines.

Moreover, for a given engine, the specific impulse I_{sp} has different values on the ground and in the vacuum of space because the ambient pressure changes with altitude and this affects the effective exhaust velocity of the gas. Therefore, we typically state specific impulse values at sea level or in a vacuum. For example, if a rocket engine with a mass flow rate of 2500 kg/s, produces a thrust of 10,000 kN at sea level, it will have a specific impulse of 408 s at the surface of the Earth, but a higher value in orbit (see Chap. 4).

In designing a rocket engine, we aim to maximize the exhaust velocity of the gas. For optimum gas expansion, the *exhaust velocity* v_{ex} is given in terms of the specific impulse I_{sp} as

$$v_{ex} = g_0 \cdot I_{sp} \tag{2.14}$$

and thus, either v_{ex} or I_{sp} can be used as the figure of merit for determining rocket propulsion performance. Typical values for chemical liquid propellant rockets are $I_{sp} \sim$ 300–500 s, and for solid rocket motors $I_{sp} < 250$ s. The reason for the difference in values is explained in Sect. 2.7.3.

The specific impulse is a measure of the total impulse per unit of propellant that is consumed, while thrust is a measure of the momentary or peak force delivered by the rocket. In most cases, propulsion systems with very high specific impulse produce low thrust. We shall address this issue in subsequent chapters, as we review a variety of rocket propulsion systems.

2.5   The Rocket Equation

Let us derive a mathematical expression that can tell us how fast a rocket-powered vehicle can move. We begin with the momentum equation, Eq. (2.3), which assumes no external forces act on the rocket vehicle. This is the ideal situation for motion in a perfect vacuum. Rearranging terms we write:

$$\frac{dv}{dt} = -\frac{dm}{dt}\frac{v_{ex}}{m}.$$

Canceling dt, we obtain a first order differential equation

$$dv = -v_{ex}\frac{dm}{m}.$$

Assuming constant exhaust velocity and optimum expansion of the gases, we can integrate the velocity between the limits v_0 and v, with the respective mass change from m_0 to m, where the subscript 0 represents the *initial* state of the rocket:

$$\int_{v_0}^{v} dv = -v_{ex} \int_{m_0}^{m} \frac{dm}{m}.$$

The solution of this integral equality gives the velocity change of the rocket:

$$(v - v_0) = \Delta v = v_{ex}\ln\left(\frac{m_0}{m}\right). \tag{2.15a}$$

This expression is known as the **ideal rocket equation**. If the rocket starts from rest, its velocity is $v = v_{ex}\ln(m_0/m)$, where m_0/m is known as the rocket mass ratio.

Another representation of the rocket equation is

$$\frac{m_0}{m} = e^{\Delta v/v_{ex}} \tag{2.15b}$$

The rocket can move faster than its exhaust gases, and this occurs when the mass ratio becomes equal to the Euler number e, the base of the natural logarithm ($e = 2.71828\ldots$). If the initial mass to final mass ratio is very large, a very high speed could be achieved. However, because of the logarithmic nature of the rocket equation, there is a limit as to what is possible. As observed in the trends of the curves in Fig. 2.3, after a certain point the gains in speed increase become smaller.

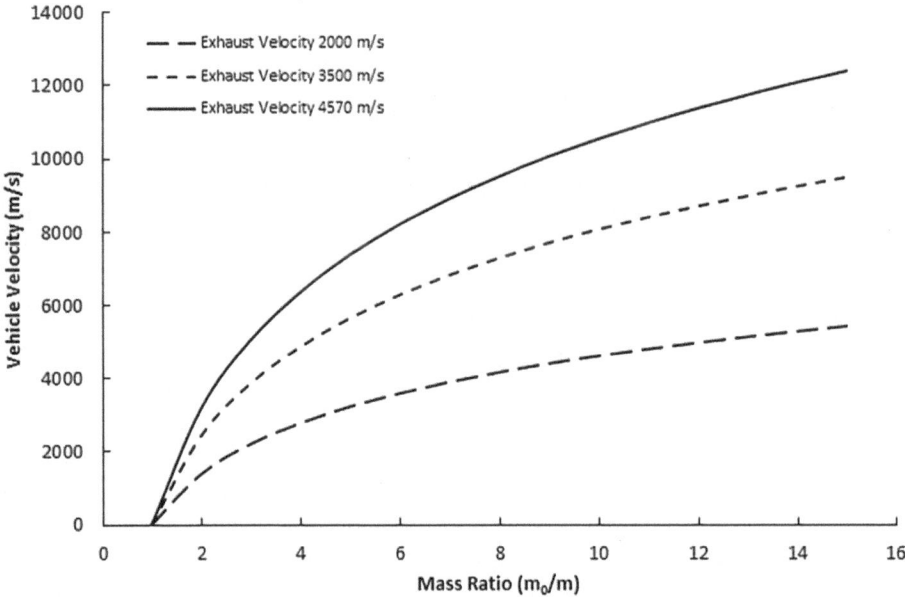

Fig. 2.3 Ideal single stage rocket velocity in dragless and gravitationless flight and required mass ratio, affected by magnitude of nozzle exhaust gas velocity

The rocket equation thus derived is ideal because it neglects effects of gravity, drag and other losses, and it assumes perfect expansion with a constant exhaust velocity v_{ex} relative to the rocket, which is valid in many cases. During ascent through an atmosphere, the pressure thrust term becomes important due to the decreasing ambient atmospheric pressure.

2.5.1   Mass Ratio

The initial mass of the rocket vehicle m_0 is the sum of the mass of propellant m_p, the mass of the vehicle's structure m_s (including the empty propulsion system mass and tanks with residual propellant plus a small mass for guidance, communications, and control devices), and the mass of the payload m_L. The final mass m (also called "dry mass") consists of the vehicle's structural mass, and the payload. Thus, the final mass m at burnout is simply the initial mass m_0 minus the mass of the propellant m_p that was exhausted, i.e., $m = m_0 - m_p$.

By defining the **mass ratio of the rocket** as,

$$\mu \equiv \frac{m_0}{m} = \frac{m_0}{m_0 - m_p} \tag{2.16}$$

the ideal rocket equation is expressed as

$$\mu = e^{\Delta v / v_{ex}} \tag{2.17}$$

The mass ratio emphasizes the sensitivity of the mass of the vehicle to the attainable effective exhaust velocity of the propulsion system, and so μ is an important parameter to compare different rockets that could be used for missions requiring a specific velocity change. Equation (2.17) indicates that a Δv of n times the exhaust velocity v_{ex} requires $\mu = e^n$. Hence, for a vehicle to achieve a Δv of 2.5 times its exhaust velocity, it would require a mass ratio of $e^{2.5}$ (approximately 12.2). Rocket mass ratios and corresponding rocket velocities are illustrated in Fig. 2.3. For example, a liquid propellant rocket that exhausts its spent propellant at 3500 m/s and has a mass ratio of 1.5 can accelerate a vehicle to 1419 m/s. Increasing the mass ratio to 3 would accelerate the vehicle to 3845 m/s.

The velocity of the rocket-powered vehicle, at any given time during its motion, only depends on the *effective* exhaust velocity v_{ex} of the gases coming out of the rocket, and on the instantaneous mass ratio m_0/m, which is always positive. For a given mass ratio, Eq. (2.15a) states that a higher exhaust velocity produces a higher vehicle velocity. As we shall find in Sect. 2.7, for chemical rockets v_{ex} is limited by the chemical nature of combustion. The most advanced liquid propellant rockets produce an exhaust velocity of less than 4600 m/s. For example, the RS-25 engine used in the core stage of NASA's SLS launch vehicle produces an exhaust velocity of 4424 m/s (at sea level) burning a mixture of liquid oxygen, LOX, and liquid hydrogen, LH$_2$, while the RL10 upper stage ejects the hot gases at almost 4565 m/s (vacuum), a value which is the theoretical limit of chemical energy extraction in a bi-propellant rocket combustion chamber.

In general, the rocket engine specific impulse I_{sp}, along with the vehicle mass ratio, determines the performance of a space vehicle. Using the definition of specific impulse from Eq. (2.9), the ideal rocket equation becomes

$$\Delta v = v_{ex}\ln\left(\frac{m_0}{m}\right) = g_0 I_{sp}\ln(\mu) \tag{2.18}$$

An efficient rocket design requires less propellant to achieve a given propulsion goal, and would therefore have a lower mass ratio; however, for any given efficiency a higher mass ratio typically permits the rocket to achieve higher Δv. For example, for a liquid propellant rocket engine with $I_{sp} = 440$s and $\mu = 12.5$, the change in velocity is $\Delta v = 10.9$ km/s. Note the strong dependence of ultimate vehicle velocity on I_{sp} (or rocket exhaust velocity): for $I_{sp} < 500$ s (limit of chemical rockets), a very high mass ratio is required to reach orbital velocity ($v \sim 9.7$ km/s). Only multi-stage launch vehicles can achieve such mass ratios.

2.5.2 Delta-v and Propellant Requirement

For spaceflight, the important measure of propulsive effort required to get from one point to another is the required *total change in velocity* Δv, also known as velocity increment, or simply "delta-v." The velocity increment appeared for the first time in the rocket equation, Eq. (2.15), indicating that the final velocity of a rocket increases with increases in the exhaust velocity of its propellant and with the mass ratio. The total delta-v is a measure of all the possible energy demands of a mission, or how much propellant is required to achieve it.

For example, to go from low Earth orbit to lunar orbit the Δv is just 4.10 km/s. However, to go from the surface of the Earth to LEO a launch vehicle requires a Δv of about 9.70 km/s, demonstrating that most of the effort of space travel near the Earth is spent in launching the spacecraft, lifting it from the surface of the planet to about a 100 km altitude due to the large gravitational hold of the planet.

The change in velocity Δv is also considered a scalar measure for the "effort" needed to carry out an orbital maneuver, i.e. to change from one orbit to another. The time-rate of change of delta-v is the magnitude of the acceleration caused by the engines, i.e. the thrust per kilogram total mass. Actual propellant usage depends more than linearly on delta-v because not only the rocket but also the propellant not yet spent must be accelerated. After propellant is exhausted through the nozzle, it reduces the mass of the rocket, and then it requires less propellant to achieve the necessary Δv for another maneuver.

We determine the amount of propellant needed to perform a maneuver with a known Δv. Knowing the empty mass of the spacecraft and the effective exhaust velocity of the rocket system we can solve for m_0 and so determine the propellant mass that was consumed, $m_p = m_0 - m$. For perfect exhaust, from Eq. (2.15):

$$m_p = m_0\left[1 - \exp(\Delta v/v_{ex})\right]$$
$$= m_0\left[1 - \exp\left(\frac{-\Delta v}{g_0 I_{sp}}\right)\right] = m\left[\exp\left(\frac{\Delta v}{g_0 I_{sp}}\right) - 1\right] \qquad (2.19)$$

This relation shows that, for a given v and initial mass m_0, increasing the velocity of the exhaust gases v_{ex} increases the final mass m and decreases m_p. That is exactly what we would like to do. Unfortunately, chemical rockets are limited to a $I_{sp} < 460$ s, and thus, they require a great amount of propellant. This is due to chemical propellants having a fixed energy per kilogram (i.e. heat of combustion) that cannot be exceeded, even with additives. Rocket engines burning cryogenic liquid propellants with low molecular mass (e.g. oxygen and hydrogen) are much more propellant efficient than rocket motors burning solid propellants, as the latter yield much lower specific impulse I_{sp} (lower exhaust velocity).

Example 2.1 Determine the propellant consumption of a rocket engine to perform a space maneuver requiring a delta-v of 3 km/s. The exhaust velocity of the rocket is 3100 m/s.

Solution: For the required maneuver Δv a certain amount of propellant must be exhausted at the velocity v_{ex} implicit in the rocket design. Assuming $m_0 = 1000$ kg, the propellant mass is, from Eq. (2.19),

$$m_p = m_0\left(1 - e^{-\frac{\Delta v}{v_{ex}}}\right)$$
$$= 1000\ \text{kg}\left[1 - \exp(-3/3.1)\right] = 620\ \text{kg}$$

The maneuver requires 620 kg of propellant. A rocket with a higher exhaust velocity would need less propellant mass to carry out the same maneuver.

It is important to emphasize the distinction between Δv and the actual velocity attained by a vehicle. The velocity increment Δv is the velocity calculated from the ideal rocket equation, Eq. (2.15), as a measure of the energy expended by the rocket. The velocity of a launch vehicle is less than this value, because of losses experienced while reaching orbital altitude. The difference between actual and ideal velocities represents the energy expended against gravity loss and potential energy. We will study this topic in Chap. 3.

The velocity budget for a given mission is the sum of all the flight velocity increments needed to attain the objective of the mission. For example, consider a mission with a reusable launch vehicle to place a spacecraft on a LEO parking orbit at 275 km altitude and bring a payload back. The ideal orbital velocity is 7.79 km/s, however, the delta-v budget for a round trip to LEO must include propellant to perform additional orbit maneuvers (see Chap. 3), and then to deorbit to return to Earth. The total velocity change required of the reusable vehicle for this LEO flight could be about $\Delta v = 9.347$ km/s.

Did you know? **The time required to reach LEO is just a few minutes, but the effort required to accelerate the vehicle to orbital velocity is very large.** On the other hand, to move from LEO to lunar orbit takes a few days but it requires less than half the propulsive effort. Explain the inverse relation that exists between travel time in space and effort to accelerate a vehicle.

Let us emphasize again that the flight velocity increment Δv is proportional to the effective exhaust velocity v_{ex} (or C) and thus to the specific impulse of the rocket. Increasing I_{sp} (with better propellants, optimized nozzle area ratio, or higher chamber pressure) results in improved spacecraft performance. Of course, such design improvement should not add to the inert mass of the rocket propulsion system. See Chaps. 4 and 5.

2.5.3 Velocity at Burnout

Consider a vehicle starting from rest $v_0 = 0$ with a launch mass $m_0 = m_p + m_s + m_L$, where m_p is the mass of the propellant, m_s is the mass of the structure, and m_L is the mass of the payload. Payload includes the mass of the instruments, crew and life-support systems, and all hardware that a spacecraft can carry over and above what is necessary to operate the vehicle in flight.

We define **burnout velocity** as the maximum velocity reached by a vehicle when all of the rocket propellant has been used, or as the velocity that a rocket would attain in gravity-free and empty space if its initial mass m_0 is decreased to its burnout or cutoff mass m by exhausting gases with a velocity v_{ex}. The terms cut-off and burnout are used interchangeably in some texts. Formally, burnout means the end of burning in a chemical rocket due to the exhaustion of propellant. In propulsion operations, cut-off refers to the termination of burning in a rocket engine brought about by an intentional command; also known as shutdown.

Mass at burnout or at the cut-off point is when powered flight ends and the rocket moves without propellant energy, or when the rocket is injected into an orbital trajectory and begins to coast onward. The mass at burnout is $m = m_0 - m_p = m_s + m_L$. If the rocket burns a propellant mass m_p and ejects it with v_{ex}, then the velocity of the rocket at burn-out, v_b, is,

$$v_b = v_{ex}\ln\left(\frac{m_0}{m}\right) = v_{ex}\ln\left(\frac{m_p + m_L + m_s}{m_L + m_s}\right) \tag{2.20}$$

This relation states that the velocity of the vehicle, at any given time during the burn, only depends on the rocket's exhaust velocity and the instantaneous mass ratio. If the amount of propellant carried is very small compared to the mass of the rocket vehicle at launch, then the mass ratio has a value very close to 1 and the burnout velocity approaches zero. This makes sense, as a rocket with very little propellant cannot go very far. Another limiting case is a rocket with a large propellant mass, $m_0 \gg m$, so v is many times the rocket's exhaust velocity, because a long burn will continuously accelerate the rocket to a high velocity $v > v_{ex}$. See Chaps. 6 and 7.

2.5.4 Burn Time and Distance Travelled During a Burn

Burn time determines the duration of a rocket engine operation required to achieve a specific change in velocity Δv. As noted in Sect. 2.3, propellant mass flow rate $\dot{m}$ multiplied by thrusting time is the propellant mass m_p at the beginning of a flight segment, so we write

$$t_b = \frac{m_p}{\dot{m}} = \frac{m_0 - m}{\dot{m}}$$

where t_b represents the *burnout time*.

Dividing through by m_0 yields an expression for the burn time in terms of the mass ratio:

$$t_b = \frac{m_0}{\dot{m}}\left(1 - \frac{m}{m_0}\right) \tag{2.21}$$

Typical burn times for solid rocket motors are about 2 min, while liquid propellant rockets have burn times up to 10 min. Strap-on rocket boosters are used to improve the performance of a launch vehicle helping the first stage produce sufficient thrust at low altitudes to lift the SLV off the launch pad. The SRBs are ignited at lift-off and burn until exhausted. For example, the new NASA SLS system is configured with the largest reusable solid propellant rocket motors ever flown, each producing a thrust of about 16.01 MN, with a burn time of 132.8 s.

With the mass flow rate constant, $\dot{m} = dm/dt = \text{const}$, we write $m(t) = m_0 - \dot{m}t$, where the negative sign implies that the mass of the vehicle decreases with time. Substitute this relation into Eq. (2.15) to obtain the vehicle's velocity as a function of time,

$$v(t) = -v_{ex}\ln\left(1 - \frac{\dot{m}}{m_o}t\right) \tag{2.22}$$

We can assume a constant mass flow rate if the rocket engine is not throttled. However, if this is not the case, a more sophisticated analysis is required to determine $v(t)$.

We can also calculate the *time required to achieve a specified change in velocity* Δv simply by rearranging terms in Eq. (2.22),

$$t = \frac{m_0}{\dot{m}}\left(1 - e^{-\Delta v/v_{ex}}\right) \tag{2.23}$$

which is equivalent to Eq. (2.21), but this expression is more general, as we can use it to calculate operation time of a non-chemical rocket during a thrusting maneuver.

During an engine burn the rocket moves a certain distance, which we expect to depend on the thrust. The range of a vehicle is the distance traveled along its trajectory during a burn. Since the velocity varies with time, we can determine the distance traveled s as

$$s = \int_0^{t_b} v(t)dt = \int_0^{t_b} v_{ex}\ln\left(\frac{m_0}{m_0 - \dot{m}t}\right)dt$$

After integration, it yields

$$s = v_{ex}\frac{m_0}{\dot{m}}\left[\frac{m_0 - \dot{m}t_b}{m_0}\left(\ln\frac{m_0 - \dot{m}t_b}{m_0} - 1\right)\right]$$

Substituting the expression for t_b from Eq. (2.23) we obtain the *range* or *maximum distance travelled* during a rocket burn as a function of the mass ratio:

$$s = v_{ex}\frac{m_0}{\dot{m}}\left[1 - \frac{m}{m_0}\left(\ln\frac{m_0}{m} + 1\right)\right]$$
$$= v_{ex}\frac{m_0}{\dot{m}}\left[1 - \frac{1}{\mu}(\ln\mu + 1)\right] \tag{2.24a}$$

A high mass flow rate $\dot{m}$ leads to a shorter distance traveled, but a high exhaust velocity v_{ex} allows the rocket vehicle to travel farther for a given amount of propellant burnt. Equation (2.24a) applies to any instant during the engine burn, not just at burnout. However, this expression assumes a zero-initial velocity. Thus, if the vehicle is already moving at a given velocity, then this must be included in the computation of the total distance traveled during the burn. For example, if the vehicle starts burning a second stage when its velocity is v_i, the additional distance is $v_i t$, or

$$s_i = v_i \frac{m_0}{\dot{m}}\left(1 - \frac{m}{m_0}\right)$$

Therefore, the total distance the vehicle travels (neglecting gravitational effects and other losses) is represented by

$$s = v_{ex}\frac{m_0}{\dot{m}}\left[1 - \frac{m}{m_0}\left(\ln\frac{m_0}{m} + 1\right)\right] + v_i\frac{m_0}{\dot{m}}\left(1 - \frac{m}{m_0}\right) \qquad (2.24b)$$

Equation (2.24b) is adequate to calculate the total range of an upper stage to inject a spacecraft into a predetermined orbit.

Example 2.2 Calculate the burn time and range of a spacecraft with an initial mass of 4700 kg and after burning its propellant the final mass is 700 kg. The rocket has an exhaust velocity of 2930 m/s (vacuum) and a mass flow rate of 100 kg/s.

Solution: The mass ratio is $4.7/0.70 = 6.71$. From Eq. (2.15), the change in velocity is

$$\Delta v = v_{ex}\ln(\mu) = 5577.5 \text{ m/s}$$

The burn time, from Eq. (2.23) is

$$t = \frac{m_0}{\dot{m}}\left(1 - e^{-\Delta v/v_{ex}}\right) = 34 \text{ s}$$

During the 34 s the rocket travels a distance given by Eq. (2.24a),

$$s = v_{ex}\frac{m_0}{\dot{m}}\left[1 - \frac{1}{\mu}(\ln\mu + 1)\right] = 66,484 \text{ m}$$

2.6   Single Stage Rocket Payload Capability

The purpose of a rocket is to deliver a payload (e.g. astronauts, satellite, cargo, spacecraft) to a given destination in space. We wish determine the relationship between the propulsion performance affected by the relative masses of the structure m_s, the propellant m_p, and the payload m_L. Recall, the mass of the fully loaded rocket is the mass at the beginning of the burn and is equal to the sum of all masses, i.e. payload, propellant, and structure:$m_0 = m_L + m_p + m_s = m + m_p$. The rocket empty mass at the end of a burn is the sum of the payload and structural mass, assuming all the propellant has been consumed: $m = m_L + m_s = m_0 - m_p$.

We need a measure of a launch vehicle capability to place a payload in orbit. We define the payload mass ratio as

$$\textbf{Payload mass ratio} : \lambda = \frac{m_L}{m_0 - m_L} = \frac{m_L}{m_p + m_s} \qquad (2.25)$$

This expression suggests that the structural mass m_s limits the payload mass that a launch vehicle can carry. In practice, λ depends on the orbit altitude and speed, and on the propellant energy, the design of the rockets powering the launch vehicle, and the configuration of the vehicle itself. Designers strive to achieve a very high payload ratio; that is, for a given vehicle liftoff mass, the objective is to deliver a payload as high as possible.

Let us explore further the relationship between launch vehicle mass distribution by defining the structural coefficient ϵ, a design parameter indicative of the dead structural mass:

$$\textbf{Structural coefficient} : \epsilon = \frac{m_s}{m_p + m_s} = \frac{m_s}{m_0 - m_L} \qquad (2.26)$$

Heavy rocket engines and heavy propellant thanks increase the structural coefficient, and thus reduce the payload capability, and decrease the overall mass ratio, as shown below.

Since the mass ratio is given by

$$\mu = \frac{m_0}{m} = \frac{m_L + m_s + m_p}{m_L + m_s}$$

the three mass ratios relate as

$$\mu(1 + \lambda) = \frac{m_0}{m_0 - m_p} \frac{m_s + m_p + m_L}{m_s + m_p} = \frac{m_s + m_L}{m_s + m_p} = \epsilon + \lambda$$

Thus, the mass ratio μ written in terms of nondimensional parameters is

$$\mu = \frac{1 + \lambda}{\epsilon + \lambda} \qquad (2.27)$$

In the design of a launch vehicle, the objective is to reduce the structural coefficient ϵ and increase the payload ratio λ. This is very challenging for a single stage configuration. Rewrite the rocket equation with Eq. (2.27) as follows

$$\frac{\Delta v}{v_{ex}} = \ln(\mu) = \ln\left(\frac{1 + \lambda}{\epsilon + \lambda}\right)$$

or

$$\left(\frac{1 + \lambda}{\epsilon + \lambda}\right) = e^{\Delta v/v_{ex}} \qquad (2.28)$$

Now solve for the payload ratio to assess the vehicle's payload capability:

$$\lambda = \frac{\epsilon e^{\Delta v/v_{ex}} - 1}{1 - e^{\Delta v/v_{ex}}} = \frac{e^{-\Delta v/v_{ex}} - \epsilon}{1 - e^{-\Delta v/v_{ex}}} \qquad (2.29)$$

This expression is plotted in Fig. 2.4 to illustrate the tradeoff between payload mass ratio, structural coefficient, and propulsive performance for a single stage rocket vehicle. For example, if $\epsilon = 0.05$, which represent the lower limit of the structural mass, and

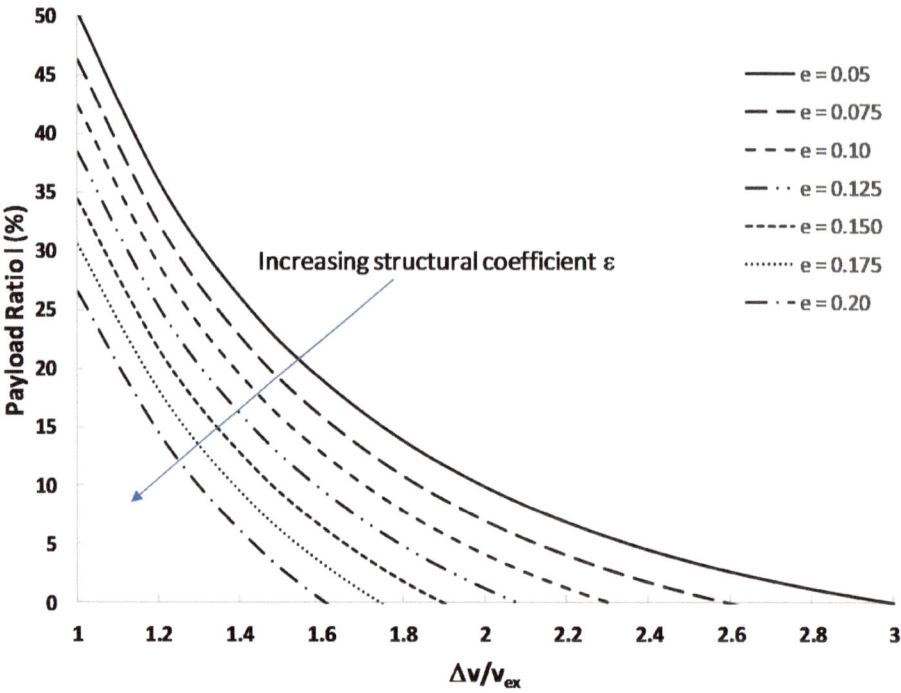

Fig. 2.4 Payload performance for a one-stage rocket vehicle with different structural coefficients

$\Delta v = 9.6$ km/s (typical value required to reach LEO), with the most advanced rocket stage having $v_{ex} = 4.0$ km/s, the payload ratio is

$$\lambda = \frac{\epsilon e^{\Delta v/v_{ex}} - 1}{1 - e^{\Delta v/v_{ex}}} = \frac{(0.05)e^{2.4} - 1}{1 - e^{2.4}} = 0.04478$$

This value represents almost 4.5% of the total mass of the rocket. If the structural coefficient were instead $\epsilon = 0.075$ for the same stage rocket, the payload capability would be reduced to 1.72% of the total rocket mass.

Adding an extra rocket stage allows a significant increase in the payload capability of a launch vehicle such that it can carry more payload, e.g. an extra astronaut, additional life-support systems, or more scientific instruments for exploration.

The NASA SLS first variant has a payload capability to LEO of 95 metric tons. Of course, the SLS can only achieve such capability by utilizing 2.5 stages, which includes strap-on solid rockets to boost the vehicle's thrust at lift-off (see Figs. 1.4 and 3.4).

Another important nondimensional parameter that is very useful to designers of launch vehicles or rocket stages is the ***propellant mass fraction***. Denoted by ζ, the propellant mass fraction is defined as the mass of the propellant in the rocket m_p divided by the total or initial mass of the rocket m_0:

$$\zeta \equiv \frac{m_p}{m_0} = \frac{m_p}{m_p + m} = 1 - \epsilon \qquad (2.30)$$

Applied to launch vehicle stages, the propellant mass fraction ζ describes the ratio of propellant in a stage to the total stage mass. Observe that a small structural coefficient yields a large propellant fraction. A high value of ζ is desirable. This is because the propellant mass fraction ζ indicates how good the design of the rocket is; if it has $\zeta = 0.92$ it means that only 9.2% of its mass is inert (hardware) and this small fraction contains, feeds, and burns a substantially larger mass of propellant.

Since $m_p = m_0 - m$, the propellant mass fraction is written in terms of the mass ratio μ:

$$\zeta = \frac{m_p}{m_0} = \frac{m_0 - m}{m_0} = 1 - \mu^{-1} \qquad (2.31)$$

Example 2.3 Consider a rocket stage with a specific impulse of 415 s and a propellant mass fraction of 0.90. If the propellant fraction is improved to 0.95, estimate the percentage improvement in Δv that the stage can achieve.

Solution: With the original mass fraction $\zeta = 0.90$, the stage mass ratio is $\mu = 10$, and the velocity increment is, from Eq. (2.18),

$$\Delta v = g_0 I_{sp} \ln(\mu) = (9.81 \text{ m/s}^2)(415 \text{ s})(\ln 10) = 9374 \text{ m/s}$$

By improving the mass fraction to $\zeta = 0.95$, the stage mass ratio is $\mu = 20$, and its Δv is

$$\Delta v = g_0 I_{sp} \ln(\mu) = (9.81 \text{ m/s}^2)(415 \text{ s})(\ln 20) = 12,196 \text{ m/s}$$

This represents a substantial increase in the Δv (2822 m/s), about 30% more.

The rocket mass ratio μ and propellant mass fraction ζ are useful to perform back-of-the-envelope calculations. The mass ratio is an easy number to derive from either Δv mission values or from rocket and propellant mass numbers, giving a practical relationship between the two. Moreover, the velocity that a vehicle can achieve depends on the type of propellant it burns and what fraction of the total lift-off mass is propellant. Therefore, the propellant mass ratio is a useful number to give an impression of the size of a vehicle: while two rockets with propellant mass fractions of, say, 92% and 95% may appear similar, the corresponding mass ratios of 12.5 and 20 clearly indicate that the two systems are very different. Typical multistage SLVs have mass ratios in the range from 8 to 20.

There are various methods for calculating ζ, and its meaning varies across the industry. For a stage, the denominator in Eq. (2.31) represents its total mass (including propellant and structure). Therefore, when using this relation, we simply divide the total propellant mass by the gross liftoff mass of the vehicle or stage, assuming that all propellant is utilized. However, the assumption of complete propellant depletion will never be satisfied exactly. There will be some propellant residuals and quantities of propellant intentionally allocated for supporting maneuvers such as a returning reusable rocket stage back to land, or transfer of an expendable upper stage to a graveyard orbit or for controlled re-entry in the Earth atmosphere. For example, if the total propellant mass in a rocket is 262,630 kg, and its gross mass at liftoff is 286,670 kg, the mass fraction would be 0.916. However, if 9525.5 kg of propellant is allocated for off-nominal conditions, engine restarts, purges/ bleeds, and other contingencies, then the total *usable* mass of propellant is 253,104.5 kg, and the mass fraction for the stage reduces to 0.883.

Thus, we can redefine the propellant mass fraction to account for propellant residuals and reserves, such that

$$\zeta \equiv \frac{m_{pu}}{m_0} = \frac{m_p - m_{res}}{m_0}$$

where m_{pu} is the total mass of usable propellant, and m_{res} is the mass of propellant residuals.

2.7   The Ideal Thermodynamic Rocket Engine

A chemical rocket engine liberates the energy in the propellant reacting in the combustion chamber under quiescent conditions. Without retorting to complicated combustion process analysis, we can develop an ideal model of a rocket engine and use thermodynamic principles to predict the behavior of the flow in the thrust chamber and obtain rocket performance parameters that are reasonably accurate for preliminary analysis, providing performance trends to compare with measured values. An idealization of a liquid propellant rocket is sketched in Fig. 2.5, indicating the main design parameters, where p_c and T_c represent the stagnation conditions, which give the pressure and temperature of the gas in the combustion chamber. The effective exhaust velocity v_{ex} that can be achieved is governed by the nozzle expansion ratio ε, defined as ratio of the nozzle exit area and its throat area, $\varepsilon = A_e/A_t$.

As a heat thermodynamic engine, the chemical rocket is a system where propellant (fuel plus oxidizer) is injected and ignited in a chamber at the combustion pressure p_c. Such combustion process converts the propellant into high pressure, high temperature gas, which then enters the converging-diverging nozzle where it expands adiabatically as it goes through the throat, and the hot gas continues to expand until it exhausts at the end plane of the nozzle with pressure p_e. As the hot gas flows through the nozzle, its pressure and temperature drop significantly, and the gas velocity increases to values several times the local speed of sound. During expansion in the nozzle a large fraction of the thermal

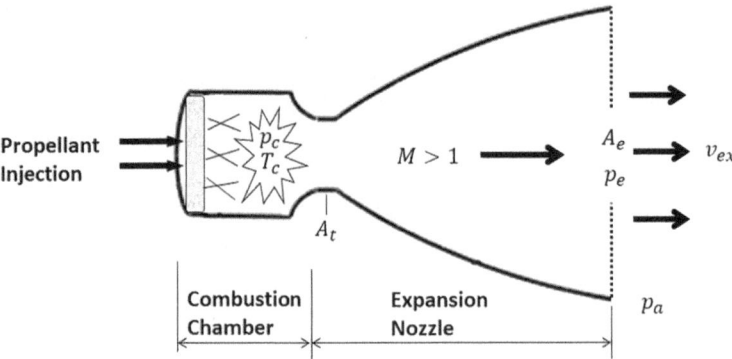

Fig. 2.5 Schematic of ideal chemical rocket engine

energy of the hot gas is converted into directed kinetic energy to produce the thrust. In this process, there exists a pressure difference between the engine and the surrounding environment.

The thrust given by Eq. (2.6) can be rewritten in terms of thermodynamic principles. Thus, we study the thermodynamic processes that control the flow in an ideal nozzle as this helps us understand what parameters determine the magnitude of the exhaust gas velocity, which in turn controls the thrust.

Although the thermodynamic concepts that we introduce in this section apply to any thermal system, in this chapter we concentrate on rocket engines that use chemical energy, and we will refer to propellant heated in the combustion chamber as the gas.

2.7.1 Assumptions

The flowfield in the nozzle is three-dimensional, i.e. the flow properties are functions of x, y and z. However, if the variation of area is gradual along the axial direction $A = A(x)$, it is convenient and sufficiently accurate to neglect the y and z flow variations, and to assume that the flow properties are functions of x only. Such a flow is defined as *quasi-one-dimensional flow*. Starting with this approximation, we derive a simple theory based on thermodynamics to represent the ideal rocket. Preliminary analysis is based on the following assumptions used to define the rocket engine theoretical performance:

(1) The working fluid (propellant) is a thermally perfect gas—it obeys the perfect gas equation of state $p = \rho RT$. This is a valid assumption since the hot propellant gas exhausting through the rocket nozzle approaches perfect gas behavior.
(2) Chemical equilibrium is established within the rocket chamber, and the propellant gas composition is homogeneous and invariant throughout the nozzle. This assumption

defines frozen equilibrium flow conditions, which require good fuel/oxidant mixing and rapid completion of combustion reactions.

(3) There is no heat transfer across the rocket engine walls; i.e. the flow expansion process through the nozzle is adiabatic, with no exchange of energy between the engine and the ambient. In a real rocket engine, there is heat loss through the walls, but for the thermodynamic analysis it is neglected.

(4) All exhaust gases leaving the rocket have an axially directed velocity. The gas velocity, pressure, temperature, and density are all uniform across any section normal to the rocket axis.

(5) The nozzle flow is frictionless, that is, there is no friction at the walls and, thus, we do not consider a boundary layer.

(6) The flow is steady and constant, the expansion is uniform and steady. Transient effects (i.e. start up and shut down) are of very short duration and are neglected.

2.7.2   Nozzle Isentropic Conditions

Three parameters are needed to model the rocket engine: the effective exhaust velocity of the heated gas (v_{ex}), the mass flow rate ($\dot{m}$), and the nozzle pressure ratio (p_c/p_e) across the thrust chamber. To determine these parameters, we need to use the ideal gas law, relations for isentropic process, all based on the conservation of mass and the conservation of energy principles. Rigorous derivations of these relations are found in many texts devoted to thermodynamics and fluid dynamics, and the reader is encouraged to consult them as needed.

First, we wish to express the exhaust velocity of a rocket engine in terms of parameters that identify the state of the thrust chamber, which is modeled as a thermal engine. In the following expressions, the subscripts c, t and e denote conditions at the combustion chamber, the nozzle throat, and nozzle exit plane, respectively.

2.7.3   Exhaust Velocity

The exhaust velocity v_{ex} is the most important performance parameter for the rocket engine, since it determines the thrust and the specific impulse that it can deliver. To obtain a representation of v_{ex} we begin with the notion that the expansion process in the ideal rocket nozzle is adiabatic. With this assumption, the energy conservation principle requires that the kinetic energy of the exhaust gas be equal to the change in enthalpy (or internal energy) of the gas as it cools and expands through the nozzle: $v_{ex}^2 = 2c_p(T_c - T_e) + v_c^2$. When the area of the chamber section is large compared with the nozzle throat area ($A_c/A_t > 4$), the velocity of the gas in the chamber is relatively small and consequently the term v_c^2 is neglected. Hence, the energy equation for the ideal rocket nozzle becomes

$$v_{ex}^2 = 2c_p T_c \left(1 - \frac{T_e}{T_c}\right)$$

Since the gas expands from stagnation chamber pressure p_c and temperature T_c via an adiabatic process, we use the isentropic relation: $Tp^{\gamma/(\gamma-1)} = $ const to rewrite the term in the bracket as

$$\left[1 - \left(\frac{p_e}{p_c}\right)^{(\gamma-1)/\gamma}\right]$$

where γ denotes the specific heat ratio ($\gamma = c_p/c_v$) of the gas. For rocket exhaust gases at high temperature the value of γ depends on the composition of the combustion products (chemical species and concentration). The specific heat at constant pressure is related to γ, to the gas constant $R = \mathcal{R}/\mathcal{M}$, and thus to the molecular mass $\mathcal{M}$ of the gas:

$$c_p = \frac{\gamma}{\gamma - 1} R = \frac{\gamma}{\gamma - 1} \frac{\mathcal{R}}{\mathcal{M}} \tag{2.32}$$

where $\mathcal{R}$ is the universal gas constant with a value of

$$\mathcal{R} = 8.3146 \frac{\text{kJ}}{\text{kmol} \cdot \text{K}} = 1545.4 \frac{\text{ft} \cdot \text{lbf}}{\text{lbm} \cdot \text{mol} \cdot {}^\circ \text{R}}$$

Therefore, we can show that the ideal rocket engine has exhaust velocity expressed as

$$v_{ex} = \sqrt{\frac{2\gamma}{\gamma - 1} R T_c \left[1 - \left(\frac{p_e}{p_c}\right)^{\gamma - 1/\gamma}\right]}$$

$$= \sqrt{\frac{2\gamma}{\gamma - 1} \frac{\mathcal{R} T_c}{\mathcal{M}} \left[1 - \left(\frac{p_e}{p_c}\right)^{\gamma - 1/\gamma}\right]} \tag{2.33}$$

where T_c denotes the maximum tolerable chamber temperature.

This equation for the rocket *exhaust jet velocity* provides some insight into the physical parameters that control the magnitude of the velocity at the exit plane of the nozzle and thus of the rocket engine performance. The parameter $\gamma/(\gamma - 1)$ does not vary appreciably for most gases, and therefore the rocket exhaust velocity is mostly a function of two parameters: (1) the ratio of exhaust to chamber pressures p_e/p_c, and (2) the ratio of the chamber temperature to the molecular mass $\mathcal{M}$ of the gas in the chamber, $T_c/\mathcal{M}$. The pressure ratio, which depends on the design of the nozzle, represents the temperature difference across the nozzle. The temperature in the combustion chamber T_c is a function of the propellant mixture, and since thrust is directly proportional to the exhaust velocity, the ratio $T_c/\mathcal{M}$ plays an important role in optimizing the oxidizer-to-fuel mixture ratio in chemical rockets. Mixture ratio is the ratio of oxidizer to fuel flow rate present in a combustion chamber. We will elaborate on this topic in Chap. 4.

The requirements for increased exhaust velocity are satisfied by reducing the pressure ratio p_e/p_c, i.e. increasing the chamber pressure or reducing the exhaust pressure by further expansion, and by increasing the ratio $T_c/\mathcal{M}$, which can be achieved by increasing the combustion temperature, or by selecting a propellant with a small molecular mass. For example, the bipropellant $O_2 + H_2$ mixture has molecular mass of $\mathcal{M} = 8.9$ kg/mol, and the bipropellant $O_2 + RP1$ has a higher molecular mass of $\mathcal{M} = 21.9$ kg/mol.

Figure 2.6 illustrates the effect of pressure ratio on gas flow velocity as the gas expands through the nozzle. Increasing chamber pressure has a favorable effect on exhaust velocity. However, the design of a rocket chamber limits p_c because a high pressure increases the structural weight requirement, and in the case of liquid chemical propellant engines, this also means increased propellant feed system weight. In practice, as p_c/p_e increases, the nozzle-area ratio must increase.

If $p_e = 0$, it would result in an infinite expansion, reaching the maximum exhaust flow velocity. The maximum theoretical value of the exhaust velocity is finite, even though the pressure ratio is infinite, as it represents the finite thermal energy content of the propellant.

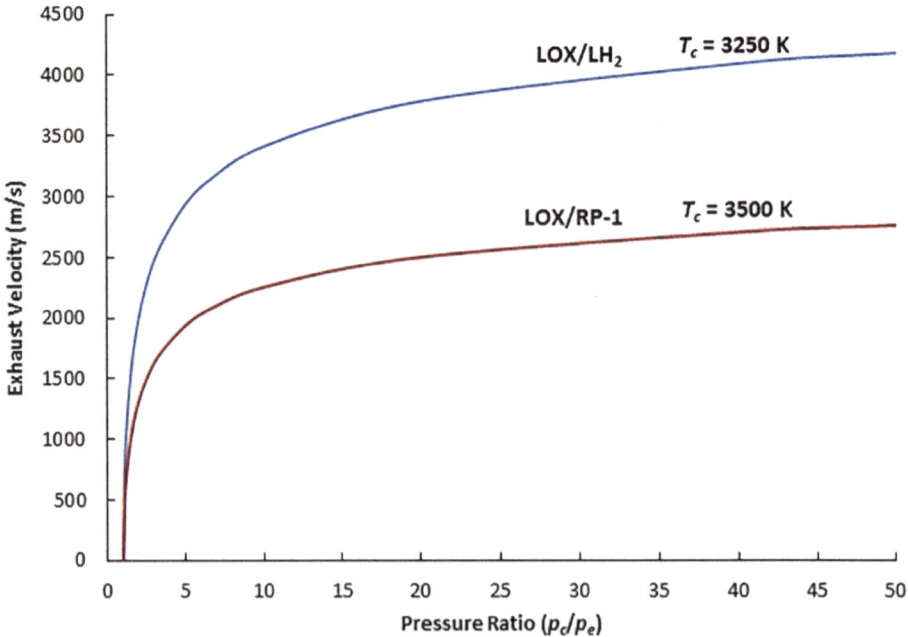

Fig. 2.6 Gas exhaust velocity as a function of the nozzle pressure ratio

The ratio of the exit velocity to the maximum possible exhaust velocity in a rocket is thus

$$\frac{v_{ex}}{(v_{ex})_{\max}} = \sqrt{1 - \left(\frac{p_e}{p_c}\right)^{\gamma - 1/\gamma}} \qquad (2.34)$$

This expression emphasizes that the pressure ratio in the rocket engine determines what fraction of the maximum possible velocity is ideally attained.

To get a perspective of the combustion chamber conditions, consider the Space Shuttle Main Engine (SSME), also known as the Rocketdyne RS-25. In its chamber, this engine mixes and burns the hydrogen fuel and oxygen oxidizer at high pressure (2×10^7 Pa or 2970 psia) and at temperatures exceeding 3588 K (3315 °C or 6000°F) to produce a sea-level thrust of 1668 kN (375,000 lbf) and a vacuum thrust of 2091 kN (470,000 lbf). With a nozzle area ratio of 77.7:1, the SSME yields a specific impulse of 455.2 s (in vacuum) and 363.2 s (at sea level).

The correlation of $T_c/\mathcal{M}$ and v_{ex} for several rocket propellants is discerned with the data in Table 2.1 adapted from that published by Brown [1]. Those values were obtained for vacuum conditions, optimum expansion, and rocket nozzle with an area ratio of 30. The results help us explain why rocket engines combusting LOX/LH$_2$ have higher specific impulse compared with engines burning other bipropellant mixtures.

In practice, it is found that the molecular mass of the hot gases increases slightly when flowing through a nozzle, thereby changing the gas density. In addition, some heat is transferred to the nozzle walls, lowering the temperature in the nozzle, and this increases the density and exhaust mass flow slightly.

Table 2.1   Exhaust velocity for monopropellant and bipropellant rockets

Propellant	$T_c(\text{K})$	$\mathcal{M}$	$v_{ex}(\text{m/s})$	$I_{sp}(\text{s})$ vac
LOX/LH$_2$	3553	10	4426	451
LOX/RP-1	3672	23.3	3466	353
N$_2$O$_4$/MMH	3389	21.5	3298	336
Monopropellant N$_2$H$_4$	1228	13	2326	237
Cold gas H$_e$	289	4	1551	158
Cold gas N$_2$	289	28	668	68

Data adapted from *Spacecraft Propulsion*, Brown [1]. Reprinted by permission of the American Institute of Aeronautics and Astronautics, Inc.

Question: If the rocket exhaust velocity depends inversely on the square root of the gas molecular mass (low $\mathcal{M}$ yields higher v_{ex}), explain why stage rockets in some launch vehicles use propellants with a high molecular mass.

A. Thrust also depends on mass flow rate ($F \propto \dot{m}$), and a heavy propellant may yield a higher mass flow for a rocket nozzle of a given throat area compared with that obtained with a lower mass propellant. Thus, although the exhaust velocity may be lower, the overall thrust will be higher. Such result is good for a first or booster stage for which a highest thrust is crucial to lift-off and overcome gravity as the vehicle gains altitude.

In thermodynamic analysis, we use the Mach number to obtain many useful relations for the ideal nozzle. The Mach number relates the flow velocity to the local sonic velocity of the gas, $M = v/a$, where the speed of sound in ideal gases is largely dependent on temperature, and it is given by $a = \sqrt{\gamma RT}$. Assuming isentropic expansion of the gas through the convergent-divergent nozzle, the Mach number monotonically increases from near 0 in the rocket chamber to $M_t = 1$ at the nozzle throat, and to the supersonic value M_e at the exit plane of the nozzle.

Once the gas Mach number axial distribution is established, then the corresponding variation of temperature and pressure in the nozzle can be found. When the local velocity comes close to zero, the local temperature and pressure will approach the stagnation pressure and stagnation temperature. The first law of thermodynamics applied to a calorically perfect gas gives a relationship between stagnation and static temperature and the flow velocity v as $T_c = T + v^2/2c_p$, where T_c denotes the stagnation condition in the rocket chamber. This is justified since in the combustion chamber, the gas velocity is relatively small, and thus the local combustion pressure is essentially equal to the stagnation pressure.

In terms of the Mach number, the first law gives a relationship for the temperature ratio of the gas as it moves through the rocket nozzle, where the subscript x denotes the axial direction:

$$\frac{T_c}{T_x} = \left[1 + \frac{\gamma - 1}{2} M_x^2\right] \tag{2.35}$$

Since the stagnation to static pressure ratio is $p_c/p = (T_c/T)^{\gamma/(\gamma-1)}$, we can write the ratio of chamber stagnation to static pressure for the flow moving along the nozzle as

$$\frac{p_c}{p_x} = p\left[1 + \frac{\gamma - 1}{2} M_x^2\right]^{\gamma/(\gamma-1)} \tag{2.36}$$

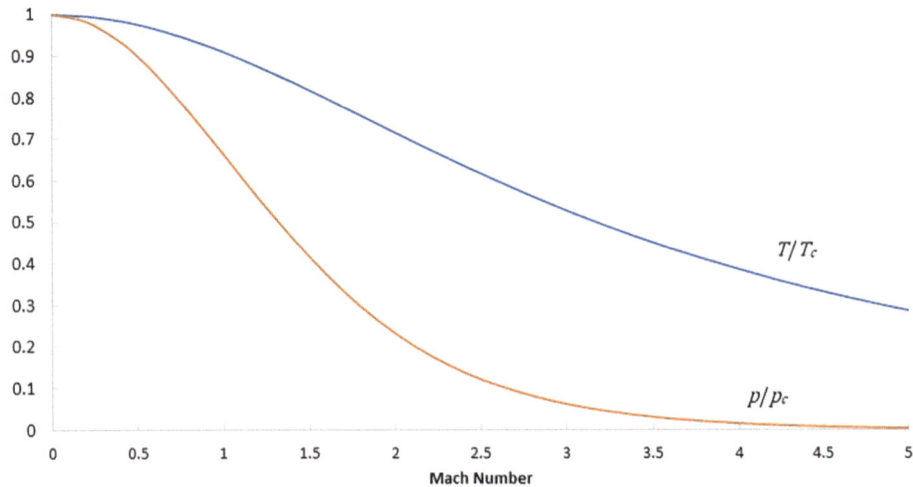

Fig. 2.7 Axial distribution of pressure ratio and temperature ratio in ideal rocket nozzle

These ratios are plotted in Fig. 2.7. As shown, the gas pressure monotonically decreases from p_c in the chamber to $0.528\,p_c$ at the throat and achieves its lower value p_e, at the exit. Similarly, the gas temperature monotonically decreases from T_c in the chamber to $0.833T_c$ at the throat and decreases to its the lower value T_e, at the nozzle exit plane.

The maximum gas flow per unit area in the nozzle occurs at the throat. The pressure for which this condition is achieved is called the *critical pressure* p_t. Therefore, for $M_t = 1$ at the throat, Eq. (2.36) reduces to the *critical gas pressure ratio* p_t/p_c:

$$\frac{p_t}{p_c} = \left(\frac{2}{\gamma + 1}\right)^{\gamma/(\gamma - 1)} \tag{2.37}$$

Typical values of the critical pressure ratio are between 0.53 and 0.57, depending on the specific heat ratio of the gas. For propellant gases the range of specific heat ratios is narrow; the lowest value is about 1.2, and the maximum possible value is 1.67 (helium).

Equations (2.35–2.37) represent important relations for stagnation properties in the rocket engine.

2.7.4   Mass Flow Rate

When we derived the equation for the velocity of the exhaust gases, we found that it is controlled to a certain extent by the pressure ratio (pressure difference between the combustion chamber and the exit plane of the nozzle, [see Eq. (2.33)], which is an important nozzle design parameter. Thus, we need to determine how this pressure difference relates to the mass flow rate. This we do starting with starting with the mass flow rate equation, $\dot{m} = \rho A v$. Use the ideal gas equation of state $p = \rho R T$, and the isentropic flow relations, we write the velocity in a form applicable to the compressible fluid flow in the nozzle.

We can use Eq. (2.33) to express the gas velocity at any point in the nozzle as

$$v = \sqrt{\frac{2\gamma}{\gamma - 1} \frac{R T_c}{M} \left[1 - \left(\frac{p}{p_c}\right)^{(\gamma - 1)/\gamma} \right]} \tag{2.38}$$

where v and p represent the local pressure and velocity of the gas (at any point in the nozzle).

An expression for the gas density is readily obtained from the equation of state in terms pressure ratio across the nozzle and chamber pressure and temperature:

$$\rho = \frac{p_c M}{R T_c} \left(\frac{p}{p_c}\right)^{1/\gamma} \tag{2.39}$$

Use Eq. (2.38) and Eq. (2.39) to obtain a relation for the mass flow rate as

$$\dot{m} = p_c A \sqrt{\frac{2\gamma}{(\gamma - 1)} \frac{M}{R T_c} \left(\frac{p}{p_c}\right)^{2/\gamma} \left[1 - \left(\frac{p}{p_c}\right)^{\gamma - 1/\gamma} \right]} \tag{2.40}$$

This equation relates the mass flow rate $\dot{m}$ to the nozzle design represented by the cross-sectional area A, and the flow properties (stagnation or chamber pressure and temperature, p_c and T_c, the ratio of specific heats of the gas γ, and the gas constant $R = R/M$).

Did you know? Each SSME on the Space Shuttle Orbiter burning LOX/LH$_2$ had a high chamber pressure of 20.477 MPa (2970 psia) (significantly higher than previous engines) and a flowrate of 467.2 kg/s (1030 lbm/s).

The maximum mass flow rate occurs when the flow is *choked*, that is when the Mach number is 1 at the nozzle throat, the section with the smallest cross-sectional area A_t.

$$\dot{m} = \frac{p_c A_t}{\sqrt{T_c}} \sqrt{\frac{\gamma}{R} \left(\frac{2}{\gamma + 1}\right)^{\gamma + 1/(\gamma - 1)}} \tag{2.41}$$

The *choked flow* equation is in important result for a rocket engine, as we can calculate the mass flow rate by simply using the conditions in the combustion chamber and the throat area.

2.7.5 Nozzle Area Ratio

Solving for the cross-sectional area of the nozzle from the mass flow rate Eq. (2.40), obtain

$$A = \frac{\dot{m}}{p_c} \left\{ \frac{2\gamma}{(\gamma-1)} \frac{\mathcal{M}}{\mathcal{R}T_c} \left(\frac{p}{p_c}\right)^{2/\gamma} \left[1 - \left(\frac{p}{p_c}\right)^{(\gamma-1)/\gamma}\right] \right\}^{-1/2} \tag{2.42}$$

Since the performance figure of merit is the nozzle exit area A_e, the expansion of the nozzle can be expressed with the dimensionless ratio relating the throat and exit areas. Denoted by ε, this is the nozzle area ratio, which is an important rocket design parameter:

$$\varepsilon \equiv \frac{A_e}{A_t} \tag{2.43}$$

The area ratio is a measure of the gas expansion process within the nozzle, and is related to the pressure ratio as follows:

$$\varepsilon = \frac{A_e}{A_t} = \left(\frac{p_c}{p_e}\right)^{1/\gamma} \left[1 - \left(\frac{p_e}{p_c}\right)^{(\gamma-1)/\gamma}\right]^{-1/2} \sqrt{\frac{\gamma-1}{2} \left[\frac{2}{\gamma+1}\right]^{(\gamma+1)/(\gamma-1)}} \tag{2.44}$$

An optimum area ratio results when the expanding flow at the exit plane pressure equals the local ambient pressure. In the propulsion of a launch vehicle, the first stage engines are designed for optimum area ratio at the average back pressure expected during the mission; that is, the value of ε is selected near midpoint of the ascent flight, having a typical value of $\varepsilon \approx 10$. For an upper stage or a spacecraft rocket the optimum area ratio is infinite. Obviously, the largest area ratio allowed by weight and volume available in the vehicle will be selected.

When the gas exhaust pressure is less than the surrounding ambient pressure ($p_e < p_a$), the pressure thrust term is negative. This is an undesirable condition because it lowers the overall thrust of a rocket engine. The rocket nozzle design that permits expansion of the exhaust gases to the pressure of the external field has an *optimum expansion ratio*. The optimum value involves a trade-off in the overall performance. For a launch vehicle, the rocket thrust will vary as it ascends through the atmosphere. The change in pressure-area nozzle thrust with altitude can be from 10 to 30% of the overall thrust. Once in space, the effect of the atmosphere vanishes. For example, since the SSME operated mostly in space where ambient pressure is zero, it needed a large area ratio (77.5:1) to fully expand the exhaust gas. For the NASA SLS, engineers report $\varepsilon = 69$ for the RS-25 engine (core stage), and $\varepsilon = 285$ for the RL10B-2 (second stage).

The above thermodynamic equations describe the nozzle expanding flow, an idealization and simplification of the rocket engine's full two- or three-dimensional equations and its real aerothermochemical behavior. However, the assumptions and simplifications we make here are adequate for obtaining useful performance trends applicable to rocket propulsion. For preliminary analysis, we assume an average specific ratio for the combustion gases, which ranges from 1.1 to 1.4 for most propellants. Mach number is small in the combustion chamber. A typical value is $M_c = 0.31$ for a thrust chamber with a contraction area ratio of $A_c/A_t = 2$, and $\gamma = 1.2$ for conventional gas mixtures. For real chemical rockets the measured performance is less than 6% below the calculated ideal value.

2.8   Thermal Rocket Engine Performance

The performance of a rocket engine is described by its thrust (F) and specific impulse (I_{sp}). Two additional parameters are very helpful in determining the performance: the *thrust coefficient* C_F, and the *characteristic velocity* C^*. We can use the previous thermodynamic concepts to write very useful expressions for these parameters.

Equation (2.6) is a general thrust expression applicable to all rocket engines. Now we can write it in a form that applies to an ideal rocket based on thermodynamic principles, where the propellant gas undergoes an adiabatic isentropic expansion process. By substituting (2.33) and (2.40) into (2.6), we obtain the ideal thermodynamic rocket thrust with a gas represented by constant γ throughout the expansion process as:

$$F = A_t p_c \sqrt{\frac{2\gamma^2}{\gamma - 1}\left(\frac{2}{\gamma + 1}\right)^{(\gamma+1)/(\gamma-1)}\left[1 - \left(\frac{p_e}{p_c}\right)^{(\gamma-1)/\gamma}\right]} + p_e A_e - p_a A_e \qquad (2.45)$$

This relation is the full thermodynamic thrust equation of an ideal rocket engine, which assumes frozen flow (constant γ). It shows that the momentum thrust is proportional to the nozzle throat area A_t and the chamber pressure p_c (nozzle inlet pressure), and is a function of the pressure ratio across the nozzle p_c/p_e, and the specific heat ratio γ. The pressure thrust term $[A_e(p_e - p_a)]$ may be positive or negative, depending on whether the exhaust pressure is higher or lower than the ambient pressure.

Operating in the vacuum of space, an ideal nozzle with infinite exit area A_e can achieve the maximum theoretical performance. Such nozzle expands the high-pressure gases from the thrust chamber to zero pressure $(p_e = 0)$, thereby attaining the maximum gas velocity. In a vacuum, $I_{sp} = v_{ex}/g_0$ is independent of chamber pressure. You can verify this is true because the exhaust velocity is solely determined by the expansion ratio of the nozzle and the properties of the propellant (see Eq. 2.33), not the absolute chamber pressure in a vacuum environment. That is why spacecraft rocket engines can (and do) have low chamber pressures compared to launch vehicle first stages. Monopropellant thrusters do not adhere to this rule, as they exhibit a second-order chamber pressure dependence due to the non-ideal dissociation of the gas.

2.8.1 Thrust Coefficient

The thrust coefficient C_F represents nozzle performance, giving a measure of the efficiency with which the rocket extracts energy from the hot gas within. It is defined as the ratio of the thrust F to the product of combustion pressure and the throat area:

$$C_F \equiv \frac{F}{p_c A_t} \tag{2.46}$$

The thrust coefficient is typically given in terms of the pressure ratio across the engine. Hence, C_F depends on the nozzle design. Using the thermodynamic equations, it can be shown

$$C_F = \left\{ \frac{2\gamma^2}{\gamma - 1} \left(\frac{2}{\gamma + 1} \right)^{(\gamma+1)/(\gamma-1)} \left[1 - \left(\frac{p_e}{p_c} \right)^{(\gamma-1)/\gamma} \right] \right\}^{1/2} + \left(\frac{p_e}{p_c} - \frac{p_a}{p_c} \right) \frac{A_e}{A_t} \tag{2.47}$$

The first term (the square root brackets) represents the contribution to thrust by the exhaust velocity or momentum; this is equivalent to $(\dot{m} v_{ex})/(p_c A_t)$. The last term, which is equivalent to $[(p_e - p_a)A_e]/(p_c A_t)$, represents the contribution to thrust by the gas exit pressure.

For vacuum operation, $p_a = 0$, and if we assume perfect gas expansion in the nozzle, $p_e = 0$, Eq. (2.47) reduces to

$$C_F = \left\{ \frac{2\gamma^2}{\gamma - 1} \left(\frac{2}{\gamma + 1} \right)^{(\gamma+1)/(\gamma-1)} \right\}^{1/2} \tag{2.48}$$

Numerical values of C_F can range between 0.6 and 2.245, with the larger values for vacuum operation conditions.

2.8.2 Characteristic Velocity

The characteristic velocity measures the efficiency of thermal energy conversion (in the combustion chamber for chemical rockets) into kinetic energy of the flow. Denoted as C^*, the characteristic velocity is defined as

$$C^* \equiv \frac{p_c A_t}{\dot{m}} \tag{2.49}$$

where p_c is the pressure in the chamber, A_t is the cross-sectional area of the nozzle throat, and $\dot{m}$ is the flow rate of the hot gases.

In terms of thermodynamic parameters, the characteristic velocity is written as

$$C^* = \frac{\sqrt{\gamma R T_c}}{\gamma \sqrt{\left(\frac{2}{\gamma+1}\right)^{(\gamma+1)/(\gamma-1)}}} \tag{2.50}$$

where γ and the gas constant $R = \mathcal{R}/\mathcal{M}$ are indicators of propellant composition, whose value is determined by the propellant combustion process. Thus, in this form, the characteristic velocity provides insight into the performance of propellants and the combustion process of a chemical rocket. The composition of the combustion gases establishes γ and the molecular mass of the exhaust.

The value of C^* for chemical rockets depends on the flame temperature in the combustion chamber. Adiabatic flame temperature is the temperature achieved by a combustion reaction that takes place adiabatically (with no heat loss, in or out of the system). Being the maximum temperature that can be achieved for the given reactants, adiabatic flame temperature is used to estimate the combustion chamber temperature (T_c) in a rocket engine.

A well-designed combustion chamber should give a value of C^* greater than 90% of the theoretical value found by modeling thermochemical combustion reactions with computational fluid dynamics (CFD) codes. Typical values of the characteristic velocity are $914 \text{ m/s} < C^* < 2134 \text{ m/s}$. If the SSME has a vacuum $I_{sp} = 455$ s, sea level $I_{sp} = 363$ s, and $C^* = 2330$ m/s, we should expect that during ascent to LEO the thrust coefficient varies as $1.53 \leq C_F \leq 1.91$.

By defining a combustion parameter in terms of the chamber combustion temperature and exhaust gas molecular mass as $\sqrt{T_c/\mathcal{M}}$, we can easily conclude that high exhaust velocities result from maximizing this parameter. Hence, we use the characteristic velocity C^* to compare the relative performance of different chemical rocket propulsion designs and propellants, as it relates to the efficiency of the combustion.

Since the exhaust velocity can also be written as $v_{ex} = C = C_F C^*$, the rocket thrust becomes

$$F = \dot{m} C_F C^* \tag{2.51}$$

Finally, we rewrite the specific impulse as follows:

$$I_{sp} = C_F C^* / g_0 \tag{2.52}$$

This expression can be used to optimize the specific impulse of a rocket engine, knowing the thrust coefficient, which is related to the nozzle shape and the external pressure, and the characteristic velocity, which is related to the propellant characteristics. The latter is clear from Eq. (2.52). Hence, a high specific impulse can be obtained if the average molecular mass of the reaction products of combustion is low (formulation rich in hydrogen), or if the available chemical energy (heat of reaction) is large, which means high combustion temperatures.

Example 2.4 Consider a second stage rocket that will ignite and begin operation at high altitude. From a preliminary design analysis, it was determined an ideal thrust of 445 kN, and the following data: $T_c = 3200$ K, $I_{sp} = 400$ s, $A_t = 0.046$ m^2, $\varepsilon = 94$.

Assume the combustion gas behaves as calorically perfect with $R = 0.375\frac{\text{kJ}}{\text{kg·K}}$, $\gamma = 1.26$, determine the following: (a) Propellant Mass flow rate, $\dot{m}_p$; (b) Characteristic velocity C^*; (c) Combustion chamber pressure p_c; (d) Effective exhaust velocity C at the design altitude; (e) Ideal thrust coefficient C_{Fi} at the design altitude; (f) Nozzle exit diameter D_e; (h) Since 445 kN is the ideal thrust, how do you estimate the actual thrust? Discuss rocket performance losses and explain how they could be quantified.

Solution:

(a) Propellant Mass flow rate, $\dot{m}_p$. From Eq. (2.13),

$$\dot{m}_p = \frac{F}{I_{sp}g_0} = 113.44 \text{ kg/s}$$

(b) Characteristic velocity C^* is found from Eq. (2.50) with $\gamma = 1.26$, and $RT_c = 1200$ kJ/kg,

$$C^* = 1660 \text{ m/s}$$

(c) Combustion chamber pressure p_c from Eq. (2.49)

$$p_c = \frac{\dot{m}_p C^*}{A_t} = 4.093 \text{ MPa}$$

(d) Effective exhaust velocity C at the design altitude is simply

$$C = g_0 I_{sp} = 3922.64 \text{ m/s}$$

(e) Ideal thrust coefficient C_{Fi} at the design altitude is

$$C_{Fi} = \frac{C}{C^*} = 2.36$$

(f) Nozzle exit diameter D_e is obtained with $\varepsilon = 94$ and with the throat area A_t known:

$$\varepsilon = \frac{A_e}{A_t} \rightarrow A_e = \varepsilon A_t \rightarrow A_e = \pi D_e^2/4$$

$$D_e = 2\sqrt{\frac{\varepsilon A_t}{\pi}} = 2.346 \text{ m}$$

(g) We expect the rocket nozzle to suffer some flow divergence losses, in addition to heat transfer losses, incomplete combustion, and other inefficiencies in the expansion process to make the actual thrust perhaps reduced by 2 or 3% of the ideal value. We could quantify those losses by carrying out a thermophysical analysis to account for reduced combustion efficiency, less than ideal expansion, and then determine the actual value of v_{ex} to calculate F_{actual}.

Final Exercise: Verify the thrust coefficient for the engines in the following table, and estimate the missing values.

Engine	Propellants	Thrust (sl) (kN)	I_{sp}(sl) (s)	p_c(MPa)	D_e(m)	C_F(sl)
SSME (RS-25)	LOX/LH$_2$	1853	363	20.4	2.4	1.53
RD-253	N$_2$O$_4$/UDMH	1410	267	14.7	1.5	1.37
Vulcain 2	LOX/LH$_2$	1350	318	11.6	2.15	1.44
RD-170	LOX/Kerosene	1925	309	24.5	4.2	
		Thrust (vac) (kN)	I_{sp}(vac) (s)			C_F(vac)
RL 10A-4-1	LOX/LH$_2$	99	451	3.9	1.53	
Aestus	N$_2$O$_4$/MMH	29	324	1.1	1.0	1.87
Vinci	LOX/LH$_2$	180	465	6.0	2.15	
J-2	LOX/LH$_2$	1052	425	3.0	2.1	

The ideal rocket parameters introduced in this chapter can be corrected by applying appropriate measures. For example, we can use empirical correction factors based on experimental data, or we develop a more detailed physical model that includes more accurate theoretical analysis and simulation of the thermochemical process in the rocket engine, including injection, propellant mixing, combustion kinetics, turbulence, energy losses. Such analysis requires a sophisticated computational approach to model and simulate the complex flow phenomena in a real rocket engine.

Notation and Definitions for Rocket Propulsion

Notation	Name	Definition
A_e/A_t	Area ratio	Ratio of the nozzle exit area to the throat area
C	Effective exhaust velocity	Average equivalent velocity at which propellant is ejected from the rocket engine. Adjusted velocity to compensate for the small pressure-area thrust produced by a real vacuum engine, a value that is readily determined on a test stand by measuring thrust and propellant flow rate

(continued)

(continued)

Notation	Name	Definition
C^*	Characteristic velocity	Empirical rocket parameter used to measure combustion performance by indicating how much mass of propellant must be burned to maintain chamber pressure. It depends on propellant combusted but is independent of nozzle performance and does not vary with ambient pressure
C_F	Thrust coefficient	Thrust divided by the chamber pressure p_c and the throat area A_t. Rocket parameter for assessing the effects of chamber pressure or altitude variations in a nozzle configuration, or to correct sea-level results for flight altitude conditions; $0.8 < C_F < 1.9$
Δv	Delta-v or Velocity Increment	Velocity change imparted to a space vehicle caused by the reaction to a given thrust applied for a given time (thrust times time or "impulse"). See also the rocket equation
F	Thrust	Force produced by a rocket propulsion system acting upon the vehicle that it propels. Force generated by momentum exchange between the exhaust and the rocket and by the pressure imbalance at the nozzle exit
F/w_0	Thrust to weight ratio	Acceleration (in multiples of Earth's surface acceleration of gravity g_0) that an engine can give to its own loaded propulsion system mass. Parameter used to compare different types of rocket engines
I	Total Impulse	Thrust force F (which can vary with time) integrated over the burning time t. Change in momentum caused by a force thrust acting over time
I_{sp}	Specific Impulse	Thrust per unit weight flow rate of propellant spent. Total impulse per unit weight of propellant consumed
p_c/p_t	Critical pressure ratio	Pressure ratio required for sonic, or choked, flow at the nozzle throat
v_{ex}	Exhaust Velocity	Velocity of the exhaust gas at the nozzle exit plane relative to the rocket or vehicle
μ	Mass Ratio	Final mass m (after rocket has consumed all usable propellant) divided by initial mass m_0 (before rocket operation). It can apply to a vehicle or to a particular propulsion stage
ζ	Propellant mass fraction	Fraction of propellant mass m_p in an initial mass m_0. It can apply to a vehicle, a rocket stage, or to an entire propulsion system. In this work we use it for a rocket engine

Recommended Reading and References

1. Brown, C. D. (1996). *Spacecraft Propulsion*. American Institute of Aeronautics and Astronautics, AIAA education series.
2. Gordon, S. and McBride, B.J. (1994).Computer Program for Calculation of Complex Chemical Equilibrium Compositions and Applications. NASA Reference Publication 1311, Oct. 1994. http://www.grc.nasa.gov/WWW/CEAWeb/RP-1311.htm.
3. Griffin, M. D. and French, J.R. (2004). *Space Vehicle Design*. 2nd Ed, AIAA Education Series, 2004.
4. Hagemann, G., Immich, H., Nguyen, T.V, and Dumnov, G.E. (1998). "Advanced Rocket Nozzles," J. of Propulsion and Power, Vol. 14, No. 5, September–October 1998.
5. Haidn, O.J. (2008), *Advanced Rocket Engines*. In Advances on Propulsion Technology for High-Speed Aircraft (pp. 6-1–6-40). Educational Notes RTO-EN-AVT-150, Paper 6. Neuilly-sur-Seine, France : RTO. Available from: http://www.rto.nato.int.
6. Isakovic, S.J., Hopkins Jr., J.P., Hopkins, J.B., *International Reference Guide to Space Launch Systems*, 4th Ed. AIAA, 2004.
7. "JANNAF Rocket Engine Performance Prediction and Evaluation," CPIA 246, April 1975.
8. Kehtarnavaz, H., Coats, D.E., and Dang, A.L. (1990). "Viscous Loss Assessment in Rocket Engines", Journal of Propulsion and Power, Vol. 6, No. 6, pp 713–717, Nov.-Dec., 1990.
9. Kliegel, J.R., Nickerson, G.R., Frey, H.M., Quan, V., and Melde, J.E. (1968). "Two-Dimensional Kinetics Nozzle Analysis Computer Program-TDK." Prepared for the ICRPG Performance Standardization Working Group, July 1968.
10. Kuentzmann, P. (2002). *Introduction to Solid Rocket Propulsion*, Paper presented at the RTO/VKI Special Course on "Internal Aerodynamics in Solid Rocket Propulsion," held in Rhode-Saint-Genèse, Belgium, 27–31 May 2002, and published in RTO-EN-023.
11. Kushida, R., Hermal, J., Apfel, S., and Zydowicy, M. (1987). "Performance of High-Area Ratio Nozzle for a Small Rocker Thruster," J. Propulsion and Power, Vol. 3, No. 4, 1987.
12. Levine, J.N. (1971). "Transpiration and Film Cooling Boundary Layer Computer Program." Dynamic Science, prepared for NASA, contract NAS7i- 791, June 1971.
13. NASA SP-8120. NASA Space Vehicle Design Criteria "Liquid Rocket Engine Nozzles," 1976.
14. NASA Computer program CEA2 (Chemical Equilibrium with Applications 2) http://www.grc.nasa.gov/WWW/CEAWeb.
15. Nickerson, G.R., Berker, D.R., Coats, D.E., and Dunn, S.S. (1993). "Two-Dimensional Kinetics (TDK) Nozzle Performance Computer Program Volume II, User's Manual." Prepared by Software and Engineering Associates, Inc. for George C. Marshall Space Flight Center under contract NAS8-39048, March 1993.
16. Pieper, J.L. (1968). "ICRPG Liquid Propellant Thrust Chamber Performance Evaluation Manual," CPIA 178, Sept. 1968.
17. Ponomarenko, A. (2010). RPA: Tool for Liquid Propellant Rocket Engine Analysis C++ Implementation, May 2010.
18. Sutton, G.P. and Biblarz, O. (2001). *Rocket Propulsion Elements*. Seventh Edition, John Wiley & Sons, Inc. New York, 2001.
19. Turner, M.J.L. (2006). *Rocket and Spacecraft Propulsion, Principles, Practice and New Developments*. Second Edition, Springer-Praxis Books (2006).

From Earth to Orbit and Spaceflight Maneuvering

3

> *Propulsion establishes how far and how fast we can navigate through the dark cosmic oceans.*
>
> —*Dora Musielak*

To deliver a payload from Earth to orbit, and to accomplish its space mission, both the launch vehicle and the spacecraft need propulsion with all supporting systems (e.g. propellant, controls). However, the performance requirements of the two vehicles are very distinct since the operational environment and times are quite different for the two propulsion systems: A typical multi-stage launch vehicle operates for less than 20 min, enough to transverse Earth's atmosphere to reach orbital velocity exceeding 9 km/s. Once delivery of the payload is accomplished, whatever remains of the launch vehicle (upper stage) is discarded. In contrast, a spacecraft operates in deep space, and its low thrust propulsion is used for maneuvers that require total change in velocity of less than 1 km/s; the in-space rockets must be ready to restart and carry out orbital maneuvers according to the needs of the years-long mission, including short impulse burn for orbit maintenance and trims. The launch vehicle requires very high thrust-to-weight rocket engines to lift off the ground, and achieve orbital velocity. A spacecraft requires propellant-efficient propulsion system to accelerate while in orbit and to perform in-space maneuvers until it completes its mission.

In this chapter, we want to answer questions such as these:

- What is gravity loss, and how does it affect the ascent of a space launch vehicle?
- Why does the propulsion system of launch vehicles require multi-stage rockets?

© The Author(s), under exclusive license to Springer Nature Switzerland AG 2025   83
D. Musielak, *Introduction to Rocket Propulsion for Astronautics*, Synthesis Lectures on Engineering, Science, and Technology, https://doi.org/10.1007/978-3-031-86141-3_3

- How does a spacecraft escape the gravitational influence of the Earth?
- What is the launch injection characteristic energy C_3 for an interplanetary mission?
- How much propellant is required for a spacecraft to maneuver?

3.1   Launch Dynamics

The launch trajectory is composed of 3 main parts: ascent, coasting (no rocket power), and injection burn or apogee boost. The first part of the trajectory, also known as boost phase, begins with lift-off and the launch vehicle ascends, traversing the atmosphere to a point where rocket thrust is terminated at MECO (main engines cut-off). In the coasting phase, the unpowered vehicle is in weightless flight with no aerodynamic forces, and it follows an elliptic transfer to a predetermined orbit. The last phase of launch is the injection burn, also known as *apogee boost* because the upper stage rocket is ignited to transfer the payload/spacecraft into the target orbit. In this section we are concerned with the equations of motion for the ascent phase.

The goal of the powered ascent phase (lasting about 12-min) is to position the spacecraft/payload for insertion in the appropriate orbit. With current expendable launch vehicles (ELVs), the ascent from Earth begins vertically. This is done to overcome the dense atmosphere as quickly as possible. At takeoff, the launch vehicle's propulsion system must deliver a thrust 10–20% greater than the vehicle mass. The performance of launch vehicles is improved by staging the rocket propulsion system. The empty rocket stages are dropped when the propellant in that stage is exhausted, reducing the mass of the vehicle as it ascends, and in the meantime the next stage is fired to propel the remaining part of the lighter mass vehicle. Profiles for both two- and three-stage missions are depicted in Fig. 3.1.

3.1.1   Launch Ascent Trajectory

Consider a two-stage SLV on a mission to low Earth orbit (LEO) such as to deliver a payload to the ISS, which circles the planet at a nominal altitude of 400 km (250 miles). After separation the second stage burns for approximately 340 to 420 s, reaching second stage cutoff (SECO 1). The second stage with its attached payload follows a Hohmann transfer path to the desired LEO altitude. Near apogee, the second stage is reignited and completes its burn to circularize the orbit. Spacecraft separation takes place approximately 250 s after second-stage cutoff command (SECO 2), Fig. 3.1a. After the spacecraft separates, the spent upper stage performs an orbit disposal maneuver for a controlled re-entry into Earth's atmosphere.

For a mission to Geosynchronous Transfer Orbit (GTO), a three-stage launcher uses the first burn of the second stage to place the payload into a 185-km (100-nmi) parking

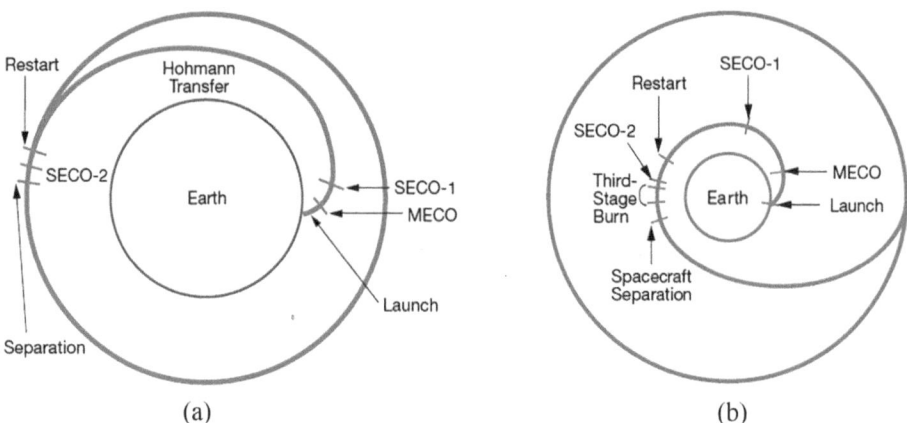

(a)   (b)

Fig. 3.1 Typical profiles for (**a**) two-stage to LEO, (**b**) and three-stage to GTO. *Note* Final circular orbit provided by spacecraft propulsion

orbit inclined at 28.7 degrees. The vehicle coasts to a position near the equator where the second stage restarts and burns until second stage cutoff (SECO 2). The third stage separates and burns, taking the payload to establish the required GTO, as depicted in Fig. 3.1b. Depending on mission requirements and spacecraft mass, the third stage can reignite to remove some inclination out of the Earth parking orbit. In Sect. 3.5 we describe the powered maneuver to achieve it.

The booster phase begins with the ignition of the first-stage main rocket engines followed by ignition of strap-on solid rocket motors (SRMs), if applicable, at liftoff. The vehicle ascends and, after about a minute, the SRMs burnout and are jettisoned. The main stage rocket engines continue to burn until main engine cutoff (MECO). By then the vehicle begins to turn in its trajectory (see Fig. 3.2). To reach orbital velocity, the launch vehicle must move fast through the densest part of the atmosphere to minimize drag and aerodynamic heating.

As shown in Fig. 3.2, the greater part of the flight ascending path is inclined with respect to the gravitational field to gain speed in the horizontal direction. The vehicle tilts by about 10° when it reaches an altitude of about 170 km. Afterwards the vehicle accelerates firing its rockets for a 5–8-min burn to gain horizontal speed. After crossing the denser air layers, the bending trajectory changes the velocity vector in a controlled manner for the upper stage carrying the payload to reach orbital velocity at perigee (point on the orbit closest to Earth). The rocket thrust vector aligns with the velocity vector, and the vertical gravity force causes the vehicle to pitch downward toward the horizon. The gradual transition from vertical to horizontal flight, illustrated in Fig. 3.2, is called **gravity turn trajectory**. On the horizontal trajectory, the vehicle is angled upwards to fight gravity and maintain altitude until orbital velocity is achieved. The exact speed required to orbit the Earth depends on the altitude of the orbit.

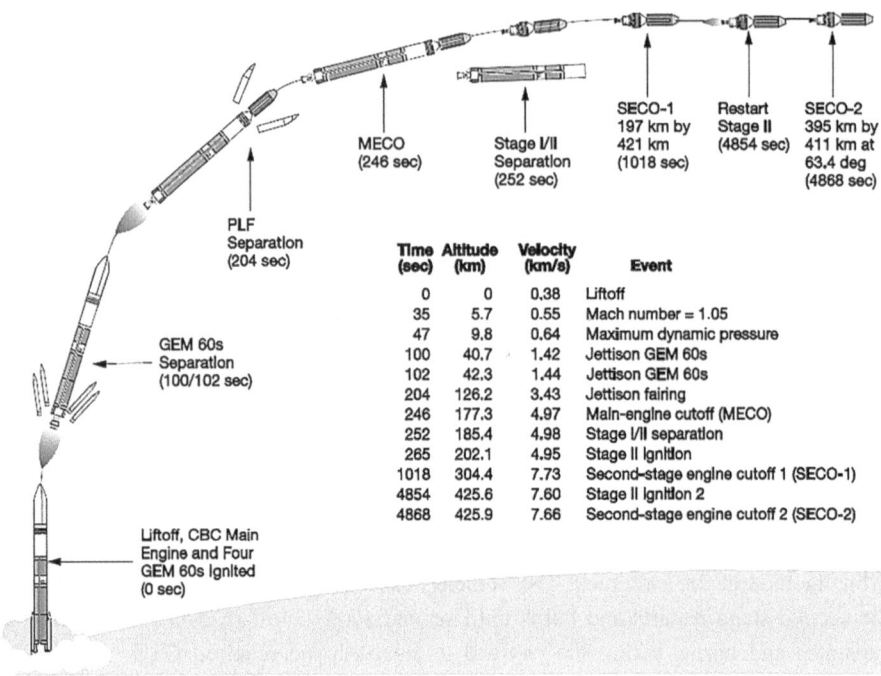

Fig. 3.2 Typical two-stage launch trajectory to LEO with four strap-on SRMs. Delta IV M+ (5,4) sequence of events for a LEO mission (western range). *Credit* ULA [16]

To intersect or achieve orbit, the rockets impart a high velocity (~ 9.5 km/sec) perpendicular to the vertical, and the optimal pitch angle must be as close to zero as possible.

The flight profile of a SLV is optimized for each mission. Typically, launch vehicle systems are designed with the flexibility of delivering payloads to different orbits, using two or three stages, and with or without strap-on SRMs. Hence, the number and arrangement of the stages depends on the mission. For example, the Atlas V-551, which was NASA's most powerful launch vehicle, was designed with a common core booster first stage, augmented by five strap-on solid rocket boosters. It included the Centaur upper stage to boost satellites into high orbit and to propel spacecraft into interplanetary space. For challenging, high energy interplanetary missions, and extra solid kick motor may be attached to the spacecraft, which in fact operates as third stage; the kick motor ensures that the spacecraft is injected with the escape velocity not only with respect to the Earth but also relative to the Sun. After payload separation, the upper stage is restarted to deplete any remaining propellants (depletion burn) and/or to move the stage to a safe distance from the spacecraft (evasive burn). Finally, after the spacecraft separates, the maneuvers of the upper stage are intended to release the spacecraft on the operational orbit or to trigger a controlled re-entry in the Earth atmosphere.

3.1.2 Equations of Motion

In Chap. 2 we derived the rocket equation assuming the vehicle moves in gravity-free rectilinear equilibrium flight, and that all control forces, lateral forces, and moments that tend to turn vehicle are zero. This idealization is not appropriate when we consider an SLV that ascends to orbit. In such situation the direction of flight may not coincide with the direction of thrust. The rocket equation must be redefined as an SLV accelerates along its flight path from lift-off until it achieves orbit.

Consider a launch vehicle moving through Earth's atmosphere. Being in the proximity of our planet, we neglect the gravitational attraction of the Sun, the Moon, and all other heavenly bodies. Assume that the vehicle moves in rectilinear equilibrium flight and that only three forces act on the vehicle: thrust F, gravitational force mg (acting from the vehicle to the center of the Earth), and drag forces F_D tangent to the flight path (opposing vehicle motion). In such situation, the trajectory is two-dimensional and is restricted to a fixed plane. Figure 3.3 illustrates the most important forces acting on the vehicle, and the path directions indicated by the tangential and normal unit vectors, $\hat{u}_t$ and $\hat{u}_n$, respectively. This model assumes that the thrust vector always points along the longitudinal axis of the vehicle, and thus F can be affected by changing the angle-of-attack α.

The equation of motion for the rocket vehicle is given by the projection along the direction of $\vec{v}$ (tangent to the path of motion). Using Newton's second law, we write the **general equation** of a rocket vehicle system in vector form as

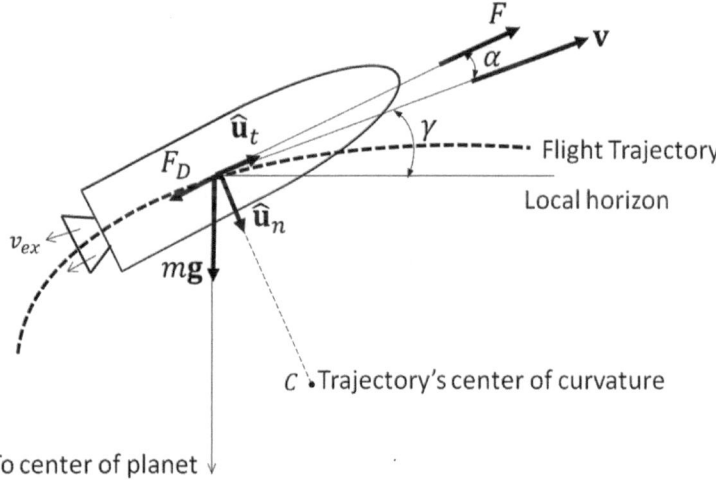

Fig. 3.3 Launch vehicle two-dimensional free-body diagram, where γ is the flight path angle, and α angle of thrust vector with the velocity vector (angle of attack)

$$\sum \vec{F}_{ext} = m\frac{d\vec{v}}{dt} + \vec{v}_{ex}\frac{dm}{dt} \qquad (3.1)$$

where the vector $\vec{F}_{ext}$ represents all external forces acting on the rocket vehicle system, including gravitational $\vec{F}_g$ and aerodynamic drag $\vec{F}_D$; the magnitudes of $\vec{v}$ and $\vec{v}_{ex}$ are v and v_{ex}, respectively, which are parallel to each other and have opposite direction.

The absolute acceleration in a rotating coordinate frame that moves with the vehicle is

$$\vec{a} = \frac{d\vec{v}}{dt} = \left(\frac{dv}{dt}\right)\vec{v}_t + \left(v\frac{d\gamma}{dt}\right)\vec{v}_n \qquad (3.2)$$

where γ is the angle of the flight measured from the local horizon to the velocity vector, dv/dt is the tangential acceleration due to the change in magnitude of the velocity vector, and $vd\gamma/dt$ is a normal acceleration component due to the rotation of the velocity vector with respect to the local horizon. For simplicity neglect all control forces, lateral forces, and moments that tend to turn the vehicle. Consider the equations of motion in two-dimensions:

The **acceleration along the same direction as the flight path** (dv/dt) is equal to the sum of propulsive, aerodynamic, and gravitational forces (tangential components), that is

$$m\frac{dv}{dt} = F\cos\alpha - F_D - mg\sin\gamma \qquad (3.3)$$

The **acceleration perpendicular to the flight path** $v(d\gamma/dt)$ is equal to the sum of propulsive, aerodynamic, and gravitational forces (normal components), that is

$$mv\frac{d\gamma}{dt} = F\sin\alpha - m\left(g - \frac{v^2}{r}\right)\cos\gamma \qquad (3.4)$$

where the term v^2/r represents the centrifugal force due to the rotation of the Earth.

The rate of change of flight path angle γ is discerned from Eq. 3.4. At lift off, $\gamma = 90°$. As the vehicle ascends and gains speed, the main engines are gimbaled to produce a small, programmed pitch-over, changing the flight path angle to less than 90°. As it gains altitude and speed, γ continues to decrease until $\gamma = 0°$ as orbital altitude/speed is reached.

The drag force F_D is a function of the vehicle's velocity v, its cross-sectional area A, the density of the atmosphere ρ, and a drag coefficient C_D, expressed as

$$F_D = AC_D\frac{1}{2}\rho v^2 \qquad (3.5)$$

The acceleration of gravity g varies inversely as the square of the distance from the Earth's center r. Thus, as a vehicle ascends to orbit, the change in g with altitude (h) can be determined with Newton's law of gravitation:

$$g = g_0\left(\frac{R_E}{r}\right)^2 = g_0\left(\frac{R_E}{R_E + h}\right)^2 \qquad (3.6)$$

where R_E denotes the mean equatorial radius of the Earth, $R_E = 6378.388$ km, and the standard acceleration due to gravity is $g_0 = 9.80665$ m/s^2.

In addition to these differential equations, two kinematic equations that define the radial velocity and down-track velocity projected along the Earth's surface are required to fully describe the motion state of the launch vehicle. A general case launch trajectory requires six equations: three for translation along each of three perpendicular axes, and three for rotation about these axes.

The equations of motion for the ascent launch trajectory have no general solution since $v, C_D, \rho, \alpha, \gamma$ and g vary independently with time, mission profile, vehicle characteristics, and altitude. For example, the value of g varies locally with the Earth's bulge at the equator, with the height of nearby mountains, and with the local difference of the Earth's density at specific regions. Hence, for more accurate analysis we use computational techniques and numerical methods of solution to compute g, velocities, and distances with respect to time during the flight.

A sophisticated iterative analysis is carried out with numerical codes to determine flight properties for actual launch trajectories. A comprehensive analysis that considers actual flight trajectories incorporating the variation of all flight parameters is outside the scope of this book. The interested reader may consult Kluever [11] and specialized literature for more details.

Nevertheless, for quick estimates and mission planning, the equations of motion may be simplified. To give some perspective here, we consider the following idealized cases.

3.1.3 Vertical Flight

Consider a rocket-powered vehicle in vertical flight, the initial part of the booster phase depicted in Fig. 3.2. The thrust, velocity, gravitational field, and drag forces are all aligned, and the motion of the rocket vehicle is one-dimensional, with $\gamma = 90°$, and $\alpha = 0°$. The vertical acceleration of the vehicle, Eq. (3.3), is written as

$$m\frac{dv}{dt} = v_{ex}\frac{dm}{dt} - F_D - mg \tag{3.7}$$

Assuming the propellant flowrate is constant with respect to time, the mass of the vehicle at any time t is expressed by:

$$m = (m_0 - \dot{m}t)$$

Substituting the above into the equation of motion, Eq. (3.7) becomes:

$$(m_0 - \dot{m}t)\frac{dv}{dt} = v_{ex}\frac{dm}{dt} - F_D - (m_0 - \dot{m}t)g \tag{3.8}$$

This is a differential equation whose general solution cannot be easily found because the velocity, the drag force, the air density, and the gravitational force are all changing as the vehicle ascends in its launch trajectory.

During the ascent segment, moving through the thick atmosphere, the vehicle is subject to a *dynamic pressure*, the component of the airflow pressure that represents fluid kinetic energy (i.e. motion). Dynamic pressure q is defined as $q = \frac{1}{2}\rho v^2$, where ρ is the air density, and v is the vehicle speed.

Dynamic pressure is a scaling factor for pressure forces experienced by flying vehicles in an atmosphere. Since the velocity of the launcher is constantly increasing as it ascends, and the air density changes with altitude, there is a point where the dynamic pressure is a maximum (q_{max}), and we call it *critical q*. It represents the point of maximum dynamic pressure that builds on the nose of a vehicle after launch, the point at which aerodynamic stress is maximized. Therefore, by observing the variation of q during ascent flight we can determine when the vehicle's structural stress will reach its critical value q_{max}.

Before reaching q_{max}, the rocket engines must be throttled down sufficiently to keep structural integrity. At some point after launch (usually about one minute later) and at a predetermined altitude (typically about 10.68 km or 35,000 ft), conditions are such that the dynamic pressure reaches q_{max}. The main core rocket engines are throttled back to about 60–70% of their rated thrust (depending on payload); combined with the propellant perforation design of the solid propellant in the SRBs (if used), which reduced the thrust at q_{max} by one third after 50 s of burn, the total stresses on the vehicle are kept to a safe level. After the critical event, the density begins to drop rapidly enough that the vehicle can be throttled to full power without fear of structural damage, and the dynamic pressure drops to zero (about 2 min after launch). The NASA SLS is predicted to reach q_{max} at an altitude of about 12.97 km (42,555 ft), when the vehicle velocity is approximately 467.15 m/s (1045 mph). During the first uncrewed maiden trip around the Moon, the SLS reached $q_{max} = 661$ lbf/ft^2, 71 s after lift-off with the core engines operating at 100 percent rated thrust level.

For the sake of illustrating different aspects of launch dynamics, we now simplify Eq. (3.7) and consider the integral of the acceleration along the flight path.

Case 1: Ascent Subject to Aerodynamic Drag and Gravity Forces

This is a particularly important case, as not only gravity but also a drag force opposes the motion of the vehicle as it ascends during the initial launch segment. This is when the vehicle moves through the densest layers of the atmosphere. We can express the integral of the acceleration along the flight path as

$$\int \frac{dv}{dt}dt = \int \frac{F}{m}\cos\alpha\, dt - \int \frac{F_D}{m}dt - \int g\sin\gamma\, dt \qquad (3.9)$$

We note the integral on the left-hand-side yields the vehicle's actual velocity increment (or, change in magnitude of velocity vector v) because it is the time integral of absolute acceleration along the flight path, which we denote as Δv_{actual}. In a similar manner, we recognize on the right side of Eq. (3.9), the integral of tangential thrust acceleration, minus the integral of drag acceleration, and minus the integral of the tangential gravitational acceleration. Thus, it is customary to express all acceleration integrals as velocity increments:

$$\Delta v_{actual} = \Delta v_{thrust} - \Delta v_{drag} - \Delta v_{grav} \qquad (3.10)$$

If the tangential thrust is aligned with the velocity vector, then Δv_{thrust} is the ideal or maximum velocity change defined by the rocket equation, that is

$$\Delta v_{thrust} = \Delta v_{ideal} = v_{ex} \ln \frac{m}{m_o} \qquad (3.11)$$

Now we focus on the two energy-loss integrals that diminish the ideal velocity increment in Eq. (3.10):
Drag loss:

$$\Delta v_{drag} = \int \frac{F_D}{m} dt \qquad (3.12)$$

Gravity loss:

$$\Delta v_{grav} = \int g \sin \gamma \, dt \qquad (3.13)$$

These integrals can be solved with numerical methods. However, we can conclude that.

- A launch vehicle that can reach horizontal flight ($\sin \gamma = 0$) as rapidly as possible minimizes gravity loss.
- A launch vehicle that accelerates along a long, shallow climb at relatively low altitudes will experience severe drag losses.

Optimization of actual launch trajectories yields a compromise between these two conflicting flight effects. However, we can estimate the relative size of the drag force and compare it with the gravity force. For a 12,000 kg vehicle with a cross-sectional area $A = 5$ m^2 and a $C_D = 0.3$, we assume conditions for maximum drag as occurring at an altitude of 13 km, where the air density is $\rho = 0.26656$ kg/m^3, and the vehicle velocity is $v = 467.15$ m/s.

The drag-to-gravity ratio is

$$\frac{F_D}{mg} = \frac{\rho A C_D v^2}{2mg} = \frac{0.26656 \times 5 \times 0.3 \times (467.15)^2}{2 \times 12{,}000 \times 9.8} = 0.37$$

At such condition, the magnitude of the drag force is about 37% of the gravity force. We must remember that the atmospheric density will continue decreasing with altitude, and thus drag will diminish accordingly. When the launch vehicle reaches an altitude of about 60 km, the density is so greatly diminished that the drag force is considered negligible.

Case 2: Gravity Loss with Negligible Aerodynamic Force

Here we consider the case when the drag force on the vehicle is much smaller than the rocket thrust and the force of gravity. Again, for the vertical flight segment, Eq. (3.7) reduces to

$$m\frac{dv}{dt} = v_{ex}\frac{dm}{dt} - mg \tag{3.14}$$

The mass of the vehicle is $m = m(t)$. If v_{ex} is constant, we can easily integrate this differential equation:

$$\int_{v_0}^{v} dv = v_{ex}\int_{m_0}^{m} \frac{dm}{m} - \int_{0}^{t} g\,dt$$

Using an average value of the gravity, $g = \bar{g}$, the integral gives the velocity of the vehicle:

$$v(t) - v_0 = v_{ex}\ln\frac{m}{m_o} - \bar{g}t \tag{3.15}$$

where m is its mass at any time t.

Substituting Eq. (2.23) for the burn time t, obtain

$$v - v_0 = \Delta v = v_{ex}\ln\frac{m}{m_o} - \bar{g}\frac{m_0}{\dot{m}}\left(1 - \frac{m}{m_0}\right) \tag{3.16a}$$

If $v_0 = 0$, the velocity at thrust termination becomes

$$v = v_{ex}\ln(\mu) - \bar{g}\frac{m_0}{\dot{m}}(1 - \mu) \tag{3.16b}$$

The second term in Eq. (3.16) represents the effect of gravity on a vehicle's motion, which we call the *gravity loss* and denote it as

$$v_{gloss} = -\bar{g}t_b = -\bar{g}\frac{m_0}{\dot{m}}(1 - \mu) \tag{3.17}$$

While the ideal velocity (in absence of gravity) is independent of the thrust itself, the gravity loss term does depend strongly on thrust. A rocket vehicle with a very short acceleration, high thrust and high mass flow rate has a small gravity loss. However, a launcher with slow acceleration, low thrust and low mass flow rate has a much higher gravity loss. This is because a rocket that ejects more propellant mass has less total mass to accelerate to higher velocity or, for a launch vehicle, less mass to lift to higher altitude. Therefore, when optimizing the performance of a vehicle, we try to reduce the gravity loss term.

For motion in Earth's gravitational field, if we neglect the variation of gravity with the geographical features and the oblate shape, the acceleration of gravity varies inversely as the square of the distance from the Earth's center (Eq. 3.9). Starting with the nominal value on the surface $g_0 = 9.80665$ m/s^2, g decreases to 8.169 m/s^2 at an altitude of 559 km over the ground. At distances as far away as the Moon, $g \approx 3.3 \times 10^{-4} g_0$. The average value in Eq. (3.17) is sufficient for a first approximation of the gravity loss.

Example 3.1 Estimate the gravity loss experienced by a core rocket stage of an SLV with burn time of 150 s. Explain if the value obtained is acceptable. If not, suggest how it could be reduced.

Solution: Assume the stage operates from sea level to an altitude where the average gravitational acceleration may be $\overline{g} = 9.7$ m/s^2. From Eq. (3.17), the gravity loss is

$$v_{gloss} = -\overline{g}t_b = -1455 \text{ m/s}$$

This is a substantial loss in velocity for the SLV. To overcome this gravity loss, we must select a rocket core stage with very high thrust, capable of providing short acceleration.

3.1.4   Launch Vehicle Velocity

The actual velocity achieved by an SLV as it ascends to orbit is less than the velocity predicted by the ideal rocket equation. The actual vehicle velocity is less than the velocity change Δv calculated with the ideal rocket equation because of gravity loss and additional energy to overcome drag and other external forces. For example, the actual velocity of a spacecraft moving on a circular orbit 500 km over the ground is 7.61 km/ s, while the velocity required to get there may be 8.7 km/s. The difference represents the energy expended to overcome gravity, drag, and other forces acting on the launch vehicle. Hence, we determine the actual velocity change from the time integral of the absolute acceleration along the flight path, which includes velocity increments due to tangential thrust, aerodynamic drag, and tangential gravitational accelerations.

We can approximate the actual velocity that a vehicle will achieve without solving the governing equations by simply defining a velocity loss factor that would reduce v_{ideal}. If the vehicle experiences velocity losses (Δv_L) due to atmospheric drag (Δv_D) and gravity (Δv_g), the orbital velocity requirement may be expressed as

$$v = v_{ideal} + \Delta v_L \tag{3.18}$$

where $v_{ideal} = v_{ex} \ln \mu = g_0 I_{sp} \ln \mu$.

Defining a velocity loss factor as $k = \Delta v_L / v_{ideal}$, we obtain the required velocity as

$$v = (1+k)^{-1} g_0 I_{sp} \ln(\mu) \tag{3.19}$$

The velocity loss may range from 20 to 30%, i.e. $k = 0.2 - 0.3$. Thus, the rocket engines must deliver a velocity of 9360 to 10,140 m/s as compared to the ideal value 7800 m/s in a drag and gravity free environment. Then the actual mass ratio μ required would be 8.33 to 9.94 as compared to the ideal case of 5.85.

The velocity vector is related to the gravity loss v_{gloss}, which depends on the pitch angle. For a rocket vehicle with constant exhaust velocity, moving at constant pitch angle, this angle has a strong effect on the velocity achieved: the smaller the pitch angle, the higher the velocity but this would be mainly horizontal velocity, and thus the vehicle would ascend sufficiently to achieve orbital velocity.

The initial speed due to the Earth's rotation is also considered in the velocity calculation. As stated in Chap. 1, Sect. 1.5, the geographical location of the launch site influences the amount of energy required to get the launch vehicle off the ground. Near the equator, where the Earth is moving at 1650 km/h (0.458 km/s), a launch vehicle takes advantage of this rotational speed to boost its own. For example, the Chinese Long March 7A vehicle is launched from Wenchang Satellite Launch Center in the northeastern part of Hainan Island. This is China's fourth and southernmost launch facility and because of its low latitude (19° north of the equator) the Wenchang allows for launching larger payloads, for which it was selected to launch the Chinese space station. The Guiana Space Center (CSG) in La Guyane française, a region of France in South America, offers ideal conditions for launching any payload to any orbit at any time, as it is located 310 miles north of the equator at a latitude of 5°. Launching from CSG, the vehicles from the European Space Agency (ESA) can gain a velocity boost of about 0.465 km/s—a slingshot effect that is greater there than at most other launch sites.

Example 3.2 Verify that a launch vehicle can gain a natural boost of about 1670 km/h if it lifts off from a spaceport very near Earth's equator.

Solution: Earth's radius at the equator is 6,378,137 m. If $\omega = 2\pi/T$ is the angular velocity of the planet's rotation, where T denotes the rotation period (23 h 56 min 4 s), and r is the radius, then the rotational velocity v is simply

$$v = r\omega = \frac{2\pi r}{T} = 465.10 \, \text{m/s} = 1674.36 \, \text{km/h}$$

This confirms that the launch pad at the Earth's equator is moving at 1674 km/h, relative to Earth's center of mass, and thus the launch vehicle is boosted by this velocity.

3.2   Rocket Staging for Launch Vehicles

A single chemical rocket cannot achieve the necessary terminal velocity to reach LEO carrying a substantial payload. This fact was recognized from the beginning of the space era, and rocket designers developed stage rockets, dividing the launch vehicle propulsion system into several stages. The stages may utilize liquid propellant rocket engines, or combinations of both liquid and solid propellant rockets, which are arranged in special geometric configurations to work in series or tandem. Each propulsion stage has its own structure, propellant tanks, instrumentation, controls, and all other supporting systems.

Once the propellant of a given stage is expended, the dead mass of that spent stage (including empty tanks, cases, instruments, and other support systems) is jettisoned, reducing the mass of the vehicle as it continues its ascent, and in the meantime the next stage is fired to propel the remaining part of the lighter-mass vehicle. This process is repeated until all the stages are fired. Staging allows for higher mass ratios; hence, a space mission can be achieved with low I_{sp} rocket engines. By reducing the mass of the upper stage carrying the payload, it produces a higher final velocity. Multistage rocket vehicles permit higher orbital velocities, more payload mass, and have improved launch performance by minimizing gravity and drag losses.

In a two-stage configuration, the first (booster) stage carries the second stage and the payload to a pre-determined altitude, after which the second stage lifts the payload to its ultimate destination. For the initial part of the launch, the vehicle must have a large powerful first or core stage. This first stage may be comprised of many liquid propellant rocket engines, such as the SpaceX Falcon 9 which incorporates nine LPRs, or it can be a combination of LPRs with strap-on SRMs to augment the lift-off thrust. Figure 3.4 illustrates typical staging configurations of launch vehicles: the 2-stage ULA Vulcan Centaur; the 2.5-stage NASA SLS; and the 3-stage ESA Ariane 6.

To replace the Atlas V and Delta IV launchers, ULA conceived the two-stage heavy-lift Vulcan Centaur to meet the requirements of the National Security Space Launch program, including capability to achieve human-rating certification. With a capability to lift 27,200 kg (60,000 lbm) payload to LEO), the new vehicle depicted in Fig. 3.4a consists of the Vulcan first stage (burning LOX/LCH4) and the Centaur second stage (burning LOX/LH$_2$).

The NASA SLS is conceived to deliver crewed payloads to a translunar injection (TLI) point in LEO. It is considered an evolvable launch system with the required power to support future human interplanetary missions. Figure 3.4b shows the first variant of the SLS (Block 1) with a booster stage comprised of one central core stage with four RS-25 liquid propellant rocket engines burning LOX/LH$_2$, and two strap-on solid rocket boosters (SRBs). The half-stage of the SLS refers to the SRBs, which operate simultaneously with the core stage during the initial phase of the launch and, after they burn out, the SRBs are separated and jettisoned before the core stage engines complete operating as the remaining SLS ascends with its payload.

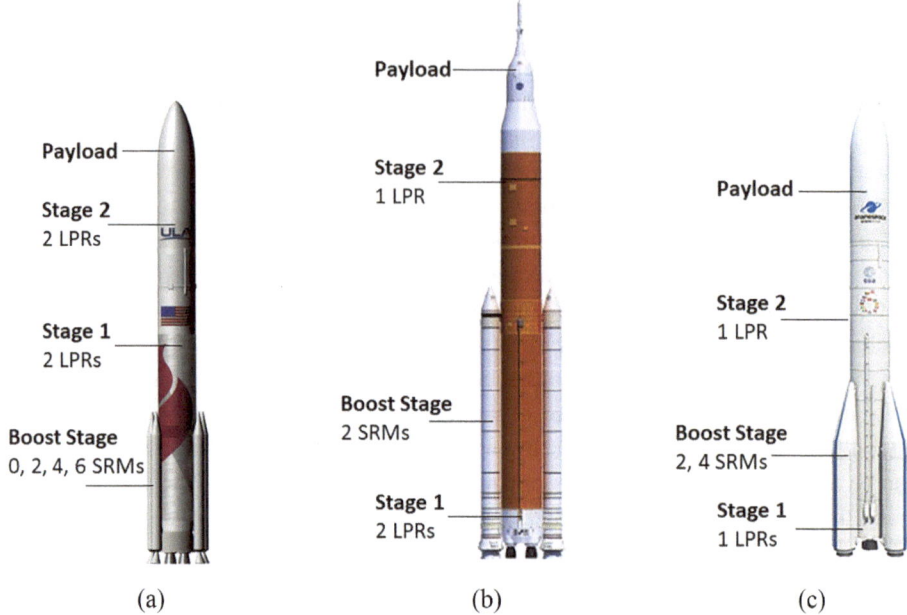

Fig. 3.4 Representative multi-stage launch vehicles; **a** 2-stage ULA Vulcan Centaur; **b** 2.5-stage NASA SLS; **c** 3-stage ESA Ariane 6 (not to scale)

To launch both heavy and light payloads to a wide range of orbits, the three-stage Ariane 6 from ESA, Fig. 3.4c, will be available in two versions depending on the performance required: Ariane 62, which is a version with two solid propellant rocket boosters, and Ariane 64 with four SRBs. The main and upper stage are known together as the central core. The main stage together with the solid rocket boosters propel Ariane 6 in the first phase of flight, while the upper stage will burn multiple times to reach the required orbits (LEO and GTO).

We illustrate the benefits of staging with a simple analysis of the payload mass that can be carried by a launch vehicle. For instance, to attain orbital speed of 7500 m/s, it requires a velocity gradient $\Delta v \approx 9000$ m/s, where the extra velocity is needed to overcome gravity and drag forces during ascent. To place a payload in orbit with a single stage rocket with a specific impulse $I_{sp} = 450$ s, the rocket equation prescribes a mass ratio

$$\frac{m}{m_0} = e^{-2.03} = 0.130$$

The final mass m at burn out must include all components of the vehicle (structure, rockets, empty propellant tanks, guidance equipment, and other subsystems), plus the payload. The mass of the propellant is $m_p = m_0 - m = (1 - 0.130)m_0 = 0.87\, m_0$.

Assume that structure, tanks plus empty engine is 10% of the propellant mass, $0.087\,m_0$. Hence, the payload mass that a one-stage vehicle can place in orbit is,

$$m_L(1 \text{ stage vehicle}) = (0.130 - 0.087)m_0 = 0.043\,m_0$$

The mass of the payload is 4.3 percent of the total mass of the one-stage launch vehicle.

Now consider a two-stage launch vehicle. Assume that the mass of each stage (empty tank and rocket engine) is 10% of the propellant mass it carries. For simplicity, the specific impulse for both stages is assumed to be $I_{sp} = 450$ s, so that $v_{ex1} = v_{ex2} = 4414.5$ m/s, and we take half of the total $\Delta v \approx 9000$ m/s for each stage operation.

First stage: let m_0 be the initial mass; the mass of the first stage after burn out, m_1, is

$$m_1 = e^{-4500/4414.5}m_0 = 0.3608\,m_0$$

The first stage propellant mass is $m_{p1} = m_0 - m_1 = (1 - 0.3608)m_0 = 0.639m_0$. Assume the mass to be discarded is 10% of the propellant mass, that is $0.0639\,m_0$; the second stage mass is

$$m_2 = 0.3608\,m_0 - 0.0639\,m_0 = 0.2969\,m_0$$

Second stage: The vehicle mass after burn out is

$$m = e^{-4500/4414.5}m_2 = e^{-1.019}(0.2969)m_0 = 0.107m_0$$

The propellant mass consumed by the second stage is $m_{p2} = m_2 - m = (0.2969 - 0.107)m_0 = 0.1897\,m_0$, and the mass of the stage is $0.0189\,m_0$. Therefore, the payload mass is

$$m_L(2 \text{ stage vehicle}) = (0.107 - 0.0189)m_0 = 0.0881\,m_0$$

The payload mass is 8.8% of the total mass of the two-stage launch vehicle. This simple analysis shows that the two-stage vehicle can deliver more than twice the payload to orbit compared with a single-stage vehicle.

In practice, a multi-stage rocket propelled SLV will attain a higher orbital velocity and will have improved payload capability compared with a single-stage SLV. However, a realistic SLV is limited to having no more than three or four stages due to the complications imposed by engineering and design, e.g. duplication of machinery, instrumentation, and other support systems. The ideal velocity gain derived from multi stages can be expressed as

$$v_{tot} = v_1 + v_2 + v_3 \ldots$$

To compute the total velocity gain requires careful consideration of the masses of the upper stages, and the burnout masses of the lower stages. This reduces the velocity gain of individual stages. In general, the number of stages for a particular mission is the result of an optimization analysis considering the mass between the stages, and the optimum flight path for each launch vehicle configuration for which a key parameter (payload mass, velocity increment) must be maximized. The optimal number of rocket stages n_{opt} can be assessed with this formula:

$$n_{opt} = 1.12 \frac{v}{\bar{v}_{ex}} \qquad (3.20)$$

where $\bar{v}_{ex}$ is the average value of the exhaust velocity of all stages.

For example, for a mission that requires a maximum velocity of 9.0 km/s, the optimum number of stages would be 3 if the stages yield an average exhaust velocity of 3.0 km/s. Thus, although this formula seems oversimplified, the result it yields is consistent with the engineering certainty that increasing the number of stages beyond three is not efficient. Four-stage launch vehicles have been conceived, but the gain in the velocity increment is not so great as the gain obtained when going from 1 to 2, and from 2 to 3 stages. The complexity of additional stages (including another rocket engine with propellant tanks, controls, etc.) would make the multi-stage SLV much more expensive to build and operate.

Staging is governed by mission design. Consider for example a vehicle that will launch a large crewed spacecraft attached to its service module (payload) on a mission to the Moon. A three-stage vehicle may be selected so that the first two stages lift the heavy payload into a LEO parking orbit, and use the third stage for the translunar injection (TLI). An ultra heavy SLV may accomplish the same mission with just two stages, if the first stage is powerful enough (using clusters of liquid propellant rockets and/or augmented with strap-on booster rockets) to lift the large payload to LEO, and the second stage takes the spacecraft to TLI.

A similar approach is considered for lifting payloads to LEO and GTO. For example, to increase the payload capability of its Falcon 9, SpaceX conceived the Falcon Heavy, a 2.5 stage derivative with a structurally strengthened Falcon 9 as the "core" component, with two additional Falcon 9 first stages with aerodynamic nose–cones mounted outboard serving as strap-on boosters. Twenty-seven SpaceX Merlin engines power the Falcon Heavy first stage for a total thrust of 22,819 kN (5,130,000 lbf) at liftoff. This capability makes it possible to lift a payload of 63,800 kg (140,700 lbm) to LEO, and 26,700 kg (58,900 lbm) to GTO.

Using the rocket equation, we estimate the velocity change using as input the payload mass, the number of stages, the propulsion performance for each stage, the structural mass, and propellant residuals for each stage. An efficient, cost-effective SLV with excellent payload capability requires a sophisticated optimization analysis. This topic is outside the scope of this book, and thus the reader is encouraged to consult specialized literature.

Table 3.1 Velocity requirements with launch from Kourou (French Guiana)[a]

Target orbit	Inclination	Velocity requirement (km/s)
LEO 200 km	5°	9.00
SSO 700 km	98°	9.70
GTO 200/36,000 km	0° (equatorial)	11.60
Moon (impact)		12.00
Earth Escape		12.50
Lunar Orbit		13.00 (approx.)
Mars Orbit		15.00 (approx.)

[a] Handbook of space technology, Eds. W. Ley, K. Wittmann, W. Hallmann (2011)

Did you know? SpaceX's two-stage Starship Super Heavy consists of two stages: the Super Heavy booster stage, comprised of 33 Raptor LREs, and the fully reusable second stage "Starship," which is the spacecraft proper, conceived to transport crews and cargo on long journeys, including trips to Mars. The Integrated Flight Test-4 (IFT-4) in the summer of 2024 completed a full-duration, 65-min mission, with both the first stage and the Starship making controlled splashdowns in the ocean.

No mission analysis is complete without introducing margins, as there are many uncertainties in the estimates. The simpler the model the more margins it needs to have. Thus, engineers usually apply margins for payload mass, velocity requirements of the mission, propulsion performance, and structural mass. Table 3.1 summarizes the velocity requirements for several target orbits including LEO, Sun-synchronous orbit (SSO), geostationary orbit (GEO) and/or transfer orbit (GTO), lunar missions or escape from Earth's gravitational field.

Numerical simulation, which combines theoretically-derived equations and empirical data and relations obtained from space launch experience has improved mission analysis and reduced many uncertainties that existed several decades ago.

3.3 Two-Body Problem (Keplerian Motion)

After the launch vehicle delivers a spacecraft to orbit, how does it move under the influence of the gravitational forces of the Sun, planets, and moons? The complexity of spacecraft motion in a gravitational field can only be determined numerically, using orbital mechanics models and sophisticated computer codes. This is a topic outside the scope of this book. However, it is important to provide a basic introduction to understand the relationships between rocket propulsion and space trajectories, to help us determine the propulsion requirements for a spacecraft to maneuver in the gravitational field of interest.

We will assume that the spacecraft is accelerated impulsively (by the rocket engine) at certain points of its trajectory, and that it moves unpowered (coasting) between those points under the sole influence of gravitational force of a celestial object (Earth, Moon, planets, Sun). For this introduction we simplify the spacecraft motion as a Keplerian motion, which considers the gravitational force of just one celestial body over the vehicle. This is an example of the classic two-body problem (2BP) whose solution yields Keplerian orbits (circles, ellipses, parabolas, hyperbolas) whose shape is determined by the eccentricity, and the focus is at the attracting body. This simple 2-body model is sufficient to provide us with the fundamental equations to evaluate the motion of the spacecraft under the gravitational influence of just one celestial body. That means that when the spacecraft moves beyond the gravitational sphere of influence, it follows a Keplerian orbit around the Sun. The sphere of influence (SOI) is a region where a celestial body's gravitational field dominates the orbits of surrounding bodies such as a spacecraft, despite the presence of the much more massive but distant parent star (the Sun). Thus, for orbital maneuvering, the spacecraft trajectory is defined by its position and velocity. By adjusting velocity, circular trajectories can be changed to ellipses, and ellipses can be modified with different eccentricities. Circular and elliptical orbits can be changed to parabolic or hyperbolic trajectories—all by adjusting the velocity of the spacecraft using a propulsion system.

In the following sections we provide the main formulae to help us understand the requirements on the propulsion system to carry out spacecraft maneuvering.

3.3.1 Equation of Relative Motion

Spacecraft always move in orbital paths and not straight lines. For example, to define the orbit of a spacecraft m about the attracting body M requires to know the minimum distance of the orbit from the fixed center of gravitational attraction – center of the Earth for geocentric orbits or center of the Sun for heliocentric orbits. Since the center of mass is not accelerating, an inertial frame (x, y, z) is at the center O of the gravitational mass M (see Fig. 3.5). The spacecraft moves under the combined effects of its momentum (by the rocket), and the attraction of gravity towards the center of the primary body. The spacecraft moves in a curved path represented by the *equation of relative motion* or Newton's gravitational law for the connecting vector $\vec{r}$ between M and m describing the joint synchronous motion of two masses about each other.

The two-body second order differential equation that governs the motion of the spacecraft relative to a single gravitational body M:

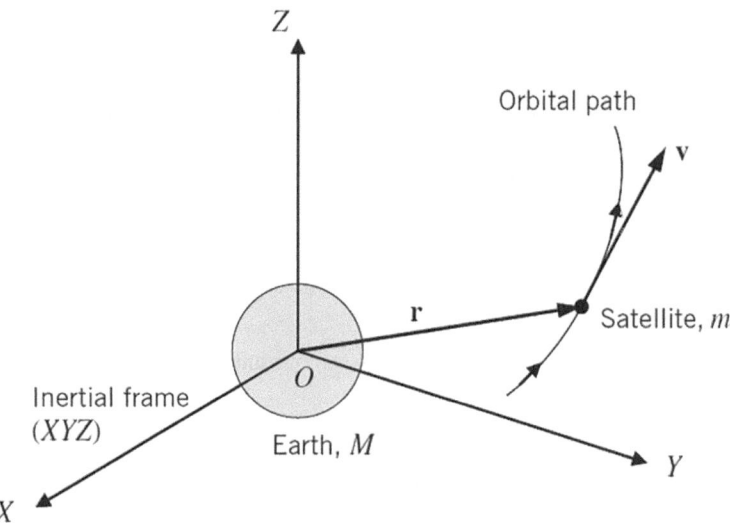

Fig. 3.5 Two-body problem representing the motion of a spacecraft of mass m under the gravitational force of the central primary mass M (e.g. Earth or any other celestial body)

$$\ddot{\vec{r}} + \frac{G(M+m)}{r^3}\vec{r} = \ddot{\vec{r}} + \frac{\mu}{r^3}\vec{r} = 0 \tag{3.21}$$

where $\ddot{\vec{r}}$ = spacecraft's acceleration; $\vec{r}$ = spacecraft's position vector; r = magnitude of the position vector, the distance from the center of the central body. This equation has two vector constants of integration, each having three scalar components, meaning that Eq. (3.21) has six constants of integration.

Note that the relative motion is practically independent of the mass of the spacecraft (secondary body), since $m \ll M$, and thus $\mu = GM$, the gravitational parameter of the primary body. The denominator is r^3 because $\vec{r}/r$ is a unit vector. Equation (3.21) is a vector acceleration equation of the relative motion for the two-body problem, which is also expressed as second-order, non-linear, differential equation. The solution gives the position and velocity vectors $[\vec{r}(t), \vec{v}(t) = \dot{\vec{r}}(t)]$ of the spacecraft of mass m relative to the central gravitational body M. The radius r of the orbit (local distance of the spacecraft from the center of M) is a function of the angle made by the radius vector to that of periapsis or closest approach (perihelion, for the case of geocentric orbits).

For geocentric orbits, the motion of spacecraft governed by Eq. (3.21) requires Earth's gravitational parameter, $\mu = 3.986 \times 10^5$ km^3/s^2. When we deal with lunar orbits, the gravitational parameter of the Moon is $\mu_{Moon} = 4903$ km^3/s^2. See Table 3.4 for the gravitational parameter of other bodies in the Solar System.

The general solution of the two-body problem yields differential equations, which once integrated produce constants of motion for the spacecraft. As a result, the solution:

(1) Confirms Kepler's laws of planetary motion, which are generalized for the spacecraft orbits to be of any conic shape.
(2) It leads to constants of motion: (a) The "specific angular momentum" or angular momentum per unit mass of a spacecraft is constant; (b) The linear momentum of a spacecraft (product of its mass m and velocity vector $\mathbf{v}$) is conserved; and (c) the specific energy (total energy per unit mass) of the spacecraft in its orbit is constant.

The energy conservation establishes that the sum of potential energy and kinetic energy of an orbiting body is a constant at all points in its orbit. This yields the energy integral:

$$\varepsilon = \frac{v^2}{2} - \frac{\mu}{r} = \text{constant} \tag{3.22}$$

where ε denotes the total mechanical energy per unit mass (specific energy) of the spacecraft about a central or major attracting body (e.g. Earth, Moon, Sun), which is a constant at all points in its orbit. The kinetic energy term is $v^2/2$, and the potential energy is $-\mu/r$, which approaches zero as $r \to \infty$. This energy integral is the most useful relation from the 2BP solution. It states that the specific mechanical energy is the same at all points of the trajectory.

Conservation of angular momentum explains the orbital motion of artificial satellites, planets, and other heavenly bodies. For a spacecraft, the total angular momentum vector ($\mathbf{H}$) is a constant, equal to the cross product of the radius and velocity vectors, $\mathbf{H} = \mathbf{r} \times \mathbf{v}$. Conservation states that a spacecraft's velocity and distance from the attracting body may change around its orbit, but the product speed $\times$ distance will not change unless some perturbation may change it. This means that the position and velocity vectors $\mathbf{r}$ and $\mathbf{v}$ always remain in the same plane. Hence, two-body trajectories are restricted to a plane that intersects the center of the attracting body.

The magnitude of the specific angular momentum h is simply

$$h = rv \cos \gamma \tag{3.23}$$

where γ is the flight path angle, which is the angle between the local horizon and the velocity vector at position r. The specific angular momentum h of the spacecraft is constant, since the only external force (gravity) acts along the radial direction and does not produce a torque.

Figure 3.6 illustrates the four types of trajectories a spacecraft may take. The circular and the elliptical orbits are known as *captured* or closed orbits because bodies on these closed orbital paths remain within the gravitational influence of the central attracting body. Thus, confined within the gravitational field of a star, planet or moon, spacecraft move in captured orbits. Once it has sufficient velocity to escape from that field, a spacecraft will escape or move in open trajectories. The parabolic and hyperbolic orbits are called *escape* or open orbits as there is no apogee and no empty focus. They are called escape orbits

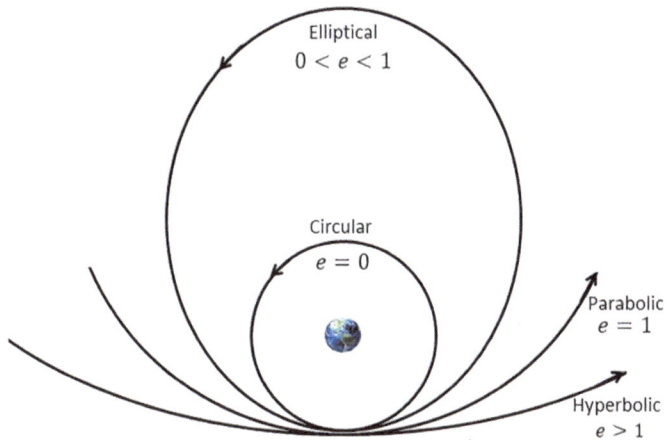

Fig. 3.6 Spacecraft closed orbits and escape trajectories

because a body in such path will escape the gravitational field of the primary attracting body. These are the typical orbits taken by a body as it leaves the gravitational field of a celestial body. Parabolic orbits describe the path of a body falling in towards another from very far away when it starts with very low velocity. Long period comets follow this type of orbit.

The orbital velocity in an elliptical orbit is always less than that required to escape from the central body, as we shall see below.

The following discussion considers spacecraft orbiting the Earth. However, the concepts apply to spacecraft orbiting the Sun, the Moon, and other planets.

3.3.2 Elliptical Orbit Relations

Let us consider a spacecraft of mass m, on an elliptical orbit with the attracting body centered at the focus M, and r is the distance from the body's center of mass, called the *radius vector*. Its direction is measured as its angle from perigee, as depicted in Fig. 3.7. The point on the orbit closest to the central attracting body is called the periapsis, and the point furthest is called the apoapsis. Collectively, these points are known as the **apsides**, and the line connecting them is the line of apsides or the major axis of the orbit. We drop the suffix "apsis" and replace it with a suffix which identifies the central attracting body of the orbit. For Earth-orbiting satellites we use the suffix "gee" from the Greek word *geo* meaning Earth, thus the terms perigee and apogee. For planetary orbits, with the Sun at the occupied focus, these points are called perihelion and aphelion, respectively. In an orbit around Mars the points are called the apoareon and the periareon, respectively.

For an elliptical orbit, the conservation of energy (Eq. 3.22) may be written as

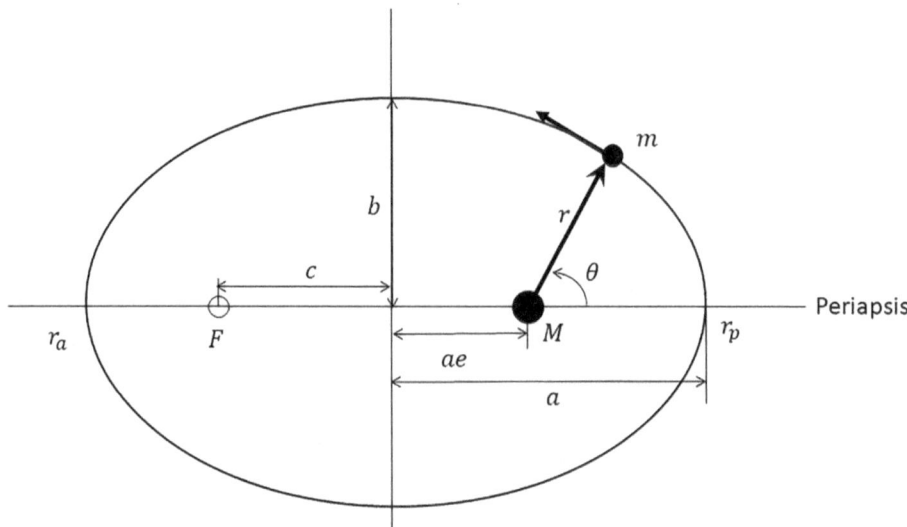

Fig. 3.7 Spacecraft of mass m in elliptical orbit about body of mass M, which is the primary focus of the ellipse. The vacant or unoccupied focus F has no physical significance

$$\varepsilon = \frac{v^2}{2} - \frac{\mu}{r} = -\frac{\mu}{2a} \tag{3.24}$$

The semi-major axis a is equal to the average of the radius of apoasis r_a and the radius of periapsis r_p:

$$a = \frac{r_p + r_a}{2} = \frac{r_p}{1 - e} \tag{3.25}$$

where e denotes the orbital eccentricity,

The eccentricity is the ratio of the distance between the foci ($2c$) to the length of the ellipse ($2a$), or the ratio of the minor and major axes:

$$e = \frac{2c}{2a} = \frac{r_a}{a} - 1 = \frac{r_a - r_p}{r_a + r_p} \tag{3.26}$$

The orbit eccentricity defines the shape or type of conic section. Because the distance between the foci in an ellipse is always less than the length of the ellipse, the numerical value of the eccentricity is between 0 and 1.

From the above orbital relationships, we can deduce convenient expressions for the radius of apoapsis r_a (maximum radius on an elliptical orbit), and the radius of periapsis (the minimum radius on an elliptical orbit):

$$r_a = a(1 + e) = 2a - r_p \tag{3.27}$$

$$r_p = a(1 - e) = 2a - r_a \qquad (3.28)$$

Formally, the fundamental equation of the two-body problem yields the shape of the orbit in polar coordinates centered in the body of mass M. This yields the position in orbit defined by the radius r and the position angle, called the true anomaly denoted θ. In orbital mechanics this is known as the **orbit equation**:

$$r = \frac{p}{1 + e\cos\theta} = \frac{h^2}{\mu} \cdot \frac{1}{1 + e\cos\theta} \qquad (3.29)$$

where e is the orbit eccentricity, the constant p is the *semi-latus rectum,* or orbit parameter as it is known in astronautics, and θ is the angle that determines the spacecraft location in the orbit, known as the true anomaly, measured from the line of apsides.

Now observe the second equality in Eq. (3.29): the orbit parameter p relates to the angular momentum of the spacecraft as $p = h^2/\mu$, confirming that one of the constants of motion (h) is related to the size of the orbit (p). The orbit equation defines the path of the spacecraft of mass m around a body M, with μ, h, and e constants. It is the solution to the restricted, two-body equation of motion (3.21). The radius attains its minimum value for $\theta = 0$ and the constant eccentricity vector is directed from the central body to the periapsis. In terms of the orbit constants, we write the distance to periapsis of the orbit as

$$r_p = \frac{p}{1 + e} = \frac{h^2}{\mu} \cdot \frac{1}{1 + e}$$

Equation (3.26) is a mathematical statement of Kepler's first law (planets follow elliptical paths around the Sun), and for that reason the two-body orbits are known as Keplerian orbits. Therefore, in the ideal two-body model, all spacecraft trajectories are described by the radius vector r.

Another more general representation of the orbit equation is:

$$r = \frac{a(1 - e^2)}{1 + e\cos\theta} \qquad (3.30)$$

Thus, given a defined orbit for the spacecraft, we can determine the true anomaly angle θ by solving from above expression. The true anomaly angle θ defines where a spacecraft is within the orbit with respect to perigee, and is the only orbital element that changes with time.

Orbital distances for a spacecraft in an elliptical orbit are, from Eq. (3.30):
At closes approach, $\theta = 0°$,

$$r = r_p = \frac{a(1 - e^2)}{1 + e\cos(0°)} = a(1 - e)$$

At apogee $\theta = 180°$,

$$r = r_a = \frac{a(1 - e^2)}{1 + e\cos(180°)} = a(1 + e)$$

which are consistent with (3.27) and (3.28).

The orbital period T is the time required for a spacecraft to complete one orbit (representing Kepler's third law of planetary motion); for an elliptical orbit the period is expressed as

$$T = 2\pi\sqrt{\frac{a^3}{\mu}} \tag{3.31}$$

where a is the semimajor axis of the elliptical orbit, measuring one-half of the long dimension of an ellipse.

The circular orbit is a special case of the elliptical orbit with $e = 0$, and $a = r$. From Eq. (3.24), the *circular speed* of the spacecraft at the orbital altitude r is

$$v = \sqrt{\frac{\mu}{r}} = v_c \tag{3.32}$$

where $\mu = GM$ is the gravitational parameter. For the Earth, $\mu = 398{,}600\ \text{km}^3/\text{s}^2$.

For spacecraft in low Earth orbits we usually give its altitude over the ground. Hence, the orbital radius is $r = R_E + H$, where R_E is Earth's mean equatorial radius equal to 6378 km, and H is the spacecraft altitude over standard sea level. Note that the greater the radius of the circular orbit, the smaller the speed required to keep the spacecraft on this orbit. Also, the circular velocity is independent of the mass of the spacecraft.

The orbital period T in a circular orbit is simply

$$T = 2\pi\sqrt{\frac{r^3}{\mu}} \tag{3.33}$$

For a geocentric orbit, spacecraft's trajectory has an altitude at perigee above 160 km. To remain in orbit at this altitude requires the spacecraft to move at greater than 7.808 km/s. The speed is slower for higher orbits, but attaining them requires greater launch energy. Examples of artificial satellites in almost circular LEO include the International Space Station (ISS) orbiting at an average altitude of 400 km (248 miles), and the Hubble Space Telescope with a nominal orbit at 515 km (320 miles) altitude.

3.3.3   Orbit Shape and Energy

The law of conservation of energy ensures that the sum of potential and kinetic energies of an orbiting body is a constant at all points in the orbit. Thus, for a spacecraft in any orbit, the total mechanical energy (per unit mass) ε is given by the energy equation or *vis-viva* (living force) equation, and it is constant.

Now let us search for a relationship between the geometry of the orbit and the energy. In this manner we can evaluate the constant values of the angular momentum and total energy at any point of the spacecraft orbit, especially at the periapsis, where the velocity is orthogonal to the radius vector, that is $v_p = h/r_p$. Rewrite Eq. (3.22) as

$$\varepsilon_p = \frac{v_p^2}{2} - \frac{\mu}{r_p} = \frac{h^2}{2r_p^2} - \frac{\mu}{r_p} \tag{3.34}$$

where the radius at periapsis is r_p is given by Eq. (3.28).

Since $p = a(1 - e^2) = h^2/\mu$, we solve for h^2 and substitute it into Eq. (3.34), to get the energy at perigee for orbits other than parabolic,

$$\varepsilon_p = \frac{h^2}{2r_p^2} - \frac{\mu}{r_p} = \frac{\mu a(1 - e^2)}{2a^2(1 - e)^2} - \frac{\mu}{a(1 - e)} = -\frac{\mu}{2a} \tag{3.35}$$

Equating Eqs. (3.22) and (3.35) and solving for v, we obtain:

$$v^2 = \mu\left(\frac{2}{r} - \frac{1}{a}\right) \tag{3.36}$$

This is the most useful result from the two-body solution, since it yields the general formula for velocity that we use to calculate velocity of spacecraft in any orbit.

3.3.4 Hyperbolic Excess Speed

Spacecraft escaping the Earth's gravity at the onset of an interplanetary mission follow hyperbolic trajectories. As shown in Fig. 3.8, the hyperbolic is an open orbit with eccentricity $e > 1$. The spacecraft velocity in a hyperbolic orbit is always greater than the velocity it has on a parabolic orbit at any radius. On a hyperbolic trajectory, the spacecraft velocity at infinity is finite (see Eq. 3.36):

$$\text{Hyperbolic Orbit, } a < 0 \quad v = \sqrt{\frac{2\mu}{r} + \frac{\mu}{a}} \tag{3.37}$$

The distance from the focus to the periapsis in the hyperbolic orbit is

$$r_p = \frac{h^2}{\mu} \cdot \frac{1}{1 + e} \tag{3.38}$$

Let v_∞ denote the spacecraft speed on a hyperbolic path arriving at infinity ($r \to \infty$). According to Eq. (3.37) this velocity at infinity takes the form

Fig. 3.8 Hyperbolic
trajectory. The length of the
major axis connecting the
extreme ends of the conic
section is $- 2a$ for a hyperbola

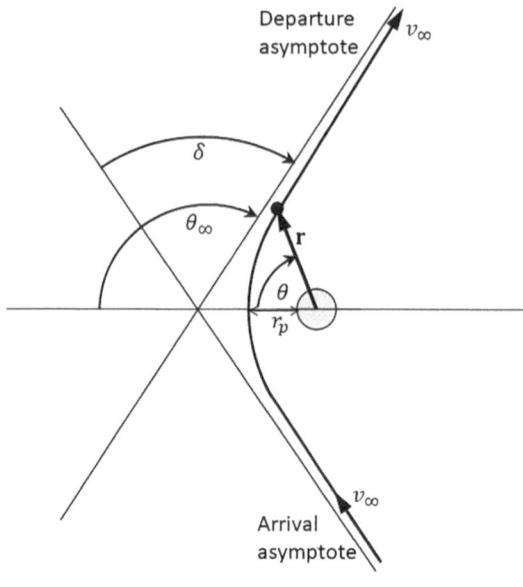

$$v_\infty = \sqrt{\frac{\mu}{a}} \qquad\qquad (3.39)$$

The speed at infinity v_∞ is called the **hyperbolic excess speed** of the spacecraft, representing its velocity when it leaves the gravitational sphere of influence (SOI) of the primary or attracting body.

The arrival and departure paths of the hyperbola are along two straight-line asymptotes (see Fig. 3.8), and the asymptotic velocity of the spacecraft at either end is computed from the energy equation. A spacecraft using slingshot or gravity assist to modify its trajectory (without use of propellant) follows hyperbolic trajectories. For example, the New Horizons spacecraft was launched directly into an Earth-and-solar escape trajectory with a speed of about 16.26 km/s (58,536 km/h; 36,373 mph). Then the spacecraft used the gravity assist from Jupiter to increase its speed (relative to the Sun) by 4 km/s (14,000 km/h; 9000 mph); this velocity gain reduced New Horizons' flight time to Pluto by 3 years.

It is important to note, v_∞ is the velocity that must be added to the Earth's velocity to achieve departure on a planetary mission. That is, v_∞ is related to the escape velocity from the primary body (v_{esc}), and the velocity of the spacecraft at perigee (v_p) by the relation $v_\infty^2 = v_p^2 - v_{esc}^2$. Table 3.2 gives values of v_∞ for spacecraft in hyperbolic trajectories with respect to the Sun. The values shown are approximate and are likely to vary as a function of time caused by perturbations from celestial bodies other than the Sun, or caused by orbital maneuvers while performing their space mission.

Due to its unusual shape, the hyperbolic orbit has a different sign convention. Because the length of the hyperbola (distance between the "ends") bends back on itself, or is

Table 3.2 Representative spacecraft solar system departure velocities

Spacecraft	Mission	v_∞ (km/s)
New horizons	First spacecraft to explore Pluto up close, flying past it and its moons. It reached the Kuiper Belt object Arrokoth, the most distant Solar System object ever explored	16.26
Pioneer 10	Mission designed for exploring outer planets. First of five probes and 11 spacecraft to achieve escape velocity needed to leave Solar System	11.87
Pioneer 11	Probe to study the asteroid belt, environment around Jupiter and Saturn, solar wind, and cosmic rays	11.39
Voyager 1	Interplanetary probe to study outer Solar System and interstellar space	17.57
Voyager 2	Probe launched on a trajectory towards Jupiter and Saturn, and had encounters with Uranus and Neptune	15.197

measured outside the conic, we define this distance, $2a$, as negative. The same convention also applies for the distance between the foci, $2c$, so $2c$ is also negative. But the magnitude of $2c$ is always larger than the magnitude of $2a$, so the eccentricity is greater than 1.0.

Did you know? Pioneer 11 was launched on a direct trajectory to Jupiter without any gravitational assists. A year later, Pioneer changed course to fly past Jupiter, facilitating a later flyby of Saturn. This maneuver consumed 7.7 kg of hydrazine, lasting 42 min and 36 s, and increased the speed of the probe by 230 km/h.

The value of the orbit specific energy ε (+ or −) in Eq. (3.22) identifies the type of trajectory the spacecraft takes. If ε is positive, the spacecraft moves along a hyperbolic orbit. If ε is zero, its path is a parabola. If ε is negative, the spacecraft orbital path is either a circle or an ellipse. For example, a spacecraft in a circular LEO at a 300 km altitude has a negative total specific energy, as the kinetic energy term is much smaller than the potential energy term, as you can easily verify. For all closed orbits specific energy is negative. On the other hand, a body on a parabolic path will continue moving away from the Earth, forever gaining altitude, increasing potential energy, and slowing down or decreasing kinetic energy. Thus, on a parabolic path, if a body would reach infinity with $v \to 0$, the total energy would approach zero, as you can verify:

$$\varepsilon = \frac{1}{2}(0 \text{ m/s})^2 - \frac{\mu}{\infty} = 0$$

Potential energy is zero at infinity ($r \to \infty$), and negative at radii less than infinity. What this means is that if it were possible for a spacecraft to reach infinity, it would have zero potential energy. But since it is impossible to reach infinity, the specific potential energy will always be negative, increasing towards zero as r increases.

Table 3.3 summarizes the characteristics for spacecraft closed orbits and escape trajectories.

Table 3.3 Orbit and trajectory characteristics in 2-body model

Orbit	Circular	Elliptic	Parabolic	Hyperbolic
Eccentricity	$e = 0$	$0 < e < 1$	$e = 1$	$e > 1$
Semimajor axis	$a = r$	$a > 1$	$a = \infty$	$a < 1$
Orbit parameter	$p = r$	$p = a\left(1 - e^2\right)$	$p = p$	$p = a\left(e^2 - 1\right)$
Specific energy ε	$\varepsilon < 0$	$\varepsilon < 0$	$\varepsilon = 0$	$\varepsilon > 0$
Spacecraft captured by gravity field of primary body?	Yes	Yes	No	No
Spacecraft velocity	$v = \sqrt{\mu/r}$	$v = \sqrt{(2\mu/r) - (\mu/a)}$	$v = \sqrt{2\mu/r}$	$v = \sqrt{(2\mu/r) + (\mu/a)}$

Table 3.4 Gravitational parameter (μ) and gravitational sphere of influence (SOI) radius for the bodies in the solar system

Body in solar system	Equatorial radius (km)	Gravitational parameter, μ (km^3/s^2)	$r_{SOI} \times 10^6$ km
Sun	695,700	$1.3271244\left(10^{11}\right)$	
Mercury	2439.7	$2.2033\left(10^4\right)$	0.111
Venus	6051.8	$3.2486\left(10^5\right)$	0.616
Earth	6378.137	$3.98600442\left(10^5\right)$	0.9246
Mars	3396.19	$4.28283\left(10^4\right)$	0.577
Jupiter	71,492	$1.266865\left(10^8\right)$	48.157
Saturn	60,268	$3.793119\left(10^7\right)$	54.796
Uranus	25,559	$5.79394\left(10^6\right)$	51.954
Neptune	24,764	$6.83653\left(10^6\right)$	80.196
Pluto	1151	$8.719\left(10^2\right)$	3.400
Earth's Moon	1737.6	$4.902801\left(10^3\right)$	0.0662

Example 3.3 Consider a spacecraft for a lunar mission. Calculate the specific energy on the transfer ellipse if the injection velocity (at perigee) is 10.88 km/s where the orbit perigee radius is 6700 km.

Solution: According to Eq. (3.34), the specific energy on the transfer ellipse is

$$\varepsilon_p = \frac{v_p^2}{2} - \frac{\mu}{r_p} = -0.305397 \text{ km}^2/\text{s}^2$$

Since the total energy at burnout is negative, it means that the spacecraft remains within Earth's gravitational field. This is because the injection velocity is lower than the velocity required to escape, v_{esc}:

$$v_{esc} = \sqrt{\frac{2\mu_E}{r_p}} = 10.91 \text{ km/s}$$

3.3.5 Gravitational Sphere of Influence

The gravitational sphere of influence (SOI) is the region around every gravitational celestial body that controls its influence on smaller bodies near them, a concept very useful to define spacecraft interplanetary trajectories. The Earth has an SOI that extends almost a million kilometers around it, so every spacecraft moving inside this region is gravitationally bound to the Earth. At the edge of the SOI the gravitational attraction of the planet and the Sun are approximately equal. Once the spacecraft leaves our planet's SOI, the main gravitational influence is the Sun, until the spacecraft enters another body's SOI. Thus, if the spacecraft has sufficient speed to escape Earth's SOI, it will fall into an orbit around the Sun, a heliocentric orbit. When the spacecraft nears a target planet on this heliocentric orbit, it will eventually enter the SOI of that planet.

According to Laplace's model, the radius of the SOI for a planet orbiting the Sun, denoted r_{SOI}, is given by

$$r_{SOI} \approx r \left(\frac{m_p}{M_s} \right)^{2/5} \tag{3.40}$$

where r = mean orbital radius of the two bodies, center-to-center distance between Sun and planet, m_p = mass of the planet, and M_s = mass of the Sun.

The SOI radius of any planet in our Solar System can be calculated by using Eq. (3.40) with the appropriate values for the mean planet–Sun distance (i.e., semimajor axis) and planetary mass. A moon has also a SOI, which is determined using Eq. (3.40) but the parent planet then takes the role of the Sun. Table 3.4 gives the radius of the SOI for the most important bodies in our Solar System, including Earth's Moon. Note our planet's

SOI extends well beyond the Moon's 384,400 km orbital radius. This means that the orbit of the Moon is determined primarily by the Earth. This is important when we define spacecraft trajectories on lunar missions.

3.4   Launch Injection Characteristic Energy

In this section, we wish to emphasize the propulsion system requirements for the powered maneuvering in space. For example, for an interplanetary mission, the spacecraft must be equipped with the right rocket engines to first escape the grip of Earth's gravity. It begins with the characteristic energy, a measure of the maximum energy a launch vehicle can impart to a spacecraft of a given mass at the injection point for interplanetary flight.

Interplanetary spacecraft sent from the Earth escape on hyperbolic trajectories relative to the Earth. A hyperbolic path is necessary if we want the spacecraft to have some speed left over after it escapes the Earth's gravitational SOI. Figure 3.9 depicts the Earth departure geometry and related velocity vector diagram.

In such interplanetary trajectory, the vector difference between the Earth heliocentric velocity and the velocity required on the transfer ellipse is called the hyperbolic excess velocity, which is the required spacecraft velocity with respect to the Earth. Assuming that the transfer ellipse and the Earth's orbit are tangent and coplanar, the hyperbolic excess velocity is v_∞:

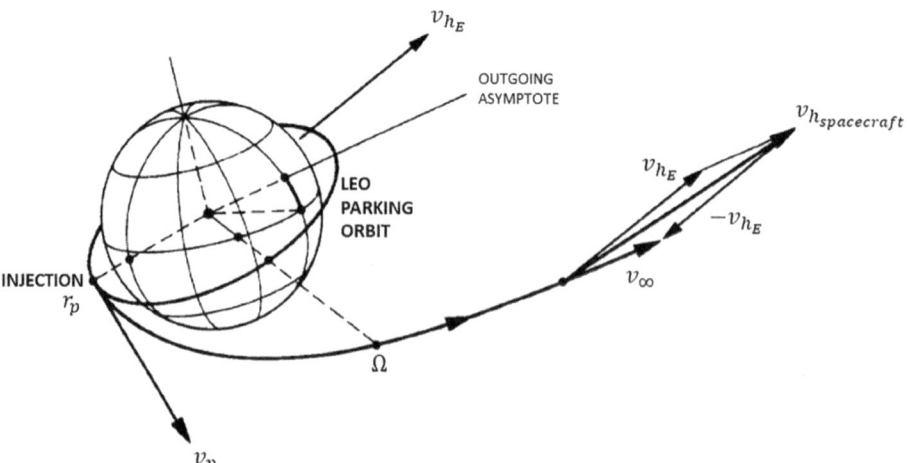

Fig. 3.9 Earth departure geometry for hyperbolic escape trajectory and velocity vector diagram. Injection velocity v_p, hyperbolic excess velocity v_∞, Earth heliocentric velocity v_{h_E}, and spacecraft heliocentric velocity $v_{h_{spacecraft}}$

$$v_\infty = v_{h_{spacecraft}} - v_{h_E} \tag{3.41}$$

where $v_{h_{spacecraft}}$ is the spacecraft heliocentric velocity on the transfer ellipse, and v_{h_E} is the Earth's heliocentric velocity, which is equal to 29.766 km/s.

Under standard assumptions a spacecraft traveling along hyperbolic trajectory will coast to infinity, arriving there with hyperbolic excess velocity v_∞ relative to the central body.

3.4.1 Earth Escape Trajectories

We can determine the hyperbolic excess velocity by stating the specific energy equation at two points on the hyperbolic escape trajectory: at periapsis, the point near the central primary body (such as the Earth) with orbit radius r_p where the escaping spacecraft has a velocity known as the injection velocity, and at a point far from the central body where the velocity of the escaping spacecraft is v_∞, as illustrated in Fig. 3.9.

Since the specific mechanical energy does not change along an orbit, the energy at the injection point (ε_p) must be equal to the energy at infinity (ε_∞). That is, $\varepsilon_p = \varepsilon_\infty$:

$$\frac{v_p^2}{2} - \frac{\mu}{r_p} = \frac{v_\infty^2}{2} - \frac{\mu}{r_\infty}$$

where v_∞ represents the velocity of the spacecraft at a great distance from the primary body where its gravitational attraction is practically negligible. This velocity at infinity is attained when the spacecraft has climbed away from the departure body, following injection at velocity v_I. From this equality we solve for the velocity at infinity,

$$v_\infty^2 = v_p^2 - \frac{2\mu}{r_p} = v_p^2 - v_{esc}^2 \tag{3.42}$$

This result confirms that, if the injection velocity is greater than the escape velocity, $v_p > v_{esc}$, the spacecraft escapes with a residual velocity v_∞ (as it has on a hyperbola trajectory). On the other hand, if $v_p = v_{esc}$, then $v_\infty = 0$ (as on a parabolic trajectory).

The **launch injection characteristic energy**, denoted C_3, is a measure of the excess specific energy over that required to just barely escape from a massive central body. The launch characteristic energy, also known as departure energy, is defined as the square of the hyperbolic excess velocity v_∞, which is the vector difference between Earth's velocity and the heliocentric orbit's velocity:

$$C_3 = v_\infty^2 \tag{3.43}$$

For mission analysis, C_3 is used rather than Δv, because C_3 is independent of the parking orbit from which the vehicle departs. In Fig. 3.9, the injection point refers to the velocity

v_p at r_p. For example, if orbit injection occurs at $r_p = 6554.3$ km with an injection velocity $v_p = 11.05$ km/s, the launch characteristic energy required for a spacecraft to escape Earth on a departure hyperbola is

$$C_3 = (0.7088 \text{ km/s})^2 = 0.5023 \text{ km}^2\text{s}^{-2}$$

To match a launch vehicle with a mission, the C_3 of the launcher must be greater than the required C_3 for a given interplanetary mission. In general, for missions beyond the Earth-Moon system, a launch vehicle system must provide high characteristic energy (C_3), enabling either shorter transit times or heavier payloads with more robust science packages for missions to the outer Solar System (see Fig. 3.10). The injection Δv from a typical LEO (200 km altitude) is evaluated from the C_3 using two-body dynamics. For interplanetary missions, C_3 is used to describe hyperbolic departure from Earth; it is not used to describe an arrival at a planet.

The capability of a space launch vehicle to deliver a payload to Earth escape is typically given as a range of characteristic energy values. For lunar mission design, the energy parameter C_3 is known as **trans-lunar injection (TLI) energy parameter**. The required C_3 for trans-lunar injection is -0.99 km^{2}s^{-2}. Note the negative value, which indicates that the launch energy required is less than the escape velocity from Earth. For conventional low energy lunar transfers, values of TLI energy parameter at least $C_3 = -2.0$ km^2/s^2 are required for realistic transfer durations between 2 and 6 days. Quicker or longer transfers are possible, but the injection C_3 value increases rapidly. For a mission to Mars, the trans-Mars injection (TMI) characteristic energy is $C_3 \sim 8.1$ km^{2}s^{-2}.

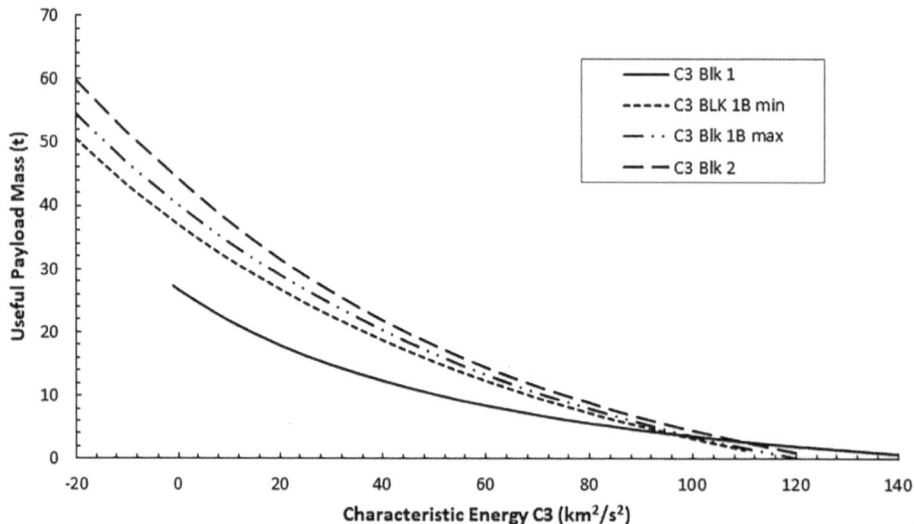

Fig. 3.10 Payload mass and launch energy C_3 capability for NASA SLS. *Source* NASA

Outer planet missions require significantly more departure energy than missions to the Moon and Mars. Interplanetary missions to outer planets are known as Ultra high C_3 missions and require heavy-lift launch vehicles to deliver the payloads to the required Earth escape conditions. For example, for a Jupiter direct transfer, C_3 is about 85 km²s⁻², while a direct transfer to Saturn is approximately 105 km²s⁻².

For interplanetary spacecraft, the value of C_3 comes from a mission design calculation, representing the minimum energy requirement needed to accomplish the mission. From the point of view of the launch vehicle, C_3 is computed as the maximum energy the launch vehicle can deliver carrying a spacecraft of a given mass. The launch vehicle C_3 capability for the expected spacecraft mass must be higher than the C_3 required for the mission, i.e. $C_{3\,launch\,vehicle} > C_{3\,mission\,requirement}$.

In general, launch vehicles can inject their payload into trajectories with different C_3, depending on their last rocket stage design and propellant. For example, the C_3 curve for the new NASA SLS vehicle is much higher than that for the Delta IV Heavy. The conventional launchers often use a kicker motor to improve payload to higher C_3, as it is done by reconfiguring the Atlas V 551 with an added Star-48 V solid motor as third stage. Figure 3.10 shows the launch energy versus payload mass capability of the SLS and its future variants projected to inject spacecraft of given mass (in tons) for interplanetary missions.

Specifying a particular target and departure and arrival dates determines the hyperbolic escape trajectory and hence the required C_3. For example, if a 10,000 kg payload requires a launch energy of $C_3 = 80$ km²/s², only the Delta IV Heavy, and Falcon Heavy (expendable) have the capability. According to ESA's published data, the Ariane Heavy can lift 4550 kg toward the Earth escape orbit: $V_\infty = 3475$ m/s($\delta = -3.8°$). This means that Ariane's launch energy capability is $C_3 = 12.07$ km²/s². Mission planners use launch performance curves provided by launch vehicle suppliers to select the best SLV configuration for the mission.

Example 3.4 Consider a 10,000 kg spacecraft on a mission to Mars. Preliminary mission estimate yields the elements of the departure hyperbola as $a = 18,849.7$ km; $e = 1.3$. Determine the characteristic energy that the launch vehicle must impart to the spacecraft at TMI, and comment on the payload capability of a given launcher.

Solution: From Eq. (3.43), the characteristic energy is

$$C_3 = V_\infty^2 = \frac{\mu}{a} = 21.146 \frac{km^2}{s^2}$$

Any heavy launch vehicle can provide this injection energy. According to Fig. 3.10, the SLS Block 1 vehicle can easily deliver it and even a higher payload mass.

The New Horizons spacecraft, with mass less than 0.5 ton, was launched toward Pluto with a $C_3 = 158\,\mathrm{km^2/s^2}$ by the Atlas V (551) AV-010 vehicle with a Star 48B third stage. The Star 48 is a SRM delivering a maximum thrust of 94.07 kN (21,150 lbf) with a specific impulse of 283.4 s. A part of NASA's New Frontiers program, the New Horizons spacecraft was launched in 2006 with the primary mission to perform a flyby study of the Pluto system in 2015, and a secondary mission to go by and study one or more other Kuiper belt objects (KBOs) in the decade to follow, which became a mission to 486,958 Arrokoth (a trans-Neptunian object located in the Kuiper belt). New Horizons is the fifth space probe to achieve the escape velocity needed to leave the Solar System (see Table 3.2).

According the NASA's Space Launch System (SLS) Mission Planner's Guide, the SLS vehicle can deliver up to 40 metric ton payloads to cis-lunar space ($C_3 = -0.99\,\mathrm{km^2/s^2}$), and deliver double payload mass or decrease flight time by half for some outer planet destinations when compared to existing SLV capabilities. For interplanetary missions, SLS is designed to provide characteristic energies up to $C_3 = 120\,\mathrm{km^2/s^2}$. Also, SLS flights may deliver the Europa Clipper spacecraft to a Jovian destination in under 3 years, compared to the 7+ years cruise time required for current capabilities. The SLS Block 1B and its 1B+ planned variant may achieve very high launch energies ($C_3 > 200\,\mathrm{km^2/s^2}$) to lift small ($\leq 2$ metric ton) payloads by adding additional upper stages (3rd and or 4th stages). (See McNutt et al. [12]).

3.5   Orbital Maneuvering

A spacecraft performs orbital maneuvers to transfer from one orbit to another. Changes to its orbit may require much propulsion effort such as when a spacecraft must move from a geocentric parking orbit to transfer into an interplanetary trajectory. Orbital maneuvers may be less energetic, such as those required for maintain a stable orbit and overcome natural perturbations (known as station-keeping maneuver). Whatever the reason for maneuvering in space, the spacecraft requires a thrust force, by firing its onboard rockets.

Orbital maneuvering is guided by fundamental principles of orbital mechanics; in simple terms, an orbit is uniquely determined by the position and velocity of the spacecraft at any point on its trajectory. Conversely, changing the velocity vector at any point instantly transforms the trajectory to correspond to the new velocity vector. Any conic section (circle, ellipse, parabola, or hyperbola) can be converted to any other conic section by adjusting velocity; a spacecraft travels on the trajectory defined by its velocity. Circular orbits can be converted to ellipses; ellipses can be changed in eccentricity; circles or ellipses can be changed to hyperbolas—all by adjusting velocity. The following overview is related to impulsive maneuvers, those that require the onboard rocket to fire in relatively short bursts to produce the required velocity change (Δv).

3.5.1   Hohmann Transfer

Perhaps the most fundamental spacecraft maneuver is the transfer between two circular nonintersecting, coplanar orbits. Such maneuver can be used to move from a low altitude orbit to one at higher altitude. A launch vehicle could place a spacecraft into a low altitude parking orbit and then the spacecraft would transfer to a higher circular orbit using an elliptical trajectory which is tangent to both circular orbits. Such is the approach conceived by German engineer Walter Hohmann in 1925.

The Hohmann transfer is used to place satellites into GTO. Starting from a stable LEO, rockets are fired to put the satellite into an elliptical orbit with the perigee at the LEO and the apogee at the geostationary orbit. Upon reaching the final altitude the rockets are fired once again to achieve the correct velocity of GTO. Another application of Hohmann transfers is for spacecraft rendezvous, the maneuver which brings two spacecraft together at the same point in orbit and at the same time.

The Hohmann transfer ellipse is a low energy maneuver, that is, it requires the least velocity increment (Δv) for a spacecraft to change its orbit via an ellipse that is tangent to both coplanar orbits, including a transfer when moving from a geocentric Earth orbit to a Mars orbit (see Sect. 3.8).

The Hohmann transfer is performed as follows. Consider the two circular orbits depicted in Fig. 3.11. A spacecraft needs to transfer from the lower orbit of radius r_1 to the higher altitude orbit of radius r_2. The semi-major axis of the transfer ellipse from point 1 to point 2 is given by Eq. (3.25), and we denote it as a_t:

$$a_t = \frac{r_1 + r_2}{2}$$

The energy of the orbit at perigee is, from Eq. (3.35)

Fig. 3.11 The Hohmann transfer between circular coplanar orbits

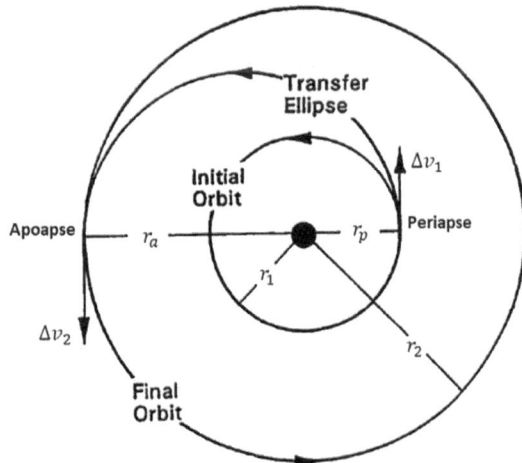

$$\varepsilon_t = -\frac{\mu}{r_1 + r_2}$$

At point 1 the spacecraft had attained a circular speed, Eq. (3.32),

$$v_{c1} = \sqrt{\mu/r_1}$$

The speed of the transfer ellipse at point 1 is given by Eq. (3.22):

$$v_1 = \sqrt{2\left(\frac{\mu}{r_1} + \varepsilon_t\right)}$$

For the spacecraft to transfer from the low circular orbit to the transfer ellipse it must accelerate, increasing its velocity from v_{c1} to v_1, that is, it needs a velocity change

$$\Delta v = v_1 - v_{c1}$$

In the same form, we can compute the delta-v to transfer from the transfer ellipse to the higher orbit. Moreover, this transfer method may be applied to transfer from higher orbits to lower orbits. The only difference is that we would require the two velocity decreases instead of two increases. The tangential Δv is the secret to the Hohmann Transfer's energy savings.

Now we can determine the time of flight for a Hohmann transfer. The time of flight (TOF) refers to the transit time between two positions in an orbit. Since the period for an elliptic orbit is given by Eq. (3.31), for the transfer ellipse we have half the period, i.e.

$$TOF = \pi\sqrt{a_t^3/\mu}$$

The Hohmann transfer is the most energy efficient, and the most economical since it requires the least Δv (least propellant). However, it also takes longer than any other possible transfer maneuver between two circular orbits. Low-energy orbit transfers are best for robotic missions. For crewed missions, other optimized transfer maneuvers are available, especially when the time of flight must be the shortest.

3.5.2 General Coplanar Maneuver

Spacecraft coplanar maneuvers are common, that is when the spacecraft orbiting a central celestial object moves into a different orbit within the same orbital plane. For example, when a spacecraft must change from a circular to an elliptical orbit. The geometry of coplanar maneuvers is simple: the spacecraft changes its initial orbit velocity v_i to an intersecting coplanar orbit with velocity v_f. The transfer can be made at any intersection of two orbits. The spacecraft velocity on the final orbit is equal to the vector sum of the

Fig. 3.12 Generalized
coplanar maneuver

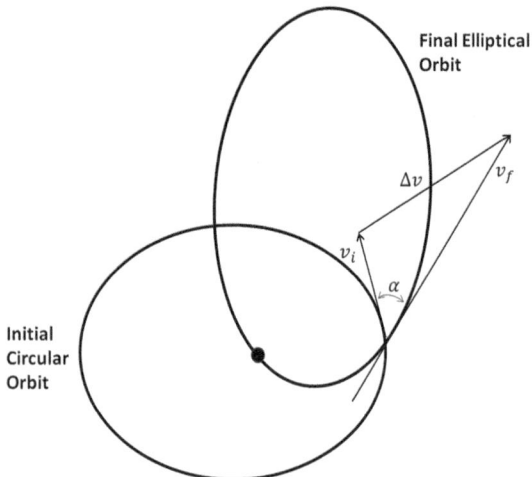

initial velocity and the velocity change vector (see Fig. 3.12). Applying the cosine law
yields the velocity change that a spacecraft must achieve to move its initial orbit to the
final orbit:

$$\Delta v = v_i^2 + v_f^2 - 2v_i v_f \cos \alpha$$

where α is the angle between vectors v_i and v_f. The least velocity change is obtained
when the orbits are tangent and α is zero.

3.5.3 Plane Changes

Changing the orbital inclination of a spacecraft orbit requires a propulsion burn. We calculate the magnitude of the velocity change to determine the amount of propellant this
maneuver requires. There are instances in which orbital plane changes are necessary. For
example, when the launch site does not match the desired orbit. Since the latitude of the
launch pad affects the orbital inclination, achieving certain orbits requires an orbital plane
change that depends on the launch site latitude.

Consider a spacecraft in a circular orbit. To change the orbital plane, the thruster
applies an impulse perpendicular to the orbit plane at the point where the initial and
final orbit planes intersect, as depicted in Fig. 3.13. Assuming the spacecraft velocity is
unchanged and that the final orbit shape is unaffected by the plane change (eccentricity,
semimajor axis, and radii remain the same), the velocity change (Δv) required to produce
the plane change maneuver is obtained from the cosine law,

$$\Delta v = 2v_i \sin(\alpha/2) \tag{3.44}$$

Fig. 3.13 Plane change
maneuver

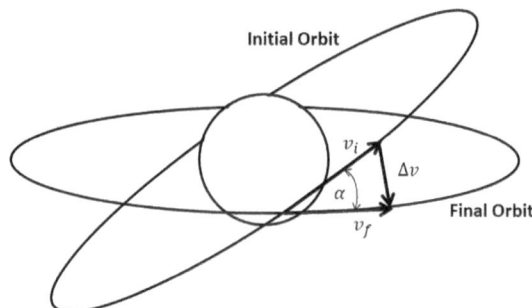

where v_i is the velocity of the spacecraft on the initial orbit at the point of intersection of the initial and final planes, and α is the angle of the plane change.

Orbital change maneuver is also known as "orbital inclination change" since it requires to tip the plane of the original orbit. Thus, Eq. 3.44 gives the Δv for pure rotation of the velocity vector. This formula also emphasizes that large plane changes are costly in terms of propellant expenditure. Consider for example a plane change of 20°. If the spacecraft has an initial velocity of 8 km/s, the resulting Δv is 2.77 km/s, which requires 610 kg of propellant if the thruster has a specific impulse of 300 s and the initial spacecraft mass is 1000 kg, as you can verify with Eq. (2.19).

Because orbital plane changes are expensive, if it is possible, plane adjustments should be made during the powered ascent phase of the mission. However, in some cases plane changes must be done in orbit. For example, this may be the case of a GEO satellite launched from a launch site far from the equator. As we know, the GEO satellite must orbit the Earth in the equatorial plane (inclination angle $i = 0°$), and so orbit plane adjustments must be made if the satellite cannot be launched directly into the equatorial orbit.

3.5.4 Station-Keeping, Repositioning, and Orbit Trim Maneuvers

Orbit trim maneuvers are required to remove any residual errors from an established orbit. The errors may be caused by the launch vehicle itself or by the spacecraft. The orbit parameters that may need correction are r, a, e, or the orbit inclination i. For each error, the thruster may burn to perform a small orbit change, a Homman transfer, or a plane change, depending on the magnitude of the required correction.

Orbit perturbations caused by the gravitational field of the Sun, the Moon, produced by the Earth's non-spherical shape, and induced by solar pressure, also require appropriate correction; the thrusting maneuvers to overcome those perturbations are part of the orbit maintenance requirements for any mission.

Orbit maintenance, or station-keeping, refers to the powered maneuvers required to overcome perturbing forces (aerodynamic drag, solar radiation, gravity gradients, gravitational pulls from the Sun, Earth, and the Moon, magnetic fields, or internal propulsion accelerations), ensuring that the spacecraft remains in its intended orbit with the correct orbital position. The thrusters that comprise the orbit maintenance propulsion for the spacecraft must be sized to correct the perturbed or altered orbit by periodically applying small burn thrusts in predetermined directions. Typically, orbit maintenance is done by a Reaction Control System (RCS) comprised of small thrusters that provide predetermined total impulses in needed directions. These corrections are performed throughout the life of any spacecraft (for 1 to 20 years and sometimes more) to overcome disturbance effects. For station-keeping, the Δv required must be fully quantified to ensure enough propellant is onboard to perform the required periodic maneuvers.

For some applications, a spacecraft may need to be repositioned. This may require increasing or decreasing the orbital velocity to change the eccentricity of the orbit. Increasing the velocity, for example, the position of the thruster impulse is at periapsis. The Δv required for repositioning a spacecraft must be accurately calculated to correlate it with the propellant mass required for this maneuver.

The thrusters on a spacecraft selected for attitude control must be re-startable, that is, designed to restart frequently. Hence, bi-propellant, monopropellant, and cold gas thrusters are considered for the attitude control system of spacecraft. Such system controls the orientation of the spacecraft with respect to an inertial frame of reference. An attitude maneuver consists of rotations about each of the spacecraft three axes. The thrusters are arranged so that the torques applied to the spacecraft are pure couples (no translational component). In a spin-stabilized spacecraft, attitude maneuvers consist of translations for station-keeping and repositioning, reorienting the spin axis, and adjusting spin velocity.

An excellent description of attitude maneuvers, repositioning, station-keeping, orbit errors, and other spacecraft maneuvering is given by Brown [5], Curtis [7], and Kluever [11].

3.6 Restricted Three-Body System

In Sect. 3.5, orbital motion analysis is simplified by assuming that a spacecraft moves under the attraction of only one body (Earth), neglecting the attraction of the Sun and of the Moon. But the spacecraft in cis-lunar space, for example, will move under the attraction exerted by at least three celestial bodies. When the spacecraft moves away from Earth towards the Moon, the attractive force exerted by the Earth becomes smaller, that of the Sun remains practically the same, but that of the Moon increases. On a trip from the Earth to the Moon, the spacecraft is subjected consecutively to the attractions exerted by the following bodies, arranged in their order of magnitude: Earth, Earth and Sun, Sun, Earth and Moon, Sun and Moon, Moon.

The analysis of spacecraft moving between the Earth and the Moon is an example of the restricted three-body problem (R3BP) since the mass of the vehicle is sufficiently small that it exerts negligible gravitational effect on the orbits of the two celestial bodies. There are no close solutions for the three-body problem; however, analytical expressions are developed to study the spacecraft motion by considering the circular restricted three-body problem (CR3BP), which is a model that assumes the two main gravitational bodies move in circular orbits about their center of mass. Thus, for lunar trajectories analysis we may ignore the Sun and only consider the motion of the spacecraft as influenced by just the Earth and Moon. Although this assumption causes a loss of accuracy for translunar trajectories (the Moon's orbit about the Earth has an eccentricity of ~0.055), the CR3BP provides a useful method for developing *preliminary* lunar and circumlunar free return trajectories.

3.6.1   Euler–Lagrange Equilibrium Points in the Sun-Earth System

The Euler–Lagrange equilibrium points, which result from the sum of the gravitational forces from two primary masses, plus the centrifugal force of its rotation around the center of mass, are very important for space exploration. The equilibrium solutions of the R3BP yield the coordinates of points in the rotating two-body system where a small third mass can be in equilibrium with respect to the massive bodies. In general, the equilibrium points, also known as libration points, are the locations in space where the third body of small mass m would have zero velocity and zero acceleration, i.e. where m would appear permanently at rest relative to the two massive bodies m_1 and m_2 (and therefore appear to an inertial observer to move in circular orbits around m_1 and m_2). Once placed at an equilibrium point, the tiny body will remain there. Thus, the R3BP model can consider three-body systems such as Sun-Earth-spacecraft, or Earth-Moon-spacecraft, or any other three-body system that meets the criteria of the R3BP model. In each instance, five equilibrium points are found due to the balance between the bodies gravitational and centrifugal forces. Figure 3.14 shows the equilibrium points for the Sun-Earth system, denoted SEL_1, SEL_2, SEL_3, SEL_4, and SEL_5.

The collinear points SEL_1, SEL_2 and SEL_3, also known as Eulerian equilibrium points, are very important for space exploration and are used for spacecraft to conduct astronomical observations. Most importantly, if certain initial conditions are satisfied, it is possible for a spacecraft to follow a periodic orbit about any of those points. Depending on the application, those initial conditions are chosen so that only the oscillatory mode of the xy-motion is excited, and the result is a halo orbit. A spacecraft can execute a periodic orbit (e.g. Halo or Lissajous) about one of these locations with small expenditure of energy.

Halo and Lissajous are orbits around the collinear equilibrium points of the Restricted Three-Body Problem (R3BP). A halo orbit is a periodic, three-dimensional orbit, while a Lissajous orbit is quasi-periodic path, both used for spacecraft moving around one of the Sun-Earth-Moon equilibrium points.

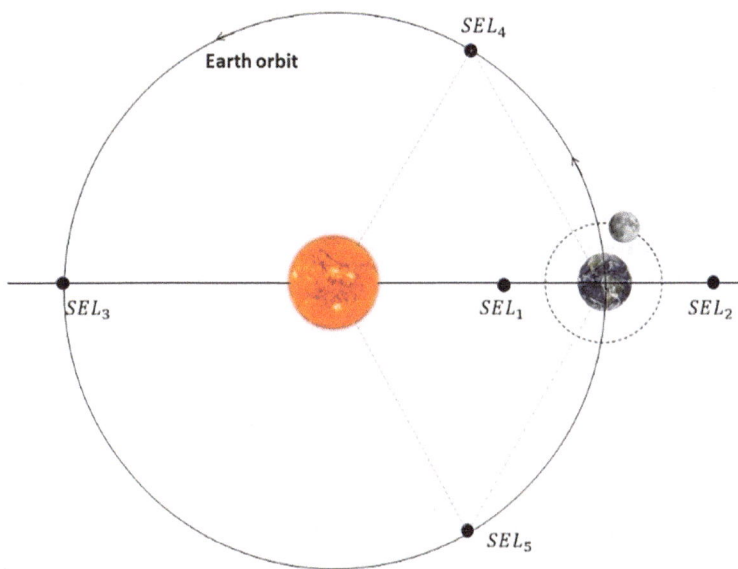

Fig. 3.14 The equilibrium points in the Sun-Earth system (not to scale)

The point SEL_1 is important for solar physics research. Solar probes orbiting this point can observe the magnetosphere over a long period and statistically investigate the characteristics of its magnetic field and plasma. From SEL_1 spacecraft can monitor the solar wind in front of the magnetosphere, and the interplanetary magnetic field, which determines the disturbance levels of the magnetosphere and ionosphere of the Earth.

Orbits about SEL_2 provide favorable survey geometry relative to Earth orbits by mitigating viewing restrictions imposed by terrestrial and lunar stray light. The average distance from Earth to SEL_2 is approximately 1.5×10^6 km. However, this distance varies over the course of a year due to the eccentricity of Earth's orbit. Being about four times farther away from the Earth than the Moon, the SEL_2 is an excellent place from which a space telescope can observe the entire universe. Otherwise, a spacecraft would have to make constant orbits around the Earth, which would require that it passes in and out of Earth's shadow, causing it to heat up and cool down, distorting its view. SEL_2 provides a much more stable viewpoint free from this restriction and far away from the heat radiated by Earth. The James Webb Space Telescope (JWST), the WMAP, Planck, and the Gaia spacecraft move around SEL_2 in halo orbits. The WMAP is a NASAprobe launched June 2001 to make fundamental measurements forcosmology research, including temperature differences in cosmic microwave background (CMB), the afterglow of the Big Bang, while Plank, Europe's first mission to study CMB, substantially improved measurements made by NASA's COBE, and WMAP.

Since periodic and quasi-periodic point trajectories are unstable, spacecraft moving on such orbits need control to remain close to their nominal orbit. For example, the James Webb Space Telescope (JWST), the WMAP, Planck, and the Gaia spacecraft, orbiting SEL_2, require special propulsion for station-keeping, ensuring that the telescopes remain in their orbit during their mission, which can be as long as 12 years.

3.7   Circumlunar Trajectories

A circumlunar trajectory, also known as trans-lunar trajectory or lunar free return, is one type of flight path for a mission to the Moon in which a spacecraft with limited propellant uses only gravity-assist, performing a swing-by maneuver once the trajectory is established at TLI, and thus it does not include landing on the Moon. The circumlunar trajectory uses the gravity of the Moon and the Earth to slingshot around and return home. An upgraded version of circumlunar trajectory is depicted in Fig. 3.15 to illustrate a proposed 10-day mission intended for lunar tourism. It does include sufficient energy and propellant for performing trajectory corrections to ensure safe return of civilian passengers.

A spacecraft using slingshot or gravity assist (a natural maneuver) modifies its trajectory without expending propellant. That is why the term free return trajectory. One benefit of such trajectory is that if something goes wrong with the propulsion or the spacecraft, it can still return to Earth without additional or expensive maneuvers. Thus, boosting a spacecraft to go on a loop around the Moon allows a mission to test new spacecraft systems and verify lunar mission plans, such as was done with the NASA's Artemis I mission in 2022.

1	Lift Off
2	Maneuvers in Geocentric Orbit
3	Trans-Lunar Injection TLI
4	Trajectory Correction Maneuvers
5	Brake and Intersect Lunar Orbit
6	Circumlunar Ride
7	Trans-Earth Injection
8	Return to Earth
9	Intersect Geocentric Orbit
10	Reentry and Landing

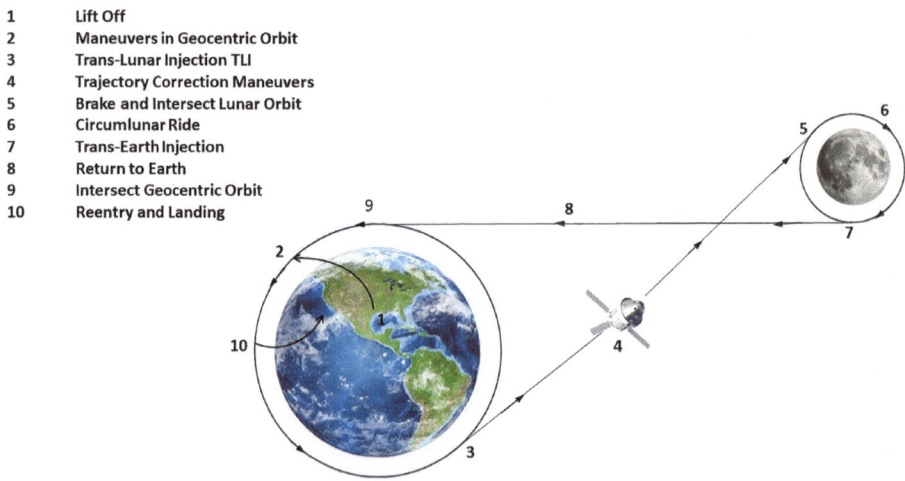

Fig. 3.15 Proposed circumlunar trajectory for space tourist trip to the Moon (not to scale)

Circumlunar trajectories are also attractive for space tourism. It is reported that SpaceX plans a one-week cruise around the Moon, taking two passengers aboard a new spacecraft now in development.

In November 2022, NASA's SLS launched for the first time, sending an uncrewed Orion spacecraft on a historical voyage around the Moon, debuting the Artemis I mission that will return humans to the Moon.

The SLS core stage and boosters inserted the interim cryogenic propulsion stage (ICPS) and Orion into an initial LEO at a velocity of 7.7967 km/s. Main engine cut off (MECO) occurred approximately eight minutes after lift-off, while the vehicle was travelling at 25,750 km/h. Orbit insertion occurred at a point into an elliptical LEO of 1564 km × 25.7 km. The low perigee was required for the spent core stage of SLS to re-enter Earth's atmosphere for safe disposal. Following orbital insertion, the ICPS performed a perigee raise maneuver to circularize the orbit.

After moving once around Earth, the upper stage fired its RL10 engine for approximately 18 min. to complete the trans-lunar injection (TLI) burn. Shortly after, Orion separated from ICPS. At the time of separation, the ICPS was moving at more than 35,406 km/h. The ICPS's final burn was done as a disposal burn to place it in a heliocentric orbit.

The uncrewed Orion spacecraft began its lunar trajectory, eventually intersecting a distant retrograde (DRO) orbit about the Moon to accomplish multiple mission objectives. The lunar trajectory required nineteen discrete translational maneuvers carried out by the ESM Propulsion Subsystem, and five burns using the Orbital Maneuvering System Engine (OMS-E), which was also use for all four large translational maneuvers (outbound powered flyby, distant retrograde orbit insertion, distant retrograde orbit departure, and return powered flyby). The OMS-E was also used for the first outbound trajectory correction burn. Orion returned safely to Earth, and separated from the service module before reentering and later landing on a controlled splashdown into the ocean.

3.8   Interplanetary Trajectories

An interplanetary mission can be achieved with minimum energy following a heliocentric Hohmann transfer ellipse trajectory between two planetary orbits, which is an ellipse with the Sun as the primary gravitational body. Such transfer requires a Δv of relatively high thrust impulse at the departure point on Earth (at time t_1) and another high thrust Δv maneuver at the arrival (at time t_2), to intersect the orbit of the destination planet. The two Δv are equal to the velocity difference between the respective circular planetary velocities

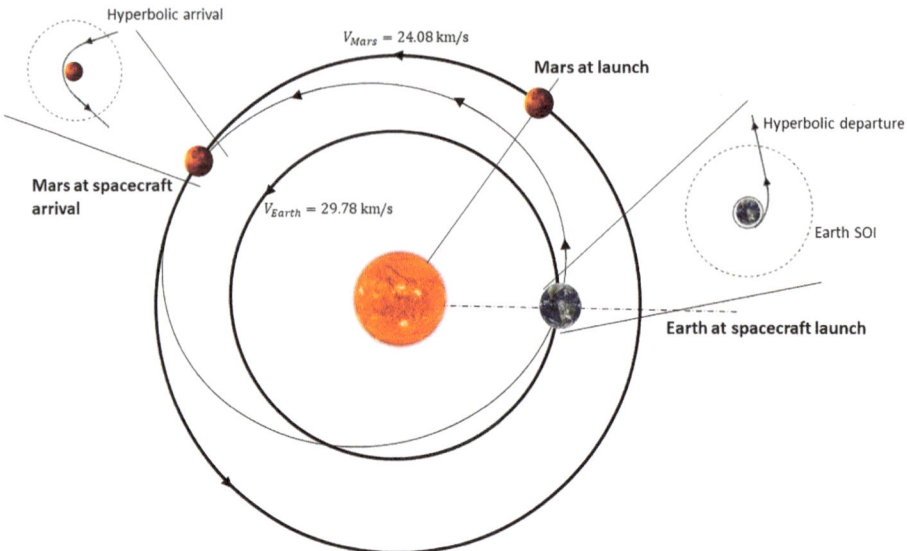

Fig. 3.16 Interplanetary trajectory based on patched-conic method, with three phases: (1) Earth departure; (2) Heliocentric transfer; (3) Target planet arrival (not drawn to scale)

and the perigee and apogee velocities which define the transfer ellipse, as described in Sect. 3.7. We illustrate such interplanetary transfer with a mission from Earth to Mars, as depicted in Fig. 3.16.

The interplanetary cruise phase for the spacecraft is the heliocentric transfer. The departure dates are determined by the relative positions of the launch planet and the target planet. This is a critical part of the mission analysis as the spacecraft must intersect the target orbit. For the Earth-Mars mission, the transfer takes about 259 days.

The interplanetary trajectory is based on the patched-conic method. It divides the interplanetary mission into a sequence of phases where each phase is a separate two-body problem. To begin the cruise, the spacecraft must first escape Earth's gravitational field, transitioning from near-Earth space to interplanetary space. As indicated in Fig. 3.16, the spacecraft departs from Earth orbit via a hyperbolic departure trajectory, leaving a LEO and crossing the Earth's sphere of influence (SOI). After moving through the transfer ellipse, the spacecraft enters Mars' gravitational field, moving on a hyperbolic arrival trajectory.

If the orbits of the two planets are not in the same plane, then the spacecraft requires additional energy to thrust and align the velocity vectors to intersect the target orbit. Moreover, an interplanetary mission may include circumnavigation, landing, and return transfers. The energy required for such maneuvers must be included in the overall Δv for the entire mission.

One of the farthest and more complex missions is NASA's Europa Clipper (launched in 2024), a probe that will take almost six years to travel 2.9 billion km to reach Jupiter. Because the spacecraft cannot carry enough propellant to power the entire voyage, it will use gravity assists from Earth and Mars's gravitational fields.

The complexities of interplanetary trajectories are best described by specialized books and articles devoted to specific mission analysis. The reader may find the fundamentals of space flight dynamics in Brown [5], Curtis [7], and Kluever [11] as excellent guides.

3.9 Mission Design, Orbital Maneuvering, and Delta-*v* Budget

Mission analysis is the process of quantifying the system parameters and evaluating the resulting performance to analyze whether the mission meets the requirements or not. Mission analysis begins with the choice of the launch system. The launch system must take a payload to its designated orbit. A simple mission analysis is performed knowing the payload mass and target orbit, so we must know the capability of the launch vehicle to provide the required velocity change (Δv), which is one of the main outcomes of the analysis.

Once in orbit, mission analysis must also determine the Δv required for the in-space maneuvers. This of course depends on the orbits (altitude, inclination, eccentricity), orbital transfers, life time of the mission, and many other parameters, but for our discussion is sufficient to concentrate on the calculation of the Δv overall budget (for orbit transfer, orbit maintenance, and return/reentry). The velocity budget for a given mission is the sum of all the flight velocity increments needed to attain the objective of the mission, and for each velocity increment a certain amount of propellant must be spent.

3.9.1 Delta-*v* for Transfer Within the Solar System

To transfer from one circular orbit around the Sun to another circular orbit of different radius but in the same plane, we wish to determine the Δv required to inject a spacecraft (initially in a circular orbit about the Earth) into a heliocentric orbit with an aphelion equal to that of a planet.

Let the radii of the two circular orbits be r_1 and r_2, as shown in Fig. 3.11. From the energy integral, Eq. (3.24),

$$\varepsilon = \frac{v^2}{2} - \frac{\mu}{r} = -\frac{\mu}{2a}$$

where $\mu = GM_S$ (M_S is the mass of the Sun) and $a = (r_1 + r_2)/2$ (the semi-major axis of the transfer orbit), we can get

$$r_1 \frac{v_1^2}{\mu} = \frac{2\sigma}{1+\sigma}$$

with $\sigma = r_2/r_1$. Since the velocity in the inner circular orbit is $\sqrt{\mu/r_1}$, the required change of velocity at the periapsis of the transfer ellipse is

$$\Delta v_1 = v_1 - \sqrt{\frac{\mu}{r_1}} = \sqrt{\mu/r_1}\left(\sqrt{\frac{2\sigma}{1+\sigma}} - 1\right)$$

As we are only interested in an encounter with the outer planet (or inner if $r_2 < r_1$), we are not going to compute the second Δv completing the Hohmann transfer.

From the energy first integral

$$v_1 = \sqrt{v_\infty^2 + 2\mu_E/r_1}$$

where now $\mu_E = GM_E$, (M_E is the mass of the Earth), and $v_\infty^2 = -\mu/a_h$.

Since the velocity in circular orbit is $v_c^2 = \mu_E/r_1$, the required increment of velocity to escape is

$$\Delta v_1 = \sqrt{v_\infty^2 + \frac{2\mu_E}{r_1}} - \sqrt{\frac{\mu_E}{r_1}}$$

This equation allows us to determine the Δv required to inject a spacecraft (initially in a circular orbit about the Earth) into a heliocentric orbit with an aphelion equal to that of a target planet, i.e. the minimum energy to encounter that planet.

The required hyperbolic escape velocity v_∞ is equivalent to the Δv_1 and the velocity increment Δv_0 to be applied from the circular Earth orbit to achieve a given v_∞. If departing from circular orbit around the Earth, at an altitude of 185 km, the velocity increments are those summarized in Table 3.5. As shown, the minimum Δv requirement for a trip to Mars is relatively low and could be achieved with available chemical propulsion technology. However, for human exploration of the outer planets, more advanced propulsion technology is required.

3.9.2   Propellant Mass

We wish to minimize the propellant (i.e. energy) requirements for a spacecraft, as this is important for the feasibility of its mission. An unreasonably high propellant requirement can render a mission infeasible.

Impulsive maneuvers are those brief firings that a rocket engine must perform to change the magnitude and direction of the spacecraft velocity vector. During the instantaneous impulse maneuver, the spacecraft position is assumed fixed, as only its current velocity

Table 3.5 Minimum Δv requirements to encounter the planets in the Solar System

Planet	v_∞ (km/s)	Δv_1 (km/s)
Mercury	7.533	5.556
Venus	2.495	3.507
Mars	2.945	3.615
Jupiter	8.793	6.306
Saturn	10.289	7.284
Uranus	11.280	7.978
Neptune	11.654	8.247
Pluto	11.813	8.363

changes. Hence, the impulsive maneuver is an ideal model that allows us to assess the motion of the spacecraft if its position changes very little during the rocket thrusting. This is the case for thrusters with burn times short compared with the coasting time of the vehicle.

The magnitude Δv for the impulsive maneuver is related to the change in mass Δm, which is the mass of propellant consumed, as we determined in Chap. 2, a result of the ideal rocket equation, Eq. 2.19:

$$m_p = m_0\left[1 - \exp(\Delta v/v_{ex})\right] = m_0\left[1 - \exp\left(\frac{-\Delta v}{g_0 I_{sp}}\right)\right]$$
$$= m\left[\exp\left(\frac{\Delta v}{g_0 I_{sp}}\right) - 1\right]$$

A mission must be carefully planned to minimize the Δv budget required for orbital maneuvering, as Δv represents rocket thrusting and the propellant mass carried aloft, which, if it is excessive, takes away from the payload mass that is required for the mission.

Example 3.6 Consider two different bi-propellant thrusters for an orbit transfer that requires a Δv of 750 m/s. One thruster has $I_{sp1} = 313$ s, and the second has $I_{sp2} = 300$ s. Calculate the amount of propellant a 1000 kg spacecraft would need to carry in its tank if you select the second thruster for the mission. Neglect the mass of all other subsystems.

Solution: The initial mass of the spacecraft is given by the rocket equation, Eq. (2.15),

$$m_0 = m e^{\Delta v/v_{ex}} = m e^{\Delta v/(g_0 I_{sp})}$$

For thruster 1: $m_{01} = 1276.67$ kg ; for thruster 2: $m_{02} = 1290.25$ kg

The mass of propellant for each case is $m_p = m_0 - m$. The second thruster is less efficient for this maneuver, as the spacecraft would need to carry an additional 13.58 kg of propellant:

$$m_{p1} = 276.67 \text{ kg}; m_{p2} = 290.25 \text{ kg}$$

3.10   Representative Example of Station-Keeping and Orbit Control

For the James Webb Space Telescope (JWST) orbiting SEL_2 (see Fig. 3.17), station-keeping maneuvers must be performed every 21 days to keep the observatory in a stable orbit, which has a period of about six months. Any change in energy from the unstable SEL_2 point will result in a departure from the orbit, either towards the Earth or in an escape direction towards the Sun-Earth-Moon regions. Since the Webb has a large sunshield, it must be repointed regularly producing significant changes in solar radiation pressure. Given the sensitivity of the instruments to stray light, by being on the other side of the Moon keeps the JWST always pointed away from the Sun, the Earth, and the Moon.

Because the JWST incorporates a passively-cooled system, the SEL_2 point was found to be the most optimum location. By orbiting the SEL_2, the telescope follows the Earth around the Sun in such a manner that the two bodies always appear to be in the same direction, and therefore a portion of JWST is continuously shielded from the Sun and Earth. Moreover, the orbit about SEL_2 was sized to ensure that the JWST never enters the Earth's shadow, thus enabling continuous generation of power with solar arrays. JWST's orbit has a period of approximately 6 months and is unstable, requiring propulsion for orbit control.

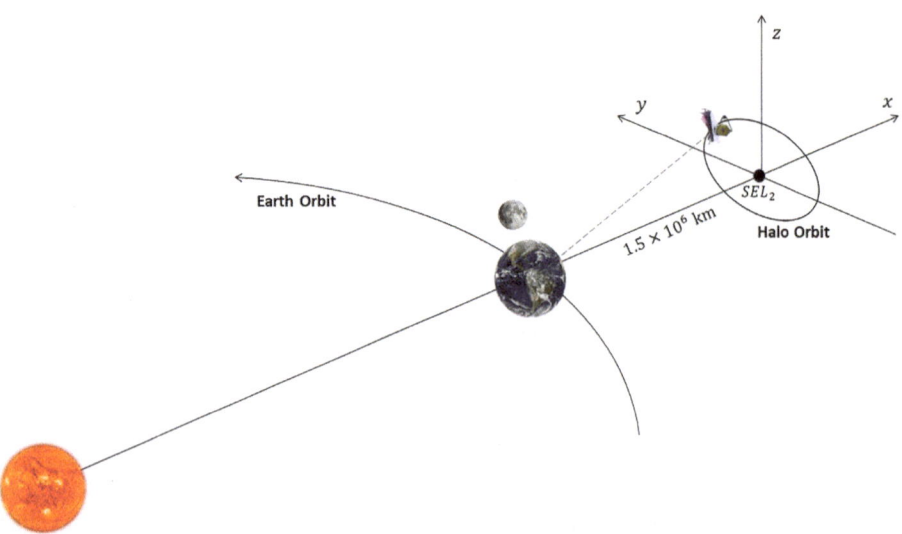

Fig. 3.17 Spacecraft orbiting the second Sun-Earth equilibrium point SEL_2

The propulsion system is one of the most critical elements of JWST and probably the system that determines the operational duration of the entire mission (10.5 years). For station-keeping and attitude control the propulsion system uses two types of thrusters: SCAT and MRE-1.

Station-keeping is required to correct for maneuver execution errors, solar radiation pressure (a natural perturbation that may result in escape trajectory from SEL_2), and other perturbations. The RCS thrusters must provide a torque to cancel out the torque produced by the change in angular momentum of the spacecraft reaction wheel (systems of orientation control of long-operating spacecraft).

The station-keeping maneuvers are performed using a set of Secondary Combustion Augmented Thrusters (SCATs) canted 37.4 deg so the thrust vector point through the center of mass during mission operations. The thrust direction is fixed in the body frame, so thrusting requires repointing the spacecraft. A typical station-keeping maneuver is estimated to be about 12 cm per second. [[8]]. The SCATs are bimodal thrust devices that can operate either as a catalytic monopropellant thruster with hydrazine, or as a high-performance bipropellant hydrazine and dinitrogen tetroxide thruster (see Chap. 4). Webb also uses eight MRE-1 hydrazine thrusters for attitude control and momentum unloading of the reaction wheels.

Recommended Reading and References

1. Ariane 6 User's Manual, Issue 2, Revision 0, Arianespace, February 2021. https://www.ariane space.com/wp-content/uploads/2021/03/Mua-6_Issue-2_Revision-0_March-2021.pdf.
2. Ariane 6 Launch Vehicle, comprises three stages: two or four boosters, and a main and upper stage—known together as the central core. https://www.esa.int/Enabling_Support/Space_Tra nsportation/Launch_vehicles/Ariane_6.
3. Ballard, R. O. (2017). "Next-Generation RS-25 Engines for the NASA Space Launch System." 7th European Conference for Aeronautics and Space Sciences (EUCASS). https://doi.org/10.13009/EUCASS2017-140.
4. Belair, M., Hennekens, M., Barsi, S., et al. (2024). "Artemis I Orion ESM Propulsion System Engine Performance." Paper SP2024_382, ESA 3AF International Conference on Space Propulsion, Glasgow, Scotland, May 20–23, 2024.
5. Brown, C. D. (1992). *Spacecraft Mission Design*, AIAA Education Series.
6. Brown, C. D. (1996). *Spacecraft Propulsion*, AIAA Education Series.
7. Curtis, H. D. (2005). *Orbital Mechanics for Engineering Students*. Elsevier Butterworth-Heinemann, Burlington, MA 01803.
8. Dichmann, D. J., Alberding, C. M., and Yu, W. H. (2014). "Stationkeeping Monte Carlo Simulation for the James Webb Space Telescope."
9. Donahue, B., Duggan, M., et al. (2023). "Interstellar Probe: Fifteen Years to the Interstellar Medium with An Enhanced NASA Space Launch System." 2023 IEEE Aerospace Conference, 04–11 March 2023.
10. Hahn, A. J. (2020. *Basic Calculus of Planetary Orbits and Interplanetary Flight, The Missions of the Voyagers, Cassini, and Juno*. April 2020, Springer, New York, NY.
11. Kluever, C. A. (2018). *Space Flight Dynamics*, First Edition. John Wiley & Sons Ltd.

12. McNutt, RL Jr, Wimmer-Schweingruber, RF, Gruntman, M, Krimingis, SM, et al. (2021), "Interstellar Probe – Destination: Universe!" International Astronautics Conference, 2021, IAC-21-D4.4.1.

13. Musielak, D. (2022). *Leonhard Euler and the Foundations of Celestial Mechanics.* Springer book series: History of Physics (HIPHY), ISBN: 978-3-031-12321-4. https://doi.org/10.1007/978-3-031-12322-1.

14. Roberts, C. (2011). "Long Term Missions at the Sun-Earth Libration Point L1: ACE, SOHO and WIND." AAS Astrodynamics Specialist Conference, 2011.

15. Space Launch System Mission Guide (2018). NASA Document No: ESD 30000. Release Date: December 19, 2018. https://ntrs.nasa.gov/api/citations/20170005323/downloads/20170005323.pdf.

16. United Launch Alliance (ULA) (2013). Delta IV Launch Services User's Guide. Cleared for public release by the Chief, Office of Security Review, Department of Defense, as stated in letter 13-S-1948, dated June 04, 2013.

17. Vallado, D.A. (2007). *Fundamentals of Astrodynamics and Applications*, third edition. McGraw Hill (2007). ISBN 978-1-881883-14-2.

18. Williams, J., Dawn, T.F., and Batcha, A. L. (2023). "A History of Orion Mission Design, Copernicus Software Development, and the Artemis I Trajectory." Paper AAS 23-241, AAS/AIAA Astrodynamics Specialist Conference, Big Sky, MT, August 13–27, 2023.

Liquid Propellant Rocket Engines

4

Four high thrust liquid propellant rocket engines power the core stage of NASA Space Launch System (SLS). The RS-25 engines fire non-stop for 8.5 min after liftoff, together with two solid rocket boosters, to take the SLS to orbit, making it a powerful launch system to propel astronauts on their voyage to the Moon.

To launch the crewed spacecraft Orion on the next lunar mission, NASA's SLS Block 1 will produce 39 MN (8.8 million lbf) of thrust during liftoff, powered by a core stage with four RS-25 rocket engines augmented with two solid rocket boosters (SRBs). The boosters will burn for about two minutes and then separate, allowing the RS-25 core stage rockets to continue lifting Orion to orbit. The fairing will be released to expose the service module, and then a jettison rocket motor will fire to separate the launch abort system, as this is no longer required because, if necessary, Orion could safely abort the flight using the propulsion system on the European Service Module. Approximately eight minutes after liftoff, the RS-25 engines will shut down and the core stage will separate from Orion, which is attached to the Interim Cryogenic Propulsion Stage (ICPS); propulsion will then be achieved with the smaller RL10 upper stage rocket, which will fire for less than one minute to position Orion on the required trans-lunar injection (TLI), the point in the orbit which determines the beginning of the trajectory to the Moon. After the RL10 engine completes the burn, the ICPS will separate from Orion, and the service module will use its own propulsion system to propel and maneuver Orion on its round-trip to the Moon, and eventually return to Earth. This scenario illustrates the rocket propulsion

© The Author(s), under exclusive license to Springer Nature Switzerland AG 2025   133
D. Musielak, *Introduction to Rocket Propulsion for Astronautics*, Synthesis Lectures on Engineering, Science, and Technology, https://doi.org/10.1007/978-3-031-86141-3_4

technology that is the focus of this chapter. In the following sections, we review the performance of modern chemical rocket designs, categorizing them according to their specific applications for both launch vehicles and spacecraft. This chapter aims to answer questions such as

- Why the booster or first stage rocket engines of a launch vehicle operate at very high pressure?
- What is the nominal performance required of first and upper stage rockets?
- What are the best propellants for launch vehicle propulsion stages and for in-space propulsion? Why?
- How do methane-fueled rockets compare with hydrogen-fueled rocket engines?
- Are monopropellant thrusters better suited for in-space propulsion?
- What are green propellants? Storable propellants? How do they compare with cryogenics?

4.1   Chemical Rocket Propulsion

Chemical rockets are propulsion systems in which the energy source is derived from the chemical reaction of propellants, and the thrust force is generated by the hot gases resulting from the chemical reaction of fuel and oxidizer propellants in the combustion chamber. More precisely, the thrust results from the conversion of the heat generated by the combustion reaction into kinetic energy. After combustion, the gases under pressure expand through a converging-diverging nozzle producing a high velocity jet of hot gases expelled by the rocket. Conceptually, a chemical rocket engine is comprised of the combustion chamber, the nozzle, the propellant injection and feed systems, control, and metering systems.

Chemical rockets can be classified according to (a) their basic function (booster stage, core and main engine, upper stage, sustainer, attitude control, orbit station-keeping), (b) the type of vehicle they power (space launch vehicle, spacecraft), (c) size (meganewton, millinewton thrust), (d) reusability (expendable or reusable), (e) type of propellant combusted (hypergolic monopropellant, cryogenic bipropellant), or (f) by their engine power cycle, a system that determines how the propellants reach the main combustion chamber.

4.1.1   Liquid Propellant Rocket Engines

A liquid propellant rocket engine (LRE) consists of a thrust chamber (combustor), the propellant injection and ignition systems, the feed system, including necessary piping, valves, and pumps, a control system, and a propellant storage system. Liquid propellant rocket engines can be classified based on the number of reactant substances injected into the

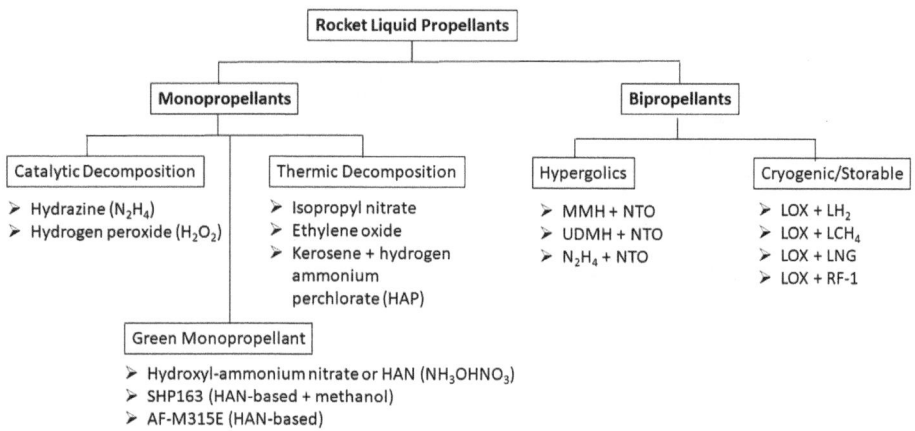

Fig. 4.1 Liquid propellants for chemical rocket propulsion

thrust chamber: bipropellant rockets, or monopropellant thrusters, as indicated in Fig. 4.1. The rocket can also be fed separate fuel and oxidizer and switch on command to operate with a single monopropellant mixture, a design known as bi-modal thruster. Thus, we classify chemical rockets as bipropellant liquid rocket engines, monopropellant rockets, and dual-mode liquid rockets. A monopropellant is a single liquid, a homogeneous substance composed of both oxidizing and fuel species which decompose into a hot gas via an exothermic reaction when it is in contact with a catalyst or due to heating. Examples of catalytic decomposable monopropellants are hydrazine (N_2H_4), and hydrogen peroxide (H_2O_2), while isopropyl nitrate, ethylene oxide and hydroxyl ammonium perchlorate are substances that decompose due to heat exposure.

4.1.2 Bipropellant Rockets

In a bipropellant liquid rocket engine, the fuel and oxidizer are stored in separate tanks and are brought together by injecting them in the combustion chamber in a manner to promote their rapid mixing and chemical reaction. Figure 4.2 illustrates the basic configuration of a liquid bi-propellant LRE.

The propellants can be ignitable or hypergolic. Ignitable propellants (e.g. LH_2 + LOX, or LCH_4 + LOX) require an external ignition system, whereas hypergolic propellant pairs react spontaneously with each other upon contact. Hypergolic bipropellant combinations include hydrazine fuel (N_2H_4), which reacts in contact with nitrogen tetroxide (NTO) (N_2O_4), and hydrazine derivative fuels such as mono-methyl-hydrazine (MMH) (CH_6N_2) and unsymmetrical dimethylhydrazine (UDMH) ($H_2NN(CH_3)_2$), which also

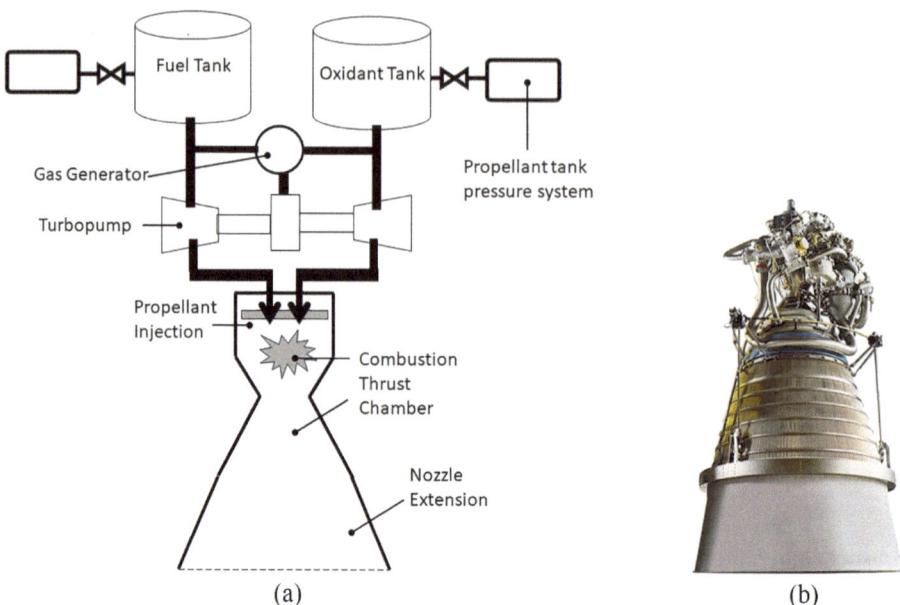

Fig. 4.2 a Graphical representation of liquid bipropellant rocket engine (LRE); **b** Photo of Pratt & Whitney RL-10A-4-1N, upper stage on Atlas launch vehicle. RL10 engine uses bipropellant LOX/ LH$_2$, expander cycle. Performance: I_{sp} = 451 s. Burn time: 740 s. Diameter: 1.53 m (5.00 ft)

spontaneously react when mixed with oxidizer NTO. When a compound is added to improve the specific impulse of a bipropellant, the mixture is known as tri-propellant.

The thrust chamber (in case of high-thrust applications) or thruster (moderate to low-thrust applications) consists of a cylindrically or spherically shaped reaction or combustion chamber, and the nozzle, a specially contoured convergent-divergent section. The fuel and oxidizer are injected, mixed, and ignited in the combustion chamber, and the chemical energy from the propellant is converted into thermal energy thereby creating a hot gas mixture. This hot gas mixture is accelerated by pressure forces to a supersonic exhaust velocity in the nozzle, thereby imparting high momentum to the vehicle. The shape or contour of the nozzle is designed so that it generates a high pressure in the thrust chamber by limiting the outlet area and the optimal conversion of the thermal energy of the hot gas mixture into kinetic energy of the jet exhaust. The nozzle extension depicted in Fig. 4.2 is a piece added to increase the expansion area ratio and thus improve the performance of a rocket engine, especially an upper stage.

The combustion chamber of a LRE is designed to be strong enough to contain the high pressure, high temperature gases resulting during the rapid mixing and turbulent chemical reaction of the propellants. The combustion temperature is limited by the energy of the chemical reaction. The rapid mixing of fuel and oxidizer substances results in the release of energy and large volumes of gaseous products. The gas generated by combustion has considerable energy in the form of heat. In typical chemical rocket engines, the temperature of the combustion gases can be higher than 3311 K (5500°F). The chamber and nozzle must be cooled to sustain the high temperature of the reaction gases and manage heat transfer through the walls. The combustion or thrust chamber must also be of sufficient length to ensure complete combustion before the gases enter the nozzle.

Did you know? **The four LREs for NASA's SLS core stage are the same RS-25 rocket main engines used for the Space Shuttle Orbiter**, where each engine provided almost 1.752 MN at 104% power, for a total of 5.255 MN of thrust at lift off.

4.1.3   Monopropellant Rockets or Thrusters

A monopropellant rocket engine, or thruster, utilizes a single liquid propellant that possesses the qualities of both the oxidizer and the fuel. Monopropellants, also called monergols, are substances or homogeneous mixtures which decompose in an exothermic reaction due to the presence of a catalyst or due to heating. Monopropellants decompose chemically, accompanied by the release of heat, providing the energy required for generating thrust. Ignition of monopropellants can be produced thermally (electrical or flame heat), or by a catalytic material.

The conventional monopropellant thruster employs a catalytic bed to ignite the propellant. In this type of thruster, the liquid monopropellant is decomposed in a chamber as it passes through a catalyst mesh giving off heat. For example, hydrazine is injected into a catalyst bed, where it decomposes into hydrogen, nitrogen, and ammonia. The temperature (in the 1204 °C or 2200°F range) and species of the decomposed monopropellant gas depend on the catalyst and other factors. After decomposition, the hot gas expands at high pressure through the nozzle and provides the thrust force (Fig. 4.3). The most common monopropellants are hydrazine (N_2H_4), and hydrogen peroxide (H_2O_2).

Shown in Fig. 4.3b is the MONARC-445 from Moog, a hydrazine spacecraft thruster with a steady state thrust of 445 N and a specific impulse of 234 s. This thruster has a feed pressure of 4.8–27.6 bar, has a nozzle expansion ratio of 50:1 and a valve power of up to 58 W. The thruster has a mass of 1.6 kg and measures 41 × 14.8 cm. Monopropellant thrusters are used in low thrust applications such as attitude and velocity control. The monopropellant must be stable (chemically and thermally) to ensure good liquid storage capability, and at the same time it must be easily decomposed and reactive to yield acceptable combustion properties. In its decomposition, the monopropellant must have excellent exothermal chemical reaction to release sufficient heat for thrust.

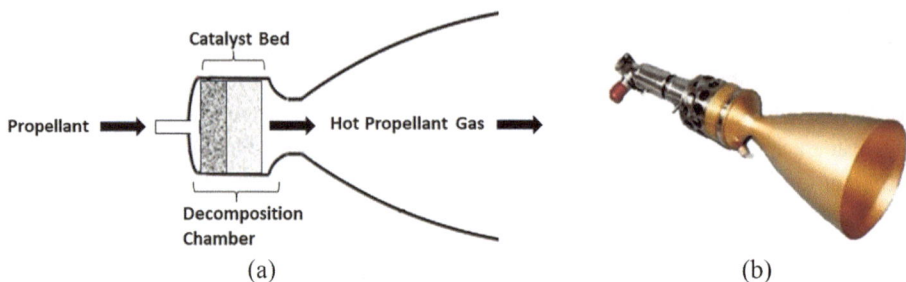

Fig. 4.3 **a** Schematic of a monopropellant thruster engine; **b** MONARC-445 hydrazine thruster for spacecraft and flight-vehicle attitude control applications. *Credit* Moog

Each of the NASA Pioneer 10 and 11 interplanetary space probes used six 4.5 N hydrazine thrusters for in-space maneuvering. At launch, the Pioneer carried 36 kg (79 lbm) of liquid hydrazine in a small, 42-cm (17 in) diameter spherical tank.

A liquid monopropellant rocket is highly reliable and easy to control. Because it consists of few moving parts, monopropellant rockets are used in many space applications that require low thrust such as for attitude control in satellites and small space probes. Monopropellant rocket thrusters are used on NASA's Orion Crew Module (CM). Twelve 160-lbf thrust MR-104G engines will provide the full complement of primary and redundant control required for critical maneuvers upon re-entry into the atmosphere. The hydrazine MR-104 engine has a proven legacy; it provided in-space propulsion for the Voyager 1 and 2 and Magellan missions. Subsequent MR-104 variants provided propulsion for Landsat and NOAA satellites.

4.1.4   Bi-Modal Thrusters

A dual-mode rocket engine is one that can operate both as a monopropellant thruster, and as a bipropellant rocket. Such design gives this type of rocket the flexibility to support several in-space thrusting maneuvers such as station-keeping, thrust-vector control, attitude control, orbit insertion, change in orbit, and propellant depletion. The Bi-Modal Thruster or Secondary Combustion Augmented Thruster (SCAT) is a thruster that operates continuously and switches between bipropellant or monopropellant modes. The SCAT is used in bipropellant mode when a spacecraft maneuver requires higher thrust. It can also operate in a monopropellant mode for lower thrust maneuvers. For example, during initial high-impulse orbit-raising maneuvers, the thruster operates in a bipropellant fashion, providing high thrust at high efficiency. When the spacecraft arrives to its operation orbit, the SCAT closes off either the fuel or the oxidizer and conducts the remainder of its mission in a simple, predictable monopropellant manner.

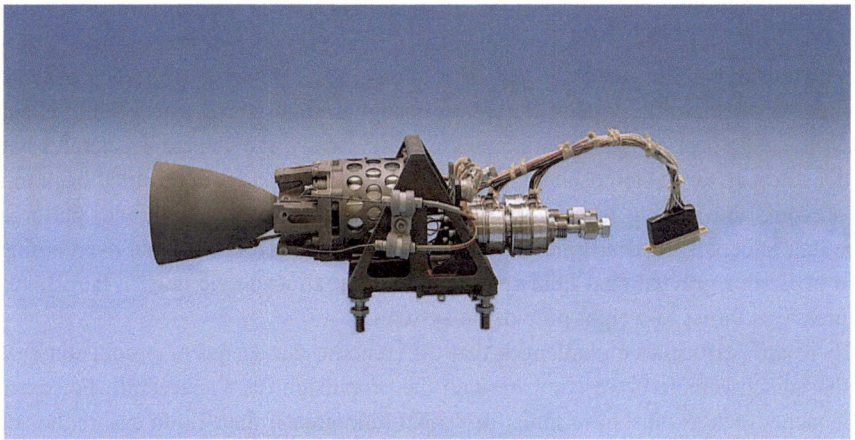

Fig. 4.4 TR501 dual-mode thruster patented by Northrop Grumman in 1995. It is 5-to-11 lbf regeneratively cooled, high-performance. Thrust level can be set to accommodate various propulsion system inlet pressures. Operates in either N_2H_4 monopropellant mode, or N_2H_4/N_2O_4 bipropellant mode. *Credit* Northrop Grumman

SCATs use integrated mono- and bipropellant feed systems with a common hydrazine fuel tank or tanks. These dual-mode thrusters use hydrazine as a fuel mixed with nitrogen tetroxide in the high-performance bipropellant thrust mode, and hydrazine as a monopropellant passing through a catalyst for the low performance mode. The hydrazine is fed from a common fuel tank. In this manner, high specific impulse is provided for long Δv burns at high thrust (e.g. apogee circularization), and reliable precision minimum impulse burns are provided from the monopropellant thrusters for attitude control. The SCAT concept was invented by the former TRW, now Northrop Grumman. Figure 4.4 depicts their TR501 Dual-Mode Thruster used by the James Webb Space Telescope for station-keeping.

4.2 Propulsion Requirements for Launch Vehicle

The first phase of spaceflight is the launch of the spacecraft. To overcome the grip of Earth's gravity, we need a powerful propulsion system producing the required thrust force to ensure the vehicle lifts off the ground and reaches orbit. If the rocket engine is too slow, then the vehicle cannot escape the gravitational field and falls back to the ground. On the other hand, if the rocket departs with sufficiently high speed, the vehicle can escape the gravity of the planet and continue its travel forever, unless it is slowed down by its own propulsion, or by the effect of gravity of a large body (such as a planet or moon) on its path.

The first and most critical phase of all space missions is the Earth-to-orbit segment. This first phase has unique requirements for the propulsion system powering the space launch vehicle. For its ascent, the vehicle requires rocket power to overcome our planet's gravitational field while passing through a changing atmosphere. The vehicle is subject to aerodynamic heating and dynamic pressures. Typically, the vehicle accelerates beyond the densest parts of the atmosphere, but then the rocket engines are throttled back to keep the pressure and heating from unsafe values. Atmospheric drag is another perturbation force that affects the ascent trajectory. That is why, to reach orbit while overcoming the force of Earth's gravitational field and high levels of atmospheric drag, a launch vehicle requires high thrust-to-weight propulsion systems.

There are performance challenges that differentiate the design of propulsion systems for the launch vehicle if we compare with the propulsion for a spacecraft. For example, the launcher rockets must have thrust-to-weight ratio greater than 1.0 to ensure the vehicle overcomes Earth's gravitational field and get off the ground. The only type of propulsion that can meet the high thrust-to-weight ratio requirement are chemical rocket engines, which can effectively convert the thermal energy of the propellants to kinetic energy. Today's heavy lift launch vehicles use liquid propellant rockets, and some also incorporate strap-on solid propellant rocket motors (SRMs) as boosters for the initial part of the ascent trajectory. Although the specific impulse of SRMs is much less than that achieved with liquid propellant, they are a cost-effective solution to generate the high thrust and high single burn total impulse required for lower stages of heavy lift launch vehicles.

4.2.1 The Multi-stage Launch Vehicle

A space launch vehicle (SLV) is a multi-stage rocket vehicle specifically designed to place payloads into orbit. All propulsion stages of a SLV are chemical rockets burning liquid or solid propellants. An SLV includes

- Propulsion systems (rocket engine stages, strap-on boosters, abort motors)
- Propellant Tanks
- Structure
- Payload (crew capsule, cargo, satellite, lunar lander, space probe, space telescope, planetary orbiter, sample return spacecraft, etc.)
- Instrumentation and Controls
- Navigation system
- The spacecraft bus to support the payload. The spacecraft is the part of a launch vehicle that carries the payload to accomplish a space mission.

Let us consider the propulsion requirements for the SLV.

4.2.2 Propulsion Requirements

The propulsion for a launch system is designed to operate in all conditions from atmospheric pressure experienced on the ground to the very low pressure in orbit. Let us consider these performance requirements, which are analyzed in greater detail throughout this book.

- **Thrust-to-weight ratio**. The propulsion system for a launch vehicle must produce a total thrust F that is greater than the vehicle's weight ($w_0 = m_0 g_0$) at liftoff to ensure the vehicle gets off the launch pad. The thrust-to-weight ratio must be $F/w_0 > 1.0$ (typical values are 1.25, 1.3). Only chemical rockets can deliver high to thrust ratios, and that is why all current launch vehicles use liquid bi-propellant rocket engines, and some also use solid propellant strap-on boosters. For example, the NASA SLS has thrust to weight ratio $F/w_0 > 1.27$ at launch by producing a combined thrust of 36,786 kN (8.27 million lbf), to lift a vehicle with a total weight of 6.5 million lbf.
- **Throttling**. The propulsion system for launch vehicles requires throttling capability. This is because the vehicle is exposed to varying high aerodynamic forces as it moves through the atmosphere. As it quickly ascends, the vehicle velocity increases while still relatively low in altitude where the atmosphere is relatively dense. Thus, within the first minute or so of launch the vehicle is subject to a *dynamic pressure*. As we saw in Chap. 3, since the air density changes with altitude, there is a point in the ascent trajectory where the dynamic pressure is a maximum, attaining a critical value q_{max} that cannot be exceeded to prevent structural failure.

Before each launch, engineers determine if current winds and other atmospheric conditions over the launch site will not exceed the vehicle's design tolerances. A specifically tailored thrust profile must be calculated for the vehicle, one that allows to decrease or "throttle down" the engines during the peak dynamic pressure. For example, each of the Aerojet Rocketdyne RS-25 cryogenic engines forming the core stage of NASA's SLS have a throttle range of 67–109%. Thus, during the critical dynamic pressure phase of the ascent flight, the RS-25 core engines can reduce the thrust from the 109% maximum to 67%. The burn profile of the boosters' solid propellant grain is also tailored to reduce thrust a similar amount to keep dynamic pressure below q_{max}.

Engine thrust throttling is also required to maintain launch vehicle acceleration below a safe level, that is especially important for the well-being of the crew. Astronauts confined to the crew capsule strapped at the top of a launch vehicle feel the thrust of lift-off as an acceleration or g-load that pushes them back into their seats. Since acceleration depends on the thrust force and the vehicle's mass, keeping the engine thrust constant means that the acceleration will gradually increase as the vehicle gets lighter due to expended propellant. Acceleration tends to increase over time. Therefore, to maintain the overall g-load on the vehicle under 3 g's, the core engines must throttle down a few minutes after lift-off to decrease the thrust in a controlled manner to match the burned propellant mass. Propulsion systems for future lunar and Mars landers will also require throttling to ensure vehicles can touch down softly on the extraneous surfaces.

- **Thrust Vector Control (TVC).** The rocket engines for launch-vehicles require a thrust vector control (TVC) system to vary the thrust direction for steering during their ascent trajectory. The TVC can gimbal the entire engine to point the thrust in the desired direction. In 2021, the NASA-designed thrust vector control (TVC) system for the SLS was tested to gimbal the RS-25 engines. "Gimbaling" refers to how the engine must move to ensure proper flight trajectory. During the test, engineers demonstrated the ability of the new TVC to move the engines on both a tight circular axis and back-and-forth on a line, pivoting around two axes of freedom with a range of $\pm 10.5°$. This gimbal range is required to correct for the pitch momentum caused by the constantly shifting vehicle's center of mass as the engines burn propellant in flight and after booster separation.
- **Nozzle Design.** The overall performance of a rocket engine depends on the external ambient pressure and the nozzle expansion ratio. Due to the wide range of operational conditions, the rocket engine nozzle in a launch system may never achieve perfect expansion. In general, at lift-off the first or core stage engines exhaust to atmospheric sea level pressure (1 bar or 14.7 psia) but this ambient quickly changes as the launch vehicle ascends, reaching near zero pressure or vacuum condition in orbit.

If it were possible, the engine nozzle would increase its expansion ratio to change the exhaust pressure as atmospheric pressure decreases throughout its trajectory. However, a variable nozzle that would be ideally expanded for all altitudes requires hardware that would add complexity and weight to the vehicle. Therefore, the nozzle design area ratio must be selected and optimized to reach ideal expansion at a predetermined design altitude that is about 2/3 of the way from the altitude of engine ignition to the altitude of engine cutoff. For example, for a rocket engine to operate from sea level to 60 km altitude it is possible to design a nozzle with an exhaust pressure matching the ambient pressure at about 40 km altitude. In this manner, the nozzle would operate over-expanded below 40 km, and under-expanded above 40 km altitude. Such nozzle yields better overall performance than one designed to be ideally expanded only at sea level. Thus, in a multi-stage SLV, each stage will have different nozzle design constraints. For example, in a three-stage vehicle, the first or core stage will have the biggest chamber and the highest thrust but the lowest nozzle area ratio, while the upper third stage will have the lower thrust but the highest nozzle area ratio.

- **Restartability.** Upper stages provide the energy to propel a spacecraft while moving from one orbit to another, or to accelerate the spacecraft to an Earth escape velocity to begin an interplanetary trajectory. The thrust and power requirements for upper stages are different than for the lower stages of the launch vehicle. Although high thrust-to-weight propulsion is not required for upper stages, this parameter may be important to meet payload injection velocity requirements. In addition, the upper stage must be capable of multiple restarts if it is intended to deliver payloads to different orbits. For example, the Vinci upper stage engine of ESA's Adriane 6 can provide four burns per mission, e.g. one burn to place the spacecraft into an elliptic orbit, and later the rocket stage can be reignited to circularize the orbit (Alliot et al. [1]).

Stabilization and orientation of the spacecraft from launch vehicle separation until end of the mission is typically provided by two engineering subsystems: attitude and articulation control (including computer, sensors, gyros, and actuators) and propulsion. For example, because of the energy required to achieve a Jupiter ballistic trajectory with an 825 kg (1819 lbm) payload, the Voyager spacecraft launched by the Titan III E/Centaur included a final propulsive stage (a Star 37E solid rocket motor) that added a final injection velocity increment of about 2 km/s (4475 mph). Approximately 11 min after solid-rocket burnout, the propulsion module was jettisoned, and the Voyager spacecraft began their interplanetary mission, each powered by a propulsion system comprised of 16 monopropellant hydrazine thrusters required for orbit trim maneuvers and attitude control; each thruster could deliver a thrust of 0.889 N (0.2 lbf).

To maximize the payload delivered to a target orbit we must use sophisticated trajectory optimization methods. Launch vehicle suppliers (e.g. NASA, SpaceX, ULA, ESA) prepare a payload user's guide, a publication that provides launch vehicle performance as a function of the orbital target, based on their own optimization analysis. The launch performance is given as a plot of payload mass versus launch injection characteristic energy (C_3), data which help us establish their launch capabilities for different missions (see Chap. 2).

4.3 Requirements for In-Space Propulsion

4.3.1 The Spacecraft

The spacecraft moves beyond our planet's atmosphere, goes into orbit or travels in deep space and/or returns to Earth. It can be a geocentric satellite, a crewed capsule or a space telescope, a planetary orbiter, or an instrumented probe for a mission of space exploration. Since energy intensive maneuvers, such as orbit injection or planetary landing, require large velocity increments, the propulsion system may be integrated with the spacecraft. For example, the NASA SLS carries the Orion crew capsule attached to its European Service Module (ESM) to the translunar injection point in LEO from where the Orion-ESM stack begins its trajectory to the Moon. The ESM provides propulsion, power, thermal control, and water and air for the astronauts, and it is ejected upon return from the mission, before Orion re-enters Earth's atmosphere.

4.3.2   Propulsion Requirements

The spacecraft must perform challenging orbital maneuvers that require thrust and therefore propellant. Because a spacecraft is always in the gravitational field of some central body (Earth, Moon, or the Sun), it is subject to orbital-motion laws, and maneuvering in space must comply with mission constrains regarding on-board propellant, hardware, and energy resources. The spacecraft propulsion is specially designed to allow reliable maneuvering in space with minimum propellant. The requirements for the rocket engines to power spacecraft include:

- **Restartability**. Thrusters for attitude control must restart frequently. Bipropellant, monopropellant, and cold-gas thrusters have this capability.
- **Steady-state and pulse mode operation**. Bipropellant thrusters can be used in pulsing or steady-state modes. The pulse duration is of milliseconds. Pulsing engines are characterized by minimum impulse bit I_{min}, expressed as $I_{min} = F(P_w)_{min}$, where $(P_w)_{min}$ denotes the minimum pulse width, a characteristic of the thruster. For example, a bipropellant MMH/N_2O_4 thruster can have a minimum impulse bit between 0.015 and 0.030 lbf-s.
- **Thrust vector control (TVC)**. Moving the net thrust vector through small angles for spacecraft stability and maneuvering is required to remove vector errors. For example, NASA's Artemis I crew capsule has an Orbital Maneuvering System Engine (OMS-E) designed to gimbal to provide two-axis (pitch and yaw) thrust vector control.
- **Reliability during a multi-year mission**. Spacecraft require propulsion to operate as need and for a long time. For example, space telescopes orbiting the unstable Sun-Earth collinear points require to perform regular station-keeping maneuvers to remain in the halo orbits around those points.
- **Propellant Storability**. If spacecraft propulsion requires to operate for long time, it must consider storable propellants that can be in a tank for the duration of the mission, stored at high temperatures.

Propulsion for in-space applications must be designed with capability for attitude, orbit control, and re-entry maneuvers, especially for heavy human-rated spacecraft, and for sample-return probes. For example, the Orion crew capsule is equipped with a Reaction Control System (RCS) comprised of twelve 160-pound-thrust MR-104 J monopropellant thrusters (designed and manufactured by Aerojet Rocketdyne) mounted in 10 pod assemblies. At the conclusion of the Artemis missions, the Crew Module separates from the European Service Module (ESM) prior to re-entering Earth's atmosphere. After this point, the RCS thrusters become the sole means for controlling the spacecraft's orientation during atmospheric reentry and altering its course in preparation for splashdown. Table 4.1 summarizes the type of propulsion used for spacecraft.

Table 4.1 Type of in-space propulsion and typical I_{sp}

Propulsion type	Orbit insertion	Orbit maintenance and maneuvering	Attitude control	Typical I_{sp} (s)
Cold gas		Yes	Yes	30–70
Solid propellant	Yes			280–300
Monopropellant		Yes	Yes	220–240
Bipropellant	Yes	Yes	Yes	305–325
Dual-mode	Yes	Yes	Yes	313–330

4.4 Liquid Propellant Characteristics

Propellants have advantages and disadvantages and selection for a given application requires compromises. The more important and desirable propellant features include: (1) High energy release per unit of propellant mass, combined with low molecular weight of the combustion or decomposition gases, for high specific impulse; (2) Ease of ignition; (3) Stable combustion; (4) High density or high density impulse to minimize the size and weight of propellant tanks and feed system; and (5) Ability to serve as an effective coolant for the thrust chamber (optimum combination of high specific heat, high thermal conductivity and high critical temperature).

For first stage rockets, we seek propellants with the highest density. Propellant density relates to propellant volume ratio, which in turn is used to determine the tank mass. The greater the propellant density, the greater the amount of propellant that can be stored in a tank and the greater the mass of propellant than can be pumped.

Liquid propellants can be conventionally storable liquids or cryogenics. Storable propellants can be stable at room temperature for a very long time (months or even years). A cryogenic propellant is a form of liquid propellant that must be kept at very low temperatures to remain liquid.

4.4.1 Cryogenic Propellants

Cryogenic propellants are liquefied gases that have a very low boiling point and must be stored at very low temperatures. Examples of cryogenic propellants are liquid hydrogen (LH$_2$) fuel, liquid methane (LCH$_4$) fuel, and liquid oxygen (LOX) oxidizer. In gaseous form, oxygen and hydrogen have such low densities that extremely large tanks are required to store them. But if these gases are cooled and compressed into liquids, their density increase greatly, making it possible to store them in smaller tanks. For hydrogen to stay in liquid state its temperature must be 20.37 K, and for oxygen to remain liquid its temperature must be 90.37 K.

Cryogenic propellants require special insulated containers and venting systems to allow gas from the evaporating liquids to escape. These propellants are pumped from the storage tanks to an expansion chamber and then are injected in the rocket combustion chamber where they are mixed and ignited by a flame or spark. Since hydrogen has a very low density, it requires a storage volume many times greater than other liquid fuels. For example, the propellant tanks in the core stage of NASA SLS carry 537,000 gallons of LH_2 cooled to -252.778 °C, and 196,000 gallons of LOX cooled to -182.778 °C. It is not possible to keep such liquids under pressure at temperatures above their boiling point. Thus, a small fraction of the cryogenic liquid is allowed to boil off. That is why the tanks are filled only a day or two before the launch. Ice forms around the vents, which then breaks off and drops down on the first moments after lift-off.

Methane is a very attractive cryogenic hydrocarbon fuel. Liquid methane (LCH_4) is thermally similar to LOX as it has a boiling point of 111.6 K (closer to the 90.37 K of LOX). Liquid methane costs less than LH_2. Having a higher density ($\rho = 1.786$ kg/m^3 at 112 K and 1 bar), LCH_4 fuel is easy to store at warmer temperatures, so the tanks do not require that much insulation, thus can be lighter and smaller, as compared with tanks for storing liquid hydrogen.

4.4.2 Hypergolic Propellants

Hypergolics are fuels and oxidizers that ignite spontaneously on contact with each other—they do not require an external ignition source. The most common hypergolic propellants are hydrazine (N_2H_4), monomethyl hydrazine (MMH) [CH_6N_2], and unsymmetrical dimethyl hydrazine (UDMH) [$(CH_3)_2NNH_2$] fuels, which are typically reacted with nitrogen tetroxide (NTO) [N_2O_4]. Mixed oxides of nitrogen (MON) are solutions of nitric oxide (NO) in dinitrogen tetroxide/nitrogen dioxide (N_2O_4 and NO_2) also used as oxidizers in some thrusters. For example, MON-3 is NTO mixed with 3% oxides of nitrogen (N_2O_4).

The easy start and restart capability of hypergolics make them ideal for spacecraft maneuvering systems and for crewed space vehicles. Also, since hypergolics remain liquid at normal temperatures, they do not pose the storage problems of cryogenic propellants. However, hypergolics are highly toxic and must be handled with extreme care. Hypergolic propellants are excellent for spacecraft and satellites that are required to start and stop their thrusters many times over their design mission life.

Hydrazine gives the best performance as a rocket fuel, but it has a high freezing point and is too unstable for use as a coolant. MMH is more stable and gives the best performance when freezing point is an issue. UDMH has the lowest freezing point and has enough thermal stability to be used in large regeneratively cooled rocket engines. Regenerative cooling refers to a cooling method that utilizes a cold propellant to circulate through a cooling jacket along the hot wall of the engine to remove excessive heat.

Hydrazine dissociates exothermically on a catalyst (iridium or platinum) to produce a hot gas mixture consisting of hydrogen (66%), nitrogen (33%) and a little ammonia. Although the combustion temperature is low (~ 800 K), the mean molecular weight is also low (~ 11), so the rocket exhaust velocity achieved with hydrazine is considerable, on the order of 1700 m/s. This is the reason why hydrazine is used as hypergolic monopropellant for attitude and orbital correction thrusters. Hydrazine is extensively used since it contains no carbon atoms and consequently will not degrade the catalyst bed. However, hydrazine may cause problems to the propellant supply system because of its high freezing point which raises the danger of the thruster propellant freezing when the spacecraft is in outer space.

For bi-propellant rocket engines, hydrazine fuel can be combusted with LOX, or with nitrogen-based oxidizers such as NTO. Nitrogen tetroxide is less corrosive than nitric acid and provides better performance, but it has a higher freezing point. Consequently, NTO is the oxidizer of choice when freezing point is not an issue. Nitrogen tetroxide has a high vapor pressure (1 bar at 294 K) and requires careful handling. MMH is a better alternative to hydrazine as a fuel, as it is safer and easier to handle. The bipropellant MMH + NTO combination is hypergolic (self-igniting), making it excellent choice for applications requiring reliable restarting such as orbital control thrusters.

4.4.3   Bipropellant Combinations

To date, four different liquid fuel/liquid oxidizer combinations have proven to work very well for most of the rocket applications in the world. Table 4.2 summarizes applications of common bipropellant combinations (Table 4.3).

Propellants with the best combinations of high energy content and low molecular mass, such as liquid Hydrogen (LH$_2$) and liquid Oxygen (LOX), can produce a specific impulse (I_{sp}) up to 467 s or exhaust velocities of 4580 m/s, about 30% higher than all other

Table 4.2 Common bipropellant combinations and their applications

Propellant combination	$I_{sp}(s)$	Application
LOX/LH$_2$	Up to 467	SLV booster, core, and upper stages
LOX/RP-1	Up to 330	SLV booster, core, and upper stages
LOX/LCH$_4$	Up to 363	SLV booster, second stage, and spacecraft or in-space maneuvering
NTO/MMH and NTO/UDMH	Up to 340	Attitude and reaction control systems for orbit change, station-keeping, reentry, or spacecraft rendezvous, planetary missions, post boost control

Table 4.3 Selected physical properties of common liquid (Fuel) propellants

Propellant	Hydrogen	Methane	RP-1	Hydrazine	MMH
Chemical formula	H_2	CH_4	$CH_{1.9-2.0}$	N_2H_4	CH_3NHNH_2
Molecular mass $\mathcal{M}$	2.016	16.03	175	32.05	46.072
Freezing point (K)	14.0	90.5		274.69	220.7
Boiling point (K)	20.4	111.6		386.66	360.6
Specific heat (kcal/kg-K)	1.75	0.835		0.736	0.698

combinations. Since the thrust developed by a chemical rocket is essentially proportional to the exhaust gas velocity, Eq. (2.5), which in turn depends on the average molecular mass, Eq. (2.33), hydrogen yields higher exhaust velocities for a given gas temperature and provides greater thrust. The Shuttle Main Engine (SSME) fueled with LOX/LH$_2$ was the highest I_{sp} bipropellant rocket engine produced in the USA in the twentieth century. This bipropellant is also used for the core and upper stages of the NASA SLS (see Tables 4.4 and 4.5).

Hydrocarbons such as kerosene and methane are also excellent rocket fuels. Rocket Propellant-1 (RP-1), a highly refined form of kerosene ($CH_{1.953}$), is typically combusted with liquid oxygen and used in the first-stage boosters of powerful SLVs, e.g. Delta, Atlas-Centaur, and Falcon V. The F-1 engine powering the first stage of the Saturn V launch vehicle that made possible the Apollo lunar missions was one of the most powerful rocket engines ever built, producing a massive thrust of 6.77 MN at sea level, fueled by RP-1/LOX. Developed by Rocketdyne, the F-1 established the foundation for modern rocket propulsion and set new standards for bi-propellant engine performance.

The combination RP-1/LOX (known as *kerolox*) is also used in the Merlin 1D engine, developed by SpaceX for the core stage of the Falcon 9 launcher. Each Merlin produces a thrust 845 kN at sea level. Kerosene delivers a specific impulse considerably less than that of cryogenic fuels (I_{sp} decreases with increasing carbon content in the fuel), but it contributed to making the Merlin 1 a reusable rocket, revolutionizing access to space through its affordability.

Did you know? **The first launch rocket engines were fueled by hydrocarbons.** The first man in space was launched into orbit by a vehicle propelled by the most powerful rocket engines developed by Russia, the RD-170 burning LOX and kerosene propellant. In the U.S., the five F-1 engines in the first stage of the Saturn V vehicle that launched the Apollo astronauts to the Moon used mixtures of LOX/RP-1 propellants.

Table 4.4 Performance characteristics of booster, main and core stage engines

Rocket engine	Nation	Engine cycle	Propellant	Thrust (MN) [sl]	Thrust (MN) [vac]	$I_{sp}(s)$ [sl]	$I_{sp}(s)$ [vac]	p_c(MPa)	Burn time, t_b (s)
RS-68	USA	GG	LOX/LH$_2$	2.95		410	412	10.26	249
RD-170	Russia	SC	LOX/RG-1	7.25	7.904	309	337	25.1	150
RD-180	Russia-USA	SC	LOX/RP-1	3.83	4.15	311	338	25.5	150
RD-107	Russia	SC	LOX/RP-1	0.81	1.00	256	313	5.9	119
F-1	USA	GG	LOX/RP-1	6.77	7.77	263	304	6.6	161
MA-5A	USA	GG	LOX/RP-1	2.10		263		4.4	263
RS-27A	USA	GG	LOX/RP-1	0.89		255		4.8	265
RD-253	Russia	SC	NTO/UDMH	1.47		285		15.2	130
YF-20	China	GG	NTO/UDMH	0.76		259		7.4	170
Viking 6	France	GG	NTO/UH25	0.68		249		5.9	142
Merlin	USA	GG	LOX/RP-1	0.845	0.914	282	311	9.7	162
Raptor	USA	SC	LOX/LCH$_4$	2.31		327	363	30	
RS-25	USA	SC	LOX/LH$_2$	1.859	2.28	366	453	20.5	510
RS-68	USA	GG	LOX/LH$_2$	2.89	3.31	360	420	9.7	249
RD-0120	Russia	SC	LOX/LH$_2$	1.51	1.96	359	455	21.8	600
LE-7A	Japan	SC	LOX/LH$_2$	0.87	1.098	338	440	12.7	390
Vulcain 2	France	GG	LOX/LH$_2$	0.94	1.359	320	434	11.6	600
Vulcain 2.1	France	GG	LOX/LH$_2$		1.324			12.08	
Merlin 1C	USA	GG	LOX/RP-1	0.512	0.569	275	304	6.77	
RD-108	Russia	SC	LOX/RP-1	0.78	1.01	252	319	5.1	290

(continued)

Table 4.4 (continued)

Rocket engine	Nation	Engine cycle	Propellant	Thrust (MN) [sl]	Thrust (MN) [vac]	$I_{sp}(s)$ [sl]	$I_{sp}(s)$ [vac]	p_c(MPa)	Burn time, t_b (s)
Viking 5C	France	GG	NTO/ UH25	0.68	0.75	249	278	5.9	142
YF-20B	China	GG	NTO/ UDMH	0.73	0.81	259	289	7.4	170
BE-4	USA	SC	LOX/ LNG	2.4				13.4	

Table 4.5 Performance characteristic of selected upper stage engines

Nation	Rocket engine	Engine cycle	Propellant	Thrust (kN) [vac]	$I_{sp}(s)$ [vac]	p_c(MPa)	Burn time (s)
USA	RL10B-2	EC	LOX/LH$_2$	110	465.5	4.3	700
USA	AJ10-118 K	PF	NTO/ Aerozine 50	43.7	319	7.9	
USA	Merlin 1D	GG	LOX/RP-1	690	310	9.7	344
France	Vinci	EC	LOX/LH$_2$	180	465	6.1	
France	HM7-B	GG	LOX/LH$_2$	70	447	3.5	731
Japan	LE-5B	EC	LOX/LH$_2$	137	447	3.6	534
China	YF-75	GG	LOX/LH$_2$	79	440	3.7	470
USA	J-2	GG	LOX/LH$_2$	890	426	4.4	
USA	Kestrel	PF	LOX/RP-1	31	325	0.93	
Russia	11D58M	SC	LOX/ Kerosene	79.5	353	7.6	680
Russia	RD-0210	SC	NTO/UDMH	582	327	14.8	230
Germany	Aestus	PF	NTO/MMH	30	324	1.0	1100
USA	Raptor Vacuum	SC	LOX/LCH$_4$	2400	380	35	281

Methane (CH$_4$) is another hydrocarbon fuel with desirable properties. When burned with LOX, methane performs very well without the volume increase penalty for propellant storage tanks common with LOX/ LH$_2$ systems. This results in an overall lower vehicle mass. Methane is clean burning, and non-toxic. SpaceX and Blue Origin have developed the Raptor and BE-4, respectively, engines that run exclusively on LOX/LCH$_4$, a bipropellant combination referred to as "methalox."

Many bi-propellant thrusters use Unsymmetrical DiMethyl Hydrazine or UDMH fuel with NTO oxidizer. UDMH fuel is easier to handle compared with hydrazine. UDMH is liquid at temperatures between 216 and 336 K. Having additional carbon in the reacting mixture result an increase in the mean molecular mass of the exhaust gas, and thus by adding a fraction of MMH or hydrazine to the mixture enhances the magnitude of the exhaust velocity. Hydrazine-based propellant mixtures have lower freezing points and thus are easier to store.

Properties of the most common liquid fuels are summarized in Table 4.3.

4.4.4 Green Propellants

Green propellants (liquid, solid, mono- or bi-propellant) are non-toxic, and environmentally friendly. For space applications, green propellants should also have excellent storability, wide material compatibility, and high propulsive performance.

In the past decades, research to develop non-toxic propellants (green) was aimed to reduce air pollution during rocket launches and to reduce hazards, as green propellants would be easier and safer to handle than hazardous propellants. Green propellants are expected to reduce the costs associated with transport and storage, both in spacecraft development and on-ground operations. These requirements make non-toxic propellants more desirable, especially when applied to low-cost micro-satellites, and to reusable launch vehicles. Formally, propellants are defined according to their toxicity. The Acute Toxicity Classification (ATC) levels are typically categorized on a 1 to 5 scale, where level one denotes the most toxic class and level five is considered the least toxic class. Propellants possessing ATC levels of three are safer and thus considered as green propellants. I expect that the toxic hypergolic propellants that are now used on spacecraft will be replaced by green propellants.

Despite being extremely toxic, carcinogenic, corrosive, flammable, and explosive, Hydrazine has a very long history as monopropellant, dating back to the development of the first rocket engines. Hydrazine became widely employed in satellite RCS and spacecraft attitude control systems. In the past several decades, green alternatives to hydrazine have been sought, resulting in propellant formulations that have decreased toxicity and are safer to store and handle. In the 1990's, the use of Hydroxyl Ammonium Nitrate (HAN) for formulating green liquid propellants became of great interest in the USA and Japan.

In 1998, engineers at Edwards Air Force invented a non-toxic monopropellant propellant blend identified as AF-M315E. Based on HAN, this blend is denser and more viscous than hydrazine, which allows more propellant to fit into the same tank volume. Hydrazine is hypergolic (it combusts when in contact with an oxidizer such as nitrogen tetroxide). The AF-M315E fuel requires hotter temperatures to ignite, making it safer to handle, but subject to more extreme operating conditions once in space. In contrast to hydrazine, AF-M315E does not freeze in space and thus it does not require heaters to stay warm enough to remain a liquid.

Monopropellant HAN-based formulations such as LMP-103S and AF-M315E have the added characteristic of improved propulsive performance, compared with hydrazine. In 2016, Masse and collaborators at Aerojet Rocketdyne reported on the specific impulse improvement using AF-M315E, obtaining $I_{sp} = 231$ s by fueling their thruster with this monopropellant, an increase of over 21% if hydrazine were used [22].

In Japan, Togo et al. [51] developed monopropellant compositions based on Hydroxyl Ammonium Nitrate (HAN). To study their combustion, the researchers used a control a mixture comprised of HAN, Ammonium nitrate (AN), plus water (H_2O) to lower the melting point of the solution. The new monopropellants (identified as SHP069 and SHP163) included different amounts of methanol fuel, added to the control blend to ensure a high specific impulse and good combustion characteristics. Among the blends studied, the Japanese identified SHP163 as the best monopropellant candidate having over 16% of methanol. They reported a specific impulse of 254 s for a thruster fueled with SHP163 when the combustion chamber had a pressure of 0.7 MPa, the nozzle expansion ratio was 50, and the thrust coefficient was 1.7. The characteristics of SHP 163 and the development of a 1 N thruster for the Green Propellant Reaction Control System (GPRCS) were published by Katsumi and Hori [19].

In recent years, Aerojet Rocketdyne, working with NASA, Ball Aerospace, and the Air Force Research Laboratory (AFRL), began developing small satellite propulsion through NASA's Green Propellant Infusion Mission (GPIM), an initiative to demonstrate a green propellant alternative to conventional chemical propulsion systems for future spacecraft.

The GPIM satellite, launched (in 2019) on the third flight of SpaceX's Falcon Heavy SLV, conducted a sequence of thruster firings to begin falling out of orbit after successfully demonstrating the effectiveness of AF-M315E, a non-toxic monopropellant that could be used on future space missions. This mission (to test the ability to control the GPIM satellite's attitude, and demonstrate how to change the spacecraft's orbital altitude) was conceived to prove the use of AF-M315E, the non-toxic propellant that could replace hydrazine for spacecraft.

Green hydrazine blends are now advanced to reduce the toxicity of conventional hydrazine. Sponsored by NASA, Aerojet-Rocketdyne has developed a hybridization of ionic liquid and conventional hydrazine constituents to form the green hydrazine propellant blend (GHPB). The green blend aims to reduce vapor toxicity and yet maintain the high-density specific impulse of ionic liquids while retaining the low combustion and preheat temperatures of conventional hydrazine. In testing reported in 2019, green hydrazine blends demonstrated long-term thermal stability/storability, low shock/impact sensitivity, and good operational stability. This form of green hydrazine has demonstrated a 100-fold reduction in vapor pressure/toxicity and a similar low-temperature start capability as compared to pure hydrazine. Equally important, the tested monopropellant has a lower freezing point, requiring less spacecraft power to maintain its temperature (see Masse and Glassy [23]).

Today, the overall objective in the space propulsion community is to develop an affordable, replaceable, non-toxic monopropellant solution to replace hydrazine. A new monopropellant requires to yield high I_{sp}, higher than that obtained with hydrazine. It must be temperature tolerant for robust, passive storage even in extreme space environments and be characterized as a good engine coolant.

A potentially revolutionary new monopropellant propulsion system is being developed in the U.S. that will burn a green monopropellant known as NOFBX. It is a non-toxic nitrous oxide/fuel/emulsifier blended propellant, that can be made from chemicals that are widely available, making it low cost and easy to produce. Researchers expect NOFBX could be widely applicable, e.g. on space stations, commercial crew vehicles, sample return missions, and human expeditions to the Moon and Mars.

Green propellants are very attractive for lunar and planetary exploration. Non-toxic propellants will not contaminate landing sites on Mars where a robot will collect samples for return to Earth. Astronauts at a lunar or Mars base will not have to worry about contaminating their habitats with toxins that have been deposited by repeated landings at the same site.

Crewed capsules RCS and landing retro-rockets have stringent selection criteria. Today the reference propellants are MMH/N_2O_4. The replacement by non-toxic propellants would offer a considerable improvement for the safety of the astronauts (the tanks are generally located inside the re-entry body aeroshell), and for post recovery operations.

Green bipropellants combinations are much safer and better for the environment. The ignition reliability which is a critical requirement for the crew safety. For mono-propellants, the critical point would be catalytic or non-catalytic ignition and for N_2O/hydrocarbons, the catalytic (N_2O decomposition) or electric ignition. Research using some candidate chemicals (e. g. hydrocarbons, HTP, nitrates) revealed that there is no perfect green propellant, and some risks should be accepted when greens are selected, as they could be mitigated by appropriate measures (e. g. cooling to reduce vapour pressure).

Continuing R&D advances to date ensure that conventional toxic propellants are replaced with safer, green propellants.

4.5   Rocket Engine Applications

Space launch vehicles today are multi-staged and are propelled by combinations of solid propellant motors and liquid propellant rockets. The rocket stages of expendable SLVs are designed to be jettisoned after burn out, soon after leaving the atmosphere to reduce the impact of the launch vehicle's mass on the payload capability to orbit.

Space launch vehicles (SLVs) can be divided into two major classes: (a) Launchers with large strap-on boosters with core and upper stages (e.g. the U.S. Space Launch

System (SLS), the European Ariane, the Japanese H-II, and the Russian Soyuz), and (2) launch vehicles with a booster, an optional sustainer, and an upper stage (e.g. the U.S. Atlas and Delta, the Russian Proton, and the Chinese Long March).

Launch vehicles can also be distinguished by their capability for human spaceflight, which must meet more stringent requirements. Human spaceflights are actively carried out by the Soyuz program conducted by the Russian Federal Space Agency, the Shenzhou program conducted by the China National Space Administration. In the U.S., SpaceX was contracted by NASA to transport crews to the ISS, for which the Falcon 9 partially reusable launch system carrying the Crew Dragon capsule was assembled in 2020.

NASA Space Launch System (SLS) was conceived as the foundation for human exploration beyond Earth's orbit, sending the Orion spacecraft to transport astronauts and cargo directly to the Moon on a single mission. Offering more payload mass, volume capability, and energy, SLS is NASA's most powerful rocket, can carry more payload to deep space than any other vehicle developed. Refer to previous chapters for full description of the SLS and its launch capability.

Although is not possible to review all modern operational rockets, the following paragraphs provide a brief overview of representative booster, main stage, upper stage engines, and thrusters, as summarized in Tables 4.4 through 4.6. Detailed descriptions and data are available from the various sources listed in the References.

4.5.1 Boosters, Main and Core Stage Engines

A booster rocket is either the first stage of a multi-stage launch vehicle, or a strap-on rocket used to augment the core launch vehicle's takeoff thrust and payload capability. Boosters are generally necessary to launch spacecraft into Earth orbit or beyond and can be solid rocket motors (SRMs) or liquid rocket engines (LREs).

Boosters or first core engines are the largest engines and operate at the highest chamber pressure. The high thrust is required to accelerate the gross liftoff weight of the entire launch vehicle, including all stages, propellant tanks, and payload, into orbit. The high chamber pressure is required to accelerate the nozzle exhaust gases against the atmospheric pressure to high Mach numbers, to maximize specific impulse performance and achieve the high thrust-to-weight ratio $F/w > 1$.

For expendable SLVs, a booster engine is jettisoned (dropping in the ocean) once it completes its burn, a point known as booster engine cut-off (BECO). The remaining parts of the launch vehicle carrying the payload (typically called the "stack") continue ascending propelled by sustainer or upper stage engines, which are optimized for vacuum operation. Main or first stage LREs operate as booster engines from the initial stage of the launch trajectory, and generally provide high thrust. Table 4.4 gives characteristic data of liquid propellant booster, main, and core engines.

Table 4.6 Thrusters for orbital maneuvering, attitude control and station-keeping

Rocket engine	Application	Propellant	Thrust (N)	I_{sp} (s) vac
Aerojet Rocketdyne AJ10	Orion primary propulsion (space maneuvers)	NTO/MMH	2670	316 s
Aerojet Rocketdyne R-4D	Orion secondary propulsion	NTO/MMH	490	312 s
Aerojet Rocketdyne MR-104	Attitude control and velocity corrections	Hydrazine	440	239
Aerojet OMS-E	OMS	NTO/MMH	26.689	313
Kaiser Marquardt R-40	orbit maneuvers, perigee kick engine	NTO/MMH	3870 N (3.114–5.338 kN)	281
Kaiser Marquardt R-1E	Vernier attitude control and orbit adjust thruster	NTO/MMH	110 (67−155.7)	280
SpaceX Draco	Attitude control and maneuvering	NTO/MMH	400	300
Northrop Gruman TR 500 SCAT	Mid-course correction maneuvers and station-keeping	NTO/N_2H_4 or Hydrazine	17.8–62 3–20 (mono mode)	325 max 230 mono

If a rocket engine considered for a first or booster stage produces insufficient thrust, it is clustered with several engines of the same characteristics to provide the required high lift-off thrust. For example, four upgraded RS-25 engines power the core stage of the SLS, each providing 1859 kN. To boost the lift off thrust, the SLS incorporates twin SRMs, which provide more than 75% of the total thrust at launch. Performance characteristics of SRBs are described in Chap. 5.

Engine clustering, using two or more engines (or the simultaneous ignition of more than one engine in a system) in a cluster, is done to provide greater thrust for first stage liftoff and acceleration of a SLV during its ascent through Earth's atmosphere. The Russian Soyuz boost system consists of an ensemble of four RD-107 engines, the YF-21 boost stage has 4 YF-20 engines, and the Ariane 4 L40 configuration was comprised of four Viking 6 engines. Except the RS-68 (first stage of the former Delta IV SLV), almost all the booster LREs are fueled with relatively high-density propellants; the Viking 6 uses the hypergolic propellant N_2O_4/UH 25 (a mixture of 75% UDMH and 25% Hydrazine). Another example of clustering is the first stage booster of SpaceX's Falcon 9, which assembles nine Merlin engines burning liquid oxygen and RP-1 fuel. Together, the nine Merlin engines in the stage produce 7.607 MN of thrust at sea level. The largest booster cluster is that of thirty-three (33) SpaceX Raptor engines that comprise the first stage of the Starship launch system now developed by SpaceX.

A high specific impulse implies that a booster engine has higher payload capability. The burn times are necessarily much longer and may reach 400 to 600 s. Core engines provide lower lift-off thrust (1000 to 2000 kN) compared to boosters but with a higher specific impulse and with operating times in the range of about 600 s. In a launch system that incorporates both core and boost engines, the contribution of core engines to the vehicle's total thrust at liftoff is minor compared to the boosters.

Note in Table 4.4, the American RS-25 and the Russian RD-0120 have similar thrust levels and burning times, as both engines were developed to meet similar requirements, namely, to yield enough thrust to launch a crewed vehicle into LEO. Other engines show comparable thrust levels when viewed in reference to their applications. For example, note the thrust and specific impulse for the Japanese LE-7A and Vulcain 2, both core engines in launchers with two large thrust boosters each, the HII and the Ariane 5, respectively. The RD-108, Viking 5C, and YF-20B have similar performance since they operate in clusters in the first stage of SLVs without solid boosters.

The Blue Origin 4 (BE-4) is a powerful, reusable liquefied natural gas (LNG) fueled rocket engine being developed to power two of the next generation American SLVs. Using an oxygen-rich staged combustion cycle, BE-4 can produce 2.4 MN (550,000 lbf) thrust with deep throttle capability. Clustering seven reusable and throttleable BE-4 LOX/LNG engines, the first stage of New Glenn can generate 17,100 kN (3.85 million lbf) thrust at sea level.

4.5.2   Upper Stage Engines

Upper stages propel spacecraft on interplanetary trajectories, or place the payload into orbits higher than could otherwise be reached using a rocket booster. Upper stage rockets are carried aloft by the core or main engine of a launch vehicle and are then fired to put the payload (crew vehicle, spacecraft, or satellite) in predetermined orbit. The thrust levels of upper stage engines vary according to the payload and have operating times up to 1100 s, depending on the mission. The key characteristic feature of an upper stage engine is the vacuum ignition capability—they are re-startable. Upper stage engines are optimized for vacuum operation and thus they utilize nozzle extensions to improve efficiency by increasing the nozzle expansion area ratio to accommodate vacuum conditions, and achieve the highest exhaust gas velocities. The overall Δv requirement is 2.2–3.0 km/s, and the mission duration ranges from hours to 2 weeks. Table 4.5 shows performance data for several upper stages.

The wide difference in propellant combination, engine cycle, and thrust level of upper stages are better understood by noting their different applications. For example, the RL10B-2 engine is unique as its main mission is to propel the Orion crew spacecraft to leave the Earth and reach the Moon. Compare this stage with the performance of any second stage used to place a small payload into LEO or GTO, such as the Merlin 1D second stage of SpaceX Falcon 9. The second stage of Blue Origin's New Glenn heavy lift launch vehicle will be powered by two re-ignitable BE-3U engines operating with LOX/LH$_2$ to deliver 1100 kN (240,000 lbf) of thrust.

As the upper stage for the SLS, known as Interim Cryogenic Propulsion Stage (ICPS), NASA selected the Aerojet Rocketdyne RL10B-2 engine, a variant of the upper stage rockets for ULA's Atlas V and Delta IV Heavy launch vehicles. The RL 10 has a heritage dating back to the earlier phase of space exploration in the United States; the first RL10 engine was built in 1959. Several versions were evolved to power the Centaur upper stage on Atlas and Titan launch vehicles. Centaur was the first rocket stage powered by the RL10 burning cryogenic LOX/LH$_2$ propellants. Through the years, the re-startable RL10 has been optimized for vacuum use and adopted by all American launch systems. There are three larger versions of the engine under development to satisfy the requirements of NASA SLS Exploration Upper Stage, and for the Centaur V on the Vulcan heavy-lift SLV developed by United Launch Alliance. Additional variants of the RL10 were also selected to provide upper stage propulsion for the OmegA vehicle developed by Northrop Grumman.

A single RL10B-2 powered the Interim Cryogenic Propulsion Stage (ICPS) during the first un-crewed test flight of SLS and Orion crew capsule, known as Artemis 1 (November 2022), and is planned for Artemis 2 and 3 as the upper stage of SLS Block 1. The Orion spacecraft is attached to the Interim Cryogenic Propulsion Stage (ICPS), as shown in Fig. 4.5. The ICPS performs all propulsive maneuvers until the ICPS/Orion separation post-trans lunar injection (TLI). The RL10 engine has a mission life of approximately 8 h.

Four RL10C-3 engines (in development) will support the more powerful Exploration Upper Stage (EUS) that is being advanced for future versions of SLS. Artemis 1 was the first uncrewed test flight mission in 2022 to demonstrate Orion's systems in a cis-lunar spaceflight environment and ensure a safe re-entry, descent, splashdown, and recovery prior to the first flight with a crew on Artemis 2. The ICPS for Artemis 1 used the RL10B-2 variant, while the ICPS for Artemis 2 and Artemis 3 will use the RL10C-2 variant.

Figure 4.6 plots the stage mass ratios (μ) after a propulsive burn calculated with Eq. (2.16), where the velocity change Δv ranges from zero (no burn) to 3.5 km/s. As

Fig. 4.5 SLS Block 1 ICPS crew configuration. The RL10B-2 engine with extendible nozzle fully deployed is shown attached to ICPS. Upstream of the engine is the LOX tank, followed by the LH$_2$ tank. *Credit* NASA

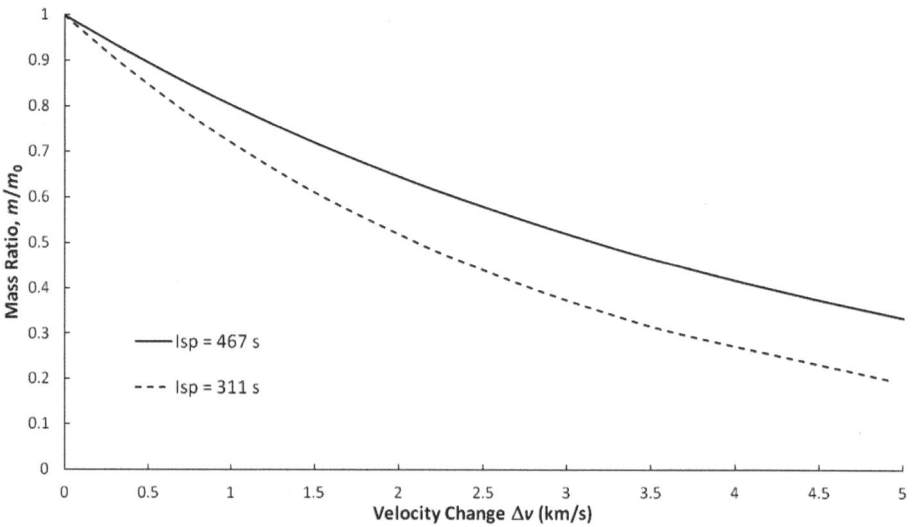

Fig. 4.6 Typical velocity increment versus mass fraction for second stages

we found in Chap. 3, a velocity change of this magnitude is enough to achieve a hyperbolic escape trajectory from low-Earth parking orbit. The Apollo lunar missions in the 1960s and 1970s utilized the Rocketdyne J-2 as upper stage for the Saturn IB and Saturn V vehicles to attain the Δv required to depart LEO and follow a high-energy trajectory to the Moon. Hence, we can conclude that J-2 engine had, after the translunar burn, a mass fraction of about 0.48 since its specific impulse was $I_{sp} = 426$ s; this implies that about half of the total spacecraft mass in the parking orbit was propellant mass, a requirement just to reach the Moon.

The J–2X is an evolved variation of two historic predecessors: the J–2 upper stage engine, and the J–2S, a simplified version of the J–2 developed and tested in the early 1970s but never flown. Each expendable J–2X is designed for four operational starts (2000s), and to restart to support missions beyond LEO. The engine will use a gas generator power cycle. At 15 feet, the J–2X will be four-feet taller than its predecessors.

In combination with the RL10 rocket, NASA's Centaur became the world's first high-energy upper stage, burning LOX/LH$_2$, designed to place payloads in geosynchronous orbits and also to provide escape velocity for interplanetary space probes. Centaur has launched over 200 payloads, most notably the Pioneer and Voyager spacecraft to the outer planets, the Viking landers, and the Curiosity rover to the surface of Mars, Cassini to the rings of Saturn, and the New Horizons probe to Pluto. Centaur has also launched dozens of communication satellites into geosynchronous orbit.

The Russian RD-0210 has a relatively high thrust but considerably shorter burn time, as its mission is to take the Block M of the Proton vehicle into GTO. All the other upper stage engines in Table 4.5 have thrust values below 200 kN and operating times between 500 and 1100 s. For example, a single 31 kN Kestrel engine (developed by SpaceX) powers the upper stage of the Falcon 1 to place small payloads in LEO. Kestrel is ablatively cooled in the chamber and throat, and the nozzle is radiatively cooled; it does not have a turbo-pump and is fed only by tank pressure.

The German Aestus rocket engine has long burn time of 1100 s to power the Ariane 5 upper stage, used for payloads insertion into LEO and GTO. Using its re-ignition capability, Aestus is also used on the ES-ATV version of Ariane 5 to place ESA's 21-ton Automated Transfer Vehicle (ATV) into LEO. Aestus is a pressure-fed engine that consumes up to 10 ton of the bipropellant combination NTO/MMH.

The Merlin 1D engine was developed by SpaceX for the upper stage of the Falcon launchers. Operating with LOX/RP-1 propellant mixtures, the Merlin 1D has an expansion ratio of 16, and chamber pressure of 9.7 MPa (1,410 psi). According to SpaceX, the Merlin 1D is now close to the sea-level thrust of the retired Rocketdyne H-1/RS-27 engines used on Saturn I, Saturn IB, and Delta II launch vehicles. A vacuum version of the Merlin 1D engine was developed for the Falcon 9 v10.1 and the Falcon Heavy second stage with the highest specific impulse ever for a hydrocarbon rocket engine made in the U.S. The increase in I_{sp} is due to the greater expansion nozzle area ratio of 165:1 obtained with a new nozzle extension. See Sect. 4.7.2 for a brief discussion on nozzle extensions and their impact on design and thermal management.

SpaceX Starship second stage will be powered by three Raptor Vacuum (RVac) engines optimized to operate in interplanetary space, aiming to achieve a specific impulse of 350 s, the highest value using LOX/LCH$_4$ bipropellant. In November 2021, SpaceX tested RVac engine's ability to burn continuously for 281 s, which is the time it may require to descend to the lunar surface.

Did you know? **Upper stage and in-space liquid rocket engines are optimized with high area ratio nozzles, incorporating nozzle extensions.** High area ratio nozzles are required to fully expand combustion gases to low exit pressures to match ambient pressures, increasing exhaust velocities and I_{sp}. The upper stages of the Atlas V, Delta IV, Falcon 9, and Ariane 5 launch vehicles incorporate metallic and composite nozzle extensions.

4.5.3   Thrusters for In-Space Propulsion

A thruster is a small rocket engine used to make alterations in spacecraft flight path or attitude. For in-space propulsion, relatively low thrust is needed but require extremely long (more than 10 years) and reliable operation and high cyclic loads during pulse operation mode. The thrust requirement for apogee engines employed on artificial satellites destined for a geostationary orbit is below 600 N, while that for satellite or attitude control thrusters may be less than 100 N. Generally, these thrusters are monopropellant engines or bipropellant LREs burning hypergolic mixtures.

To reach orbit a spacecraft is often launched into a more elliptical orbit. The spacecraft then circularizes the orbit by firing its onboard low thrust rocket engine at apogee to turn parallel to the equator. The propulsion system used for such orbital maneuvers is called liquid apogee rocket or apogee thruster, typically used as the main thrusting engine for a spacecraft. The name apogee rocket derives from the type of manoeuvre for which it is typically used, i.e. it provides the delta-v change at the apogee of an elliptical orbit to circularize it. For geostationary satellites, this orbital manoeuvre is performed to transition from a GTO and place the satellite in a circular geostationary orbit. An apogee rocket engine can also be used for a range of other manoeuvres, such as end-of-life deorbit, Earth escape injection, planetary orbit insertion and planetary descent/ascent. Typical bipropellants for apogee engines are hypergolic combinations such as NTO/Hydrazine, NTO/MMH, NTO/UDMH. These bipropellants allow the thrusters to be reignited as many times as needed.

The NASA Orion service module (SM) provides in-space propulsion for the crewed spacecraft to travel to the Moon. The propulsion system is comprised of 33 thrusters of three types: the main engine Orbital Maneuvering System Engine (OMSe), secondary Auxiliary (AUX) thrusters, and the Reaction Control System (RCS). The OMSe has a thrust of about 26,000 N, while the eight AUX thrusters combined have a thrust of about 3400 N (and a slightly lower Isp). A single AJ10 rocket engine provides the primary propulsion, while eight R-4D-11 thrusters (shown in four pairs around the bottom of the service module in Fig. 4.7) provide secondary propulsion during the mission, and act as a back-up for the main engine. The 24 reaction control thrusters in the RCS are used to steer and control Orion while in orbit. Characteristic of these engines are included in Table 4.6.

The complete propulsion system was designed to bring the crew home in a variety of emergency situations, including abort scenarios after the launch abort system has already been jettisoned. Prior to re-entering Earth's atmosphere, the Orion crew module separates from the service module. After this point, the Reaction Control System (RCS) becomes the sole means of controlling the crew module's orientation during atmospheric re-entry and altering its course in preparation for splashdown. Twelve 160-pound-thrust monopropellant thrusters provide this critical function to Orion spacecraft.

Rocket engines for lunar ascent/descent modules have an overall delta-v or total impulse requirement of 4200 m/s with mission duration of up to 2 weeks. Orbit transfer thrusters also require a Δv of about 4000 m/s. Table 4.6 includes thrusters that were used for the orbital maneuvering system (OMS) and the reaction control system (RCS) of the

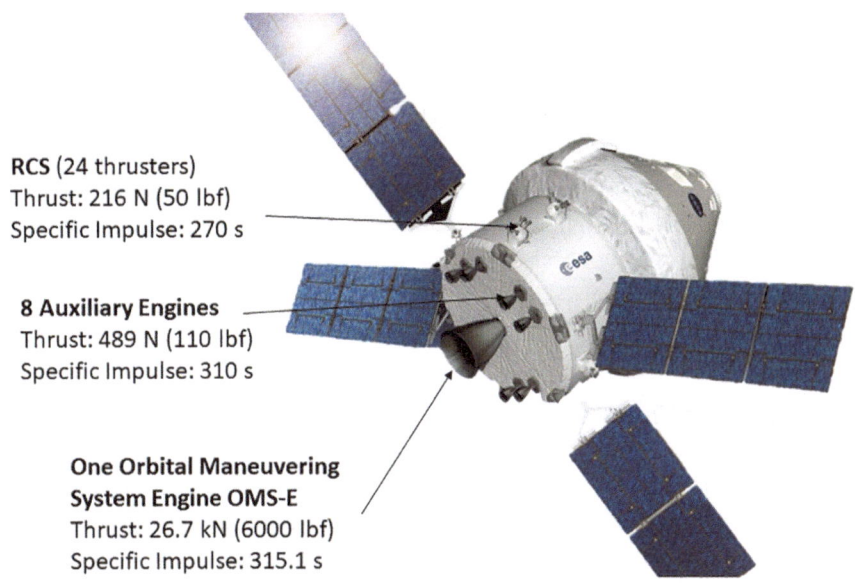

RCS (24 thrusters)
Thrust: 216 N (50 lbf)
Specific Impulse: 270 s

8 Auxiliary Engines
Thrust: 489 N (110 lbf)
Specific Impulse: 310 s

One Orbital Maneuvering
System Engine OMS-E
Thrust: 26.7 kN (6000 lbf)
Specific Impulse: 315.1 s

Fig. 4.7 NASA's Orion spacecraft. Visible are the eight R-4D-11 thrusters distributed in four pairs, and the single AJ10 rocket engine. *Credit* NASA/ESA

Space Shuttle Orbiter. The OMS provided propulsion for orbit insertion, orbit circularization, orbit transfer, rendezvous, and deorbit. The Orbiter used the OMS to provide thrust above 21 km (70,000 feet) altitude. The thruster for the RCS provided propulsion for attitude control during re-entry, for station-keeping in orbit, and for close maneuvering during docking procedures.

The original bi-propellant R-4D was manufactured by Marquardt (now Aerojet Rocketdyne) in 1962. Delivering a nominal thrust of 445 N, this thruster was used in Apollo missions and a Lunar module for reaction control system (attitude control and for orbit maneuvering), for orbital injection, and several other in-space applications. The R-4D has evolved over the last five decades, improving its performance by introducing advanced materials and better thermal management approaches. The specific impulse increased from 312 to 335 s and its operating time increased to almost 11 h.

Draco is a bi-propellant thruster designed by SpaceX for the Dragon spacecraft and for the upper stage of the Falcon 9 launch vehicle. Draco thrusters are used for attitude control and maneuvering. To maneuver in orbit and during re-entry each Draco thruster generates 400 N (90 lbf) of thrust burning MMH fuel with NTO oxidizer. The 1290 kg of propellants onboard are consumed by the engines over the course of a mission. The same type of propulsion will be used for the Launch Abort System (LAS) for the Dragon Lab Spacecraftthe crewed version of the cargo Dragon—but with upgraded performance. Named Super Draco Engine, this thruster will provide 67,000 N (15,000 lbf) of thrust.

Did you know? The JWST (the Webb) has two types of rocket thrusters. The bipropellant Secondary Combustion Augmented Thruster (SCAT) type is used for orbit correction and for orbit station-keeping. The two pairs of SCATs (paired for redundancy) use hydrazine and dinitrogen tetroxide, as fuel and oxidizer respectively. The second type of thruster on Webb is a mono-propellant rocket engine (MRE-1) fueled by hydrazine. The eight MRE-1 s are used for attitude control and momentum unloading of the reaction wheels.

4.5.4 Cold Gas Thrusters

Cold gas thrusters can also be used for in-space propulsion requiring tiny impulses. A cold gas thruster is the simplest type of rocket that relies on gas under pressure as its only source of thermodynamic energy. The cold propellant gas is stored at high pressure. In a typical design, the gas enters the thruster and remains behind a solenoid seal until it opens on command, releasing the gas through the nozzle. Gases such as air, nitrogen, or helium are good propellants. Producing low level of thrust, cold-gas thrusters have been used for attitude control propulsion, and have been proven in space flights lasting more than 10 years. These devices can be turned on and off repeatedly, producing very small, finely controlled thrust pulses (also called impulse bits)—a desirable characteristic for attitude control. The cold gas propellant is nontoxic, produce no deposit or contamination on sensitive spacecraft surfaces, such as mirrors, and they are very safe overall.

Cold-gas thrusters yield low specific impulses, which makes them only applicable to maneuvers requiring low velocity increments, compared to mono- or bipropellant systems. Nitrogen cold thrusters developed by the Moog company were used in the Spitzer Space Telescope ($F = 0.120$ N, $I_{sp} > 57$ s, $p_c = 6.9$ bar). The Kick Stage of the Electron launch vehicle developed by Rocket Lab employs a Reaction Control System (RCS) comprised of 6 cold gas thrusters to precisely point the stage to deploy the satellites.

The main disadvantage of cold gas thrusters developed in the past was due to the massive storage tanks, making the system to be relatively heavy with poor propellant mass fractions (0.02 to 0.19). However, today advances in materials and additive manufacturing techniques may improve the tank design and reduce the overall mass of the cold gas rocket system.

4.6   Rocket Engine Propellant Feed System

Liquid propellant rocket engines are also identified according to their power cycle, which determines how the propellants are fed into the main combustion chamber. The engine can use either a pressure-fed or a pump-fed system. The simplest approach is to supply the propellant by pressurization but it is limited to low chamber pressure applications. Large first and second stage engines use pump-fed systems.

To minimize dead weight, the propellant tanks are designed with thin, stiff walls, and are made of strong materials to sustain the structural loads. The propellants must be delivered to the turbo-pump at a high enough pressure to prevent cavitation. Hence, the propellant tanks are pressurized to 5–10 bar, using either a separate compressed gas supply (helium or nitrogen) stored onboard, or by using the gaseous form of the propellant to pressurize the tanks. The liquid propellant can be converted to gaseous form by absorbing the heat of the combustion chamber walls. The following paragraphs provide a brief description of the different engine cycles.

4.6.1   Gas Generator (GG) Cycle

This is an open cycle system that takes a small amount of fuel and oxidizer from the tanks (~3 to 7%) to feed a small burner, which is the gas generator proper. The hot gas from this generator passes through a turbine to generate power to run the pumps that send the propellants to the main combustion chamber. Main core engines such as the Aerojet Rocketdyne J-2X, SpaceX Merlin 1C, Rocketdyne RS-68, and the Snecma ArianeGroup Vulcain 2 use gas generator cycles.

4.6.2   Staged Combustion (SC) Cycle

This is a closed cycle system in which the propellant is burned in stages, starting with a pre-burner that generates gas to run a turbine. The Aerojet Rocketdyne RS-25 (also known as Space Shuttle Main Engine SSME) burning cryogenic LOX/LH$_2$ propellants operated on a fuel-rich dual-shaft staged combustion cycle. What it means is that most of the fuel flow (except for a small amount for cooling) and a small amount of the oxidizer flow were pre-burned. The resulting fuel-rich hot gas was used to power the turbopump turbine, and then injected into the main combustion chamber along with the remaining oxidizer and the coolant fuel, all to be finally combusted to develop thrust. The SSME engine had two high-pressure turbopumps to supply 440 kg/s (970 lbm/sec) of oxygen and 73 kg/s (162 lbm/s) of hydrogen fuel to the engine's main combustion chamber. Pre-burning the fuel resulted in a high combustion efficiency (~99%), thereby optimizing the overall performance of the reusable rocket engine.

The efficient staged-combustion cycle allowed for the SSME high chamber pressure (2970 psia, significantly higher than previous engines) and high flow rate (1030 lbm/s). Such cycle allows high turbopump propellant discharge pressures, 52.4 MPa (7600 psia) for LOX, and 42.7 MPa (6200 psia) for LH_2), which are necessary to overcome pressure losses within the engine and still maintain the 20.48 MPa (2970 psia) chamber pressure.

The Russian RS-170/180 engine operating with RP-1/LOX mixtures uses a staged combustion cycle. The Blue Origin 4 (BE-4) is a staged-combustion rocket engine, with a single oxygen-rich pre-burner, and a single turbine driving both the fuel and oxygen pumps similar to the kerosene-fueled RD-180 used on the Atlas V; the BE-4 cycle uses only a single combustion chamber and nozzle.

4.6.3 Expander Cycle

In the expander cycle the fuel is circulated through the cooling jacket of the main combustion chamber where the heat transferred vaporizes the fuel. The vaporized fuel is then passed through a turbine, which drives the propellant pumps, before being injected into the main combustion chamber to burn with the oxidizer. The expander cycle works with liquid fuels such as hydrogen and methane, which have a low boiling point and vaporize easily. The expander cycle is used by the Aerojet Rocketdyne RL10, and by the ArianeGroup Vinci upper stage engines burning LOX/LH_2 propellants.

4.6.4 Pressure-Fed Cycle

This simple cycle uses tank pressurization to feed the propellants into the main chamber. Self-pressurization is achieved by vaporization of the liquid propellant or by thermal decomposition caused by external heat addition or catalytic decomposition, and thus this approach can only be used for very low chamber pressure monopropellant rockets. Tank pressurization with a gas such as helium makes it possible to achieve up to 31 MPa chamber pressure, sufficient for upper stage applications. The RCS and OMS in the Space Shuttle Orbiter used a pressure-fed cycle. Gas-pressure fed propellant systems are not sufficiently powerful for use in first stages.

4.7   Real Rocket Engine: Key Thermophysical Processes

Even though the ideal thermodynamic analysis predicts the operating performance of any rocket propulsion system and provides tools for the determination of design parameters for a given performance requirement, this analysis must be refined to account for real flow effects. The study required to optimize the performance of a rocket engine is not trivial. It requires a multidisciplinary approach to study the complex flowfields that develop in the thrust chamber (combustion chamber plus nozzle) and its interaction with other systems.

The actual performance of a rocket engine is always lower than that predicted because of imperfections in the mixing of propellants, incomplete combustion and instability, viscous boundary layer friction, heat transfer, turbulence, and divergence or nonuniformity of the exhaust flow. The greatest losses usually result from nonuniformities of the nozzle exit flow, inefficiencies in propellant mixing and non-isentropic flow losses. In addition, to increase the nozzle area ratio and achieve optimum flow expansion one can use structural solutions such as expandable nozzle extensions, but this approach may lead to a loss in specific impulse caused by discontinuity of the nozzle contour. These and other losses will reduce the theoretical values calculated with the idealized assumptions that we made in Chap. 2.

High specific impulse is required for a successful space mission. This is because mission payload is very sensitive to a reduction in delivered specific impulse. It has been shown that there is a greater than 3% reduction in allowable payload for each second loss in I_{sp}. Therefore, the highest possible I_{sp} efficiency must be realized from a rocket engine. In designing a rocket nozzle, analysis should provide optimum values of nozzle area ratio and operating pressure, operating conditions for favorable heat transfer, and the effects of deviations from optimum conditions. Let us review briefly some of the sources of rocket specific impulse inefficiencies.

4.7.1   Combustion Process

Combustion is the sequence of exothermic chemical reactions between a fuel and an oxidant substance accompanied by the production of heat and conversion of chemical species. To characterize the chemical reaction of propellants in the combustion chamber requires a complex thermochemical analysis to determine the equilibrium state of a reacting mixture, which depends on mixture ratio (the ratio at which the oxidizer and fuel mix and react), and on the pressure and temperature conditions in the combustor. The analysis is rather complicated and requires the use of numerical methods to solve the finite-rate chemical kinetics, turbulent combustion reactions. Figure 4.8 illustrates the processes taking place in the combustion chamber, emphasizing the propellant injection through multiple injector elements. The size of the chamber considers the minimum length required for complete combustion, which is dictated by the product of the mean propellant gas in the chamber and the residence time. The residence time depends on propellant evaporation and mixing rates, and reaction timescales (function of the propellant mixture), and is directly affected by injector design. Residence times are on the order of 40 ms or less.

In the combustion chamber, the release of energy per unit mass of propellant mixture and the combustion temperature are highest at or near the stoichiometric mixture ratio. The reactants are in stoichiometric proportions when their chemical reaction goes

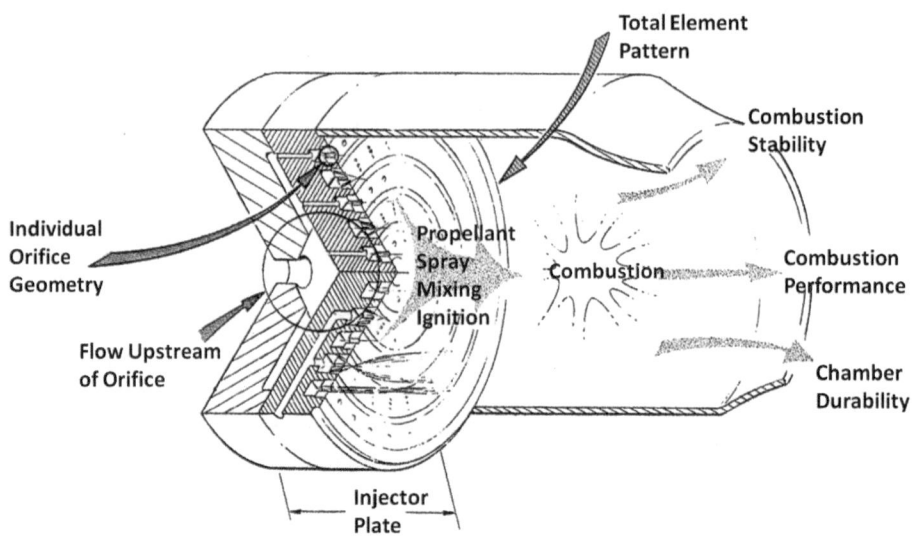

Fig. 4.8 Rocket combustion chamber. Sketch adapted from NASA SP-8089

to completion. This means that all the reactants are consumed and transformed into reaction products. To elucidate this process, consider this basic chemical reaction of hydrogen and oxygen:

$$H_2 + \frac{1}{2}O_2 \rightarrow H_2O$$

This stoichiometric reaction states that all the hydrogen and oxygen are fully consumed and form water vapor (single product) without any reactant residue of either hydrogen or oxygen. For this to occur a mixture of 1 mol of H_2 and ½ mole of O_2 will result in 1 mol of H_2O. On a mass basis this mixture requires 2 kg of H_2 and 16 kg of O_2 (half of 32). The stoichiometric mixture mass ratio (oxidizer to fuel) is simply

$$\frac{\dot{m}_o}{\dot{m}_f} = \frac{\left(\frac{1}{2}\text{kg} \cdot \text{mol}\right)\left(32\frac{\text{kg}}{\text{kg·mol}}\right)}{(1\text{kg} \cdot \text{mol})\left(2\frac{\text{kg}}{\text{kg·mol}}\right)} = 8$$

In a similar manner, methane mixed with oxygen has a stoichiometric chemical reaction

$$CH_4 + 2O_2 \rightarrow CO_2 + 2H_2O$$

which yields a mixture mass ratio of 4.

The propellant mixture ratio for a bipropellant rocket is defined by the ratio at which the oxidizer and fuel mix and react to yield the hot gases of propulsion. The mixture ratio is the ratio of the oxidizer mass flow rate $\dot{m}_o$ to the fuel mass flow rate $\dot{m}_f$:

$$\frac{O}{F} = \frac{\dot{m}_o}{\dot{m}_f}$$

The mixture ratio is usually lower than the stoichiometric mixture ratio (O/F), since the highest specific impulse (I_{sp}) is achieved for fuel rich conditions. Therefore, O/F is the parameter that gives a maximum value of $T_c/\mathfrak{M}$, where T_c denotes the combustion temperature and $\mathfrak{M}$ is the average molecular mass of the reaction gases. Each propellant combination has an optimum mixture ratio that produces maximum I_{sp}. The correlation of $T_c/\mathfrak{M}$ and v_{ex} is illustrated in Table 2.1, where the exhaust velocities were computed by Brown (1996) for vacuum conditions, optimum expansion, and a nozzle area ratio of 30.

In practice, rocket engines do not operate with stoichiometric mixtures. Typically, the combustion chamber will burn a fuel-rich mixture (having more fuel than is needed for a stoichiometric reaction), as this balances flame or chamber temperature and the final molecular mass of the gases. The following simplified fuel rich chemical reaction

$$5H_2 + O_2 \rightarrow 2H_2O + 3H_2$$

illustrates how the fuel-rich reaction allows lightweight molecules such as hydrogen to remain unreacted; such approach is taken to reduce the average molecular mass of the reaction products, which in turn increases v_{ex} and the specific impulse (see Eq. 2.33)

Burning a fuel rich mixture reduces the adiabatic flame temperature by almost 40%, but the limiting specific impulse in the extreme rich mixture case is reduced by just 5%. Thus, by burning a fuel rich mixture only a small performance penalty is paid for a large reduction in temperature that must be handled by the structure. The trade-off here is between stagnation temperature and molecular mass. The rich mixture combustion products have a high percentage of unburned hydrogen that reduces the molecular mass to about half the value in the stoichiometric mixture case. Figure 4.9 illustrates the theoretical I_{sp} of various propellant reactants in terms of the O/F ratio; Haidn [15] obtained these data with NASA's Chemical Equilibrium and Transport Properties (CET-93) computer code, limited to a chamber pressure of 100 bar (10 MPa), and gas expansion with a nozzle area ratio of 45.

For engines burning LOX/LH$_2$, the best operating mixture mass ratio O/F is between 4.5 and 6.0, which is lower than the stoichiometric value of 8.0. Optimum mixture ratios for rocket engines using kerosene fuel and oxygen (LOX/RP-1) may be 2.3 to 2.6.

The trends in Fig. 4.9 show that LOX/LH$_2$ mixtures yield the highest specific impulse for any mixture ratio, about 30–40% higher than all other propellant mixtures. This is because hydrogen provides more thermal energy than any other fuel due to its high heat of

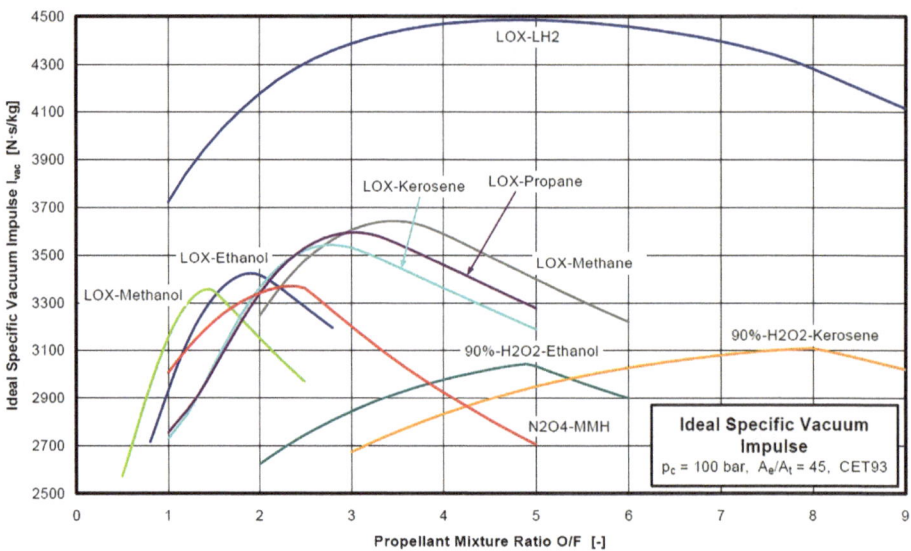

Fig. 4.9 Ideal specific impulse for propellant combinations. *Credit* Haidn [15]

reaction (with all oxidizers) and the low molecular mass of its combustion products. Since the thrust developed by a propulsion system is essentially proportional to the exhaust gas velocity, Eq. (2.47), which in turn depends on the average molecular mass, hydrogen yields higher gas velocities for a given temperature and provides greater thrust.

The next best fuels are kerosene and methane. Rocket Propellant-1 (RP-1), a highly refined form of kerosene ($CH_{1.953}$), is typically burned with liquid oxygen. This bipropellant (LOX/RP-1 is known as *kerolox*) is used in the first-stage boosters of the Delta, Atlas-Centaur, and Falcon V vehicles. Kerosene delivers a specific impulse considerably less than that of cryogenic fuels (I_{sp} decreases with increasing carbon content in the fuel), but it is much better fuel than hypergolic propellants.

Note in Fig. 4.9, for LOX/LCH$_4$ the highest specific impulse is achieved with O/F of about 3.5. For a more realistic view of methane rockets, we must consider potential soot deposition and decomposition of methane in the cooling channel. Nevertheless, since methane is the most important component of natural gas with a volume fraction of 80% to 99%, compressed natural gas (CNG) or liquified natural gas (LNG) can be used instead of pure methane, although residues such as Sulphur or phosphorus may have a negative impact on material compatibility. Nonetheless, the reusable New Glenn SLV being developed by Blue Origin will have a first stage comprised of seven BE-4 rocket engines powered by LOX/LNG to provide a combined thrust of 16.8 MN at liftoff.

From these results is seems that a LOX/LH$_2$ rocket engine is more advantageous for upper stage engines while booster engines perform well with LOX/RP-1 and LOX/LCH$_4$. Fuels which already contain oxygen atoms such as alcohols are also good, as they have an almost similar specific impulse as storable propellants. For comparison, propellant combinations which use hydrogen-peroxide (H_2O_2) as oxidizer are included in Fig. 4.9.

For realistic combustion analysis, we determine the adiabatic flame temperature. This is the temperature that would result from the combustion of fuel and oxidizer under given pressure conditions in the combustor. The adiabatic flame temperature for a given reaction is the highest temperature achievable with the reactants. The adiabatic flame temperature for the stoichiometric reaction of LOX + LH$_2$ is 5100 K. With this value, we calculate the limiting velocity for the reaction products, assuming adiabatic expansion of the gas to zero temperature,

$$V_{ex,lim} = \sqrt{2c_p T_c} = 5590 \text{ m/s}$$

Hence, the limiting specific impulse for chemical rockets is

$$I_{sp,lim} = \frac{V_{ex}}{g_0} = 560 \text{ s}$$

As we stated before, burning a fuel rich mixture reduces the adiabatic flame temperature, and therefore the limiting values above give a perspective for guidance. Computational codes accurately reproduce the thermodynamic equilibrium models in terms of flame temperature (for temperatures where dissociative affects become significant).

The coupling of propellant mixing and combustion is of great importance. For example, the processes in the immediate neighborhood of the propellant injector face are very complex, and are not well defined analytically. Also, a finite-rate chemistry model must describe the high temperature combustion of the reactants, and must be coupled to mixing models that may include multi-phase substances. For engine design and performance analysis, practical procedures demand to make calculations over a wide range of realistic conditions, using chemically frozen or chemical equilibrium models, and then use multi-reaction, finite-rate chemical kinetics for those conditions.

To achieve high rocket engine performance, many design criteria must be satisfied to assure attainment of high combustion efficiency, including complete propellant mixing and chemical reaction, complete liquid oxidizer vaporization, and gas homogeneity. Although the combustion criteria represent difficult design requirements to meet, there are both analytical and experimental techniques available to accomplish this task. Sophisticated analytical techniques include computational fluid dynamics (CFD) modeling of turbulent mixing and reaction, spray correlations, supercritical droplet vaporization models, tools that are validated with extensive cold and hot fire flow testing. To date, analysis of the combustion process depends to some extent on empirical techniques. These topics are outside the scope of this book. The interested reader is referred to the specialized literature. See Yang et al. [56] and references therein.

4.7.2   Propellant Injection and Ignition

The propellants are delivered to the combustion chamber at high pressure, with a high mass flow rate. The injection system is designed to uniformly inject the propellants at the proper mixture ratio for optimum combustion. In a liquid bipropellant rocket engine, the function of the injector is to atomize and mix the fuel with the oxidizer to produce efficient chemical reactions, yielding stable combustion that will provide the required thrust without impacting adversely the combustion chamber and related propulsion system hardware. Combustion instability is an undesired phenomenon characterized by pressure peaks and temperature spikes that are often unpredictable, and leads to hardware damage and even complete chamber destruction. This instability is driven by inefficient propellant injection and chamber flow conditions. Hence, the design of the propellant injection system is of crucial importance to ensure the optimum performance of a rocket engine.

The propellant injectors must ensure that the fuel and oxidizer (in liquid or gaseous phase) are finely atomized (with droplet size less than 100 microns in diameter) for fast evaporation and promote fast mixing. Injection systems for liquid propellant rocket engines are very complex. Different injection approaches are considered, depending on the propellants type (e.g. cryogenic, hydrocarbon, hypergolic) and their state (e.g. hot liquid, cold gas, gel), flow rate, mixture ratio, chamber pressure, chamber characteristics (e.g. length, cooling), throttling requirements, engine life (e.g. restarts, total impulse time). All these parameters directly or indirectly affect combustion performance, heat transfer, materials compatibility, and stability. Each imposes specific demands on the injector (e.g. local mixture ratio and mass gradients or spray drop size).

Four basic injection methods are considered: impinging propellant stream jets, swirl injection, parallel injection through a showerhead with coaxial passages, and shear coaxial injection; combinations of these concepts are also common incorporating mechanical means that produce maximum atomization and fuel/oxidizer mixing. The RS-25 SSME used a shear coaxial injector plate with 600 coaxial elements to inject liquid oxygen from the oxidizer manifold (Fig. 4.10). Through its annulus, each element also injects the hot, fuel-rich gas. Cold hydrogen gas that had previously migrated through the double walls of the hot gas manifold, enters the slot between the secondary plate and the lip of the primary plate. Both plates are porous and are transpiration-cooled by the cold hydrogen gas as it flows through them.

In the SSME, an augmented spark ignition system chamber was in the center of the main injector plate (see Fig. 4.10). Small quantities of hydrogen and oxygen were continuously injected into this chamber, and initially ignited by two spark igniters located therein. This flame then ignited the propellants flowing through the injector elements into the combustion chamber. The dual-redundant igniter was used during the engine start sequence to initiate combustion. See Rocketdyne Propulsion Power (1998).

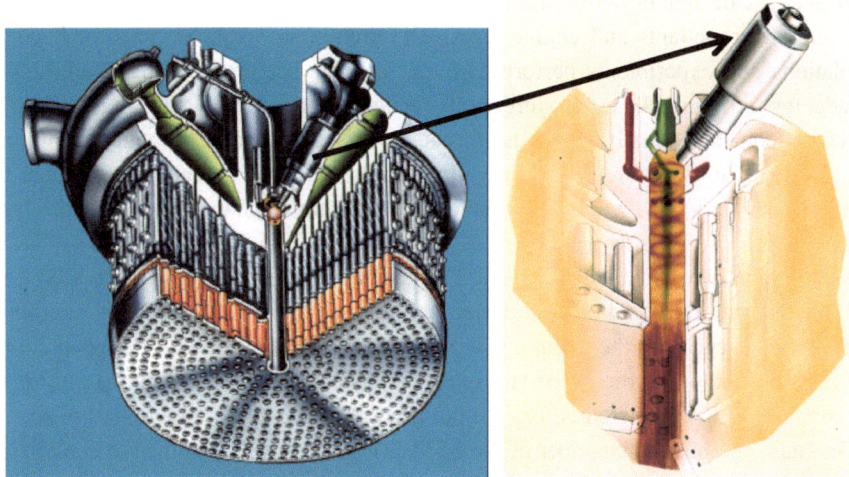

Fig. 4.10   View of SSME main injector and ignitor. *Credit* Rocketdyne Propulsion Power (1998)

> ***Did you know?*** In a bipropellant engine such as the SSME, the main injector feeds
> into the main combustion chamber a combination of hot and cold propellants: a hot
> fuel-rich gas from two pre-burners, cold hydrogen gas from the chamber cooling
> circuits, and cold liquid oxygen from the high-pressure oxidizer turbopump.

4.7.3   Combustion Chamber Size

In the combustion chamber, the liquid propellants mix and react very fast, undergoing
exothermic chemical reaction and become a hot gas at combustion temperature T_c and
chamber pressure p_c. Knowing this condition helps us assess the characteristic velocity
C^* to determine the design of the chamber, which is related to the efficiency of the combustion process in a chemical rocket. According to Eq. (2.51), C^* depends on the product
$p_c A_t$ divided by $\dot{m}$. Because this ratio is constant and the variables are propellant-specific
parameters, C^* determines the design of the chamber. For example, for a given propellant
mass flow rate $\dot{m}$ and a maximum allowable chamber pressure p_c we can calculate the
required cross section at the narrowest point of the converging section, which is the throat
cross sectional area A_t. The volume of the chamber is calculated from the characteristic
length, L^*, a parameter which is determined from a detailed combustion analysis. The
characteristic length of the chamber provides an indication of the residence time of the
reacting fuel and oxidizer.

In general, designers define the minimum characteristic length based on experience with similar propellants and engine sizes. The preliminary estimate is formalized with simulations and experiments performed under a given set of operating conditions that include type of propellant, mixture ratio, chamber pressure p_c, injector design, and chamber geometry. Finally, once the throat cross sectional area A_t and the minimum characteristic length L^* are established, the chamber volume V_c can be calculated from the relation $L_c = V_c/A_t$, where L_c is the chamber length or characteristic length. Typical values of L^* for LOX/LH$_2$ are 0.76−1.02 m.

The combustion chamber of an LRE typically includes a liner, jacket, throat ring, coolant inlet manifold, and coolant outlet manifold, as thermal management is crucial to maintain the integrity of the engine. It should be evident that in a well-designed combustion chamber, including the propellant injection system, all droplet vaporization, mixing, and combustion is completed before the hot gas enters the throat area downstream.

The main combustion chamber of the SSME (Fig. 4.11) was designed to contain internal gas pressure of 2.068×10^7 Pa (3000 psi) and sustain the hot gases at temperatures up to 3589 K (3315.556 °C or 6000°F). To operate at these extreme conditions, Rocketdyne developed NARloy-Z, a high conductivity copper-based alloy with silver and zirconium. The outer liner was made from structural nickel, applied by an electroforming process. The chamber support jacket was made from Inconel alloy 718. For thermal management, the chamber was cooled by the cryogenic hydrogen fuel, which was circulated through 430 channels machined into the liner inner wall.

Fabrication methods for LRE combustors are changing considerably with the introduction of additive manufacturing techniques. Engineers at NASA Marshall Space Flight

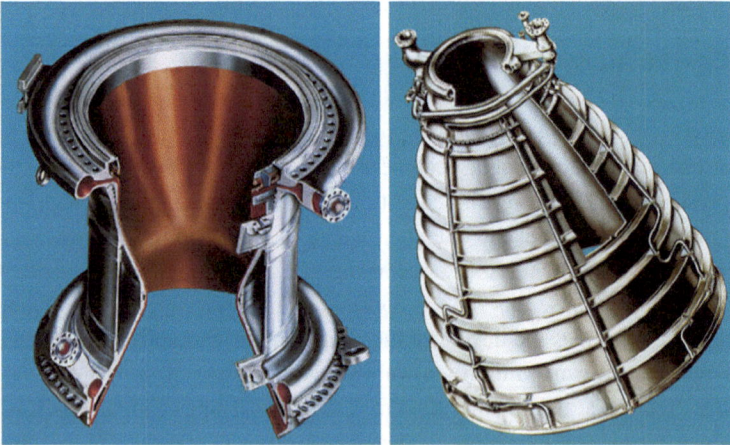

Fig. 4.11 SSME main combustion chamber and nozzle. Injector end-to-throat length 37.338 cm (14.7 in). Throat area 0.06 m^2 (93.02 in^2). Nozzle length to throat 121 in, Exit diameter 94 in (outside). *Credit* Rocketdyne Propulsion Power (1998)

Center are advancing the design for additive manufacturing, demonstrating the Laser-Powder Directed Energy Deposition (LP-DED) approach for integral channel wall nozzles and for two thrust chambers. They determined that LP-DED offers build volumes unattainable by Laser Powder-Bed Fusion (L-PBF) systems, required for large diameter thrust chambers. In a recent paper [50], the NASA engineers discuss challenges of thrust chamber design, fabrication, and analysis methodologies while considering LP-DED. This paper can guide designers toward optimal solutions, and caution them that "Not all parts [of the chamber] should be additive, and not all additive processes should be applied" [to LRE chambers].

4.7.4 Nozzle Function, Design and Geometry

The gas flow expansion process in the nozzle affects drastically the specific impulse of the rocket engine. The nozzle design parameters which determine the achievable specific impulse I_{sp} are area ratio, length, and contour. The main design requirement of a rocket engine is to achieve maximum thrust. For this, the engine requires an effective converging-diverging nozzle to expand the combustion gases. According to Eq. (2.47), we need to select the throat cross sectional area of the nozzle and the pressure ratio that maximizes the thrust F.

The engine parameter that determines its geometry is the nozzle expansion ratio, ε, which is known from the thermodynamic analysis. The layout of the engine design thus begins with the known cross-sectional areas of the nozzle at the throat A_t, and at the exit plane A_e. The design requires focus on both the chamber—where the heat addition occurs (combustion chamber in a chemical rocket)—and the nozzle where the hot gases expand and the thermal energy is converted into kinetic energy. The geometry of the divergent section extends from the throat to the nozzle exit. The function of this part of the nozzle is to accelerate the exhaust gases to a high velocity in a short distance while providing near-ideal performance. Seeking to optimize the shape of the expansion section, different nozzle geometries have been developed and studied.

The nozzle for the SSME (Fig. 4.11) was 121 in long. It was designed with 1080 stainless steel tubes brazed to themselves and to the structural jacket surrounding the nozzle. Nine hatbands were welded around the jacket for hoop strength. Coolant manifolds were welded to the top and bottom of the nozzle, along with three fuel transfer ducts and six drain lines. As the SSME operated mostly in space environment, it required the large area ratio of 69:1 to fully expand the exhaust gas.

4.7.4.1 Nozzle Shape Concepts

The best nozzle design is one that can achieve the highest exhaust velocity with a minimum structural mass and shorter overall length. The most common nozzles have conical and bell-shaped or contoured expansion sections (see Fig. 4.12).

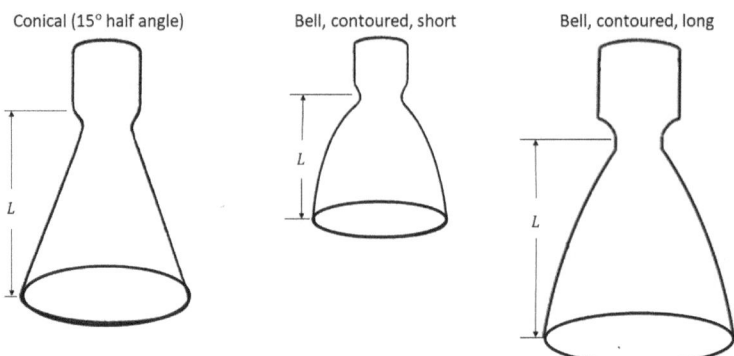

Fig. 4.12 Rocket nozzle geometry

Conical nozzle. The conical nozzle is a simple straight-wall cone shape with a typical 15°
half angle. This is the oldest and simplest approach to convert thermal combustion energy
into thrust, but it has low expansion performance. The cone half angle can be increased
to 20° to improve thrust performance with a reasonable nozzle length. The conical nozzle
is easy to fabricate and is better suited for applications where payload performance is not
a requirement such as small low-thrust attitude control engines.

Bell nozzle. The specially contoured expansion surface within a restrictive length, gives
the nozzle a characteristic bell shape. The bell nozzle can be shorter and thus less massive
than a conical nozzle, as the contouring provides for rapid early expansion, and at the
same time the exhaust flow is redirected toward the axial direction near the nozzle exit,
resulting in a more uniform exit flow field with no divergence losses. Bell nozzles are
used on the larger launch systems currently in operation. For example, the Aestus engine
on the upper stage of the Ariane 5 vehicle uses a regenerative-cooled bell-shaped nozzle
with an expansion ratio of 30 to generate an exhaust velocity of 3240 m/s in vacuum.

The ideal contoured nozzle produces a homogeneous flowfield with a constant velocity
distribution in the exit area, but it is too long for application to launch vehicles. Therefore,
most of today's rocket nozzles have a length approximately 80% of their ideal length.
Nozzles with thrust-optimized contours (TOC) are used in the Vulcain 2, SSME, and
RS-68 engines.

Parabolic nozzle. This a short bell nozzle design with a parabolic contour and small
divergence angles at the nozzle exit. Typical parabolic nozzles, known as truncated ideal
contours (TIC), are used for example in the Japanese LE-7 and the Russian RD-0120
engines.

Variable Area Nozzle. Plug nozzles, dual-bell nozzles or nozzles with an extendible
exit cone can be configured to change the expansion ratio and thus adapt to changes in
pressure. Because they are designed to operate efficiently across a wide range of altitudes,

these are known as altitude-compensating rocket nozzles, allowing a rocket engine to achieve maximum performance at more than a single altitude.

Conceptually, the dual bell nozzle is comprised of two short bell expansion contours. There exists a discontinuity in wall curvature at the interface of the two nozzles. During ascent, the dual bell nozzle operates with the lower area ratio, and this induces flow separation to develop at the inflection point. Flow separation occurs when the gas pressure exiting the nozzle is too low, causing the jet flow to separate from the nozzle surface, leading to inefficient and unstable flow expansion. As altitude increases and the gas expands further, the flow attaches itself downstream of this point, so the expanding flow fills the entire nozzle exit section and operates with the higher area ratio, thus improving propulsion performance.

Plug or aerospike and expansion-deflection nozzles, are designed with aerodynamic boundaries instead of a solid outer nozzle structure, where a center body is shaped to provide the desired flow distribution and altitude compensation. These advanced nozzles have been tested successfully but none is in a flight-ready development status to date. However, experimental studies have shown the potential of advanced nozzles for future high-performance propulsion concepts. Rocket nozzle design concepts are fully addressed in specialized literature, e.g. Hageman, et al. [13, 14]).

Regardless of the type of nozzle selected, maximum performance is achieved by optimizing the contour. Optimization methods are used for the contour of a rocket nozzle of a given length, or expansion ratio, with the objective of maximizing the thrust. Contouring the nozzle wall turns the flow closer to the nozzle axis and thus it reduces flow divergence. Calculus of variations is used to optimize the thermodynamic expansion of an ideal gas with constant specific heats. The input to start the solution includes the supersonic flow properties at the nozzle throat. The transonic flow in the throat region is calculated as a function of the upstream radius of curvature at the throat and the properties of the gas such as the specific heat ratio γ. Iteration algorithms yield an optimum nozzle contour defined to satisfy one of three initial conditions: a given length, a given expansion ratio, or a fixed envelope (exit radius and length).

Equation (2.47) expresses the ideal rocket thrust, showing the contribution of both the exit flow velocity (momentum) and the exit gas pressure. The velocity term always contributes to the rocket thrust, but the pressure term can increase or decrease it. Ideally, for a fixed chamber pressure the nozzle will expand the hot gases from the throat to the exit such that $p_e = p_a$. This condition of optimum expansion occurs only at one altitude, and a nozzle with a fixed area ratio is operating much of the time at either over-expanded ($p_e < p_a$) or under-expanded ($p_e > p_a$) conditions. In these cases, optimum expansion is not achieved.

In an over-expanded nozzle, the gas exits the nozzle with a lower pressure than the external pressure due to having an exit cross sectional area too large for optimum. On the other hand, an under-expanded nozzle discharges the gas at a pressure that is greater than the external pressure because the exit area is too small for an optimum area ratio. In this case the expansion of the gas is incomplete within the nozzle, and must take place outside.

4.7.4.2 Nozzle Area Ratio

A launch vehicle operates over a range of altitudes, which means operation with a range of nozzle pressure ratios during ascent from sea-level to orbital altitude. In practice, the nozzle shape (conical or bell), length, and area ratio (throat plane area versus exit plane area) combine with the chamber pressure to determine at what altitude optimum expansion occurs. This is the design altitude of the engine. Below this altitude, overexpansion of the available exhaust gas occurs; underexpansion occurs above. Both produce less than maximum thrust efficiency.

We also know that having very large nozzle area ratios allow a rocket engine to yield a small but significant improvement in specific impulse, particularly at very high altitudes. Thus, for upper stages designers consider increasing the nozzle exhaust cross sectional area, thereby increasing the expansion ratio. However, the extra length and extra mass necessary to house a large nozzle make this solution less optimum.

The expansion ratio ε or nozzle area ratio A_e/A_t is a measure of the gas expansion provided by a rocket engine, and is expressed in terms of the pressure ratio, Eq. (2.46):

$$\varepsilon = \frac{A_e}{A_t} = \left(\frac{p_c}{p_e}\right)^{1/\gamma}\left[1 - \left(\frac{p_e}{p_c}\right)^{(\gamma-1)/\gamma}\right]^{-1/2}\sqrt{\frac{\gamma-1}{2}\left[\frac{2}{\gamma+1}\right]^{(\gamma+1)/(\gamma-1)}}$$

The optimum area ratio provides an exit plane pressure equal to local ambient pressure. For a first-stage engine, the optimum ε is selected near midpoint of the ascent flight. For a spacecraft engine (or an upper-stage engine), the optimum area ratio is infinite; the largest area ratio allowed by volume and weight is used. Typical nozzle expansion ratios are given in Table 4.7 for current and rocket stages in development. Note the low area ratio of booster engines, which operate over a relative short time. Core stage engines, which operate for up to 10 min have nozzles a medium-range expansion ratio, whereas upper stage engines are designed with very high expansion ratios greater than 165. A special case was the SSME propelling the Orbiter. The engine was designed to operate from sea level to orbit. Thus, it needed the large area ratio of 69:1 to fully expand the exhaust gas to the conditions found in its full trajectory.

Each rocket stage has a particular set of design requirements. A main stage engine must fulfill a wider range of operation conditions during ascent of the launcher. Common main stage rocket nozzles are designed to be full flowing under sea-level conditions to avoid flow separation and undesired side loads. But these requirements limit the expansion ratio and result in performance losses as ambient pressure decreases during ascent.

The best nozzle for a space launch vehicle is not necessarily one that gives optimum gas expansion, but one that gives the greatest flight performance, e.g. highest impulse or specific impulse, longer range, or largest payload capability. And since both engine

Table 4.7 Nozzle expansion ratios of representative LREs

Engine	Chamber pressure p_c		Area ratio ε	Application
	psia	MPa		
F-1 without extension (Saturn V)	965	6.65	10	Booster
F-1 with extension	965	6.65	16	Booster
RD-170 (Russian Energia SLV)	3556	24.52	36.9	Booster
Vulcain 2 (European Ariane V SLV)	1701	11.73	58.2	Core
RS-25 (SSME American STS with 3 engines)	2994	20.64	69	Core
RS-25 (American SLS with 4 engines)	2994	20.64	78	Core
Vinci (European, developed for Ariane 6)	882	6.08	240	Upper stage
RL10B-2 (American SLS Block 1)	640	4.41	280	Upper stage
Merlin 1D Vacuum (American, under development by SpaceX)	1410	9.70	165	Upper stage

length and specific impulse strongly influence the payload capability of rocket vehicles, the design of the nozzle expansion section is crucial to ensure the maximum performance from a length commensurate with optimum vehicle payload.

4.7.4.3   Extensible Nozzle

Adding a nozzle extension can improve the performance of an engine by providing extra length to fully expand the flow to match ambient conditions and achieve the highest v_{ex} (see Fig. 4.13). Upper stages and rocket engines operating *in vacuo* improve their performance with nozzle extensions, as these increase the expansion area ratio considerably (see Table 4.7). The RL10B-2 upper stage was fitted with an extendible nozzle cone or skirt, which was placed around the engine, stowed during the ascent of the Delta III launch vehicle. The extension was then lowered into position by electromechanical devices after the vehicle was separated from the upper stage at high altitude and before firing.

Nozzle extensions can also be considered for first stage engines. For example, the Rocketdyne F-1 first stage engines of the Saturn V (launch vehicle of Apollo program) required a nozzle extension to increase its performance at higher altitudes. The nozzle extension increased the expansion ratio from 10:1 to 16:1.

Due to the large size of high area ratio nozzles and the related engine performance requirements, engineers began considering carbon-carbon (C/C) composite nozzle extensions to reduce weight impacts. In 1995, Aerojet Rocketdyne selected a translating nozzle extension assembly for the RL10B-2 engine to achieve the performance required of the

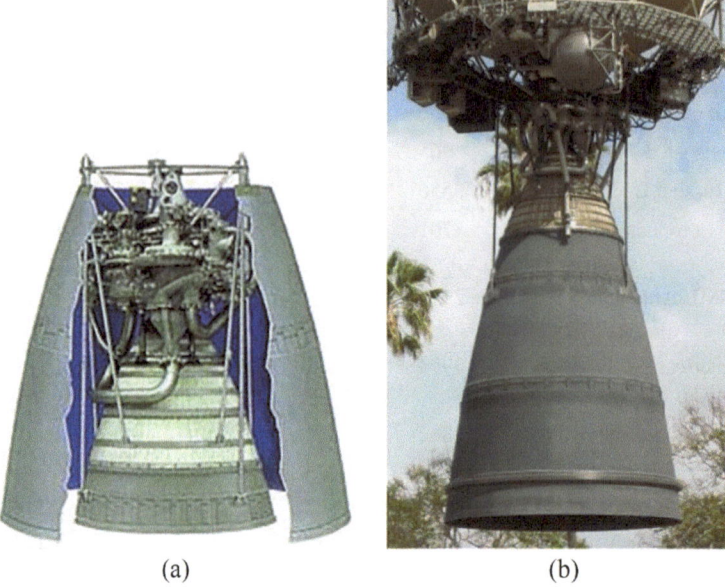

(a)   (b)

Fig. 4.13 RL1010-B2 engine with extendible nozzle cone: **a** Stowed as used in Delta IV launcher; **b** Fully deployed shown attached to ICPS of NASA SLS

upper stage for the Boeing Delta III and Delta IV launch vehicles. The three-cone extension increased the nozzle area ratio from 77:1 to 285:1. Aerojet Rocketdyne designed the mechanical deployment system and gave Airbus Safran Launchers the task of designing and manufacturing the nozzle extension proper (see Fig. 4.13). The RL10B-2 extension was comprised of three cones made of Novoltex® Sepcarb® material, a three-directionally reinforced carbon-carbon (C-C) material composed of a PolyAcryloNitrile (PAN) based carbon preform made by Safran Ceramics. See Pichon and [33].

The RL10B-2 engine with the Safran Ceramics C-C nozzle extensions performed flawlessly for the Delta IV Centaur Upper Stage. Since then, engineers concluded that substantial reductions in overall weight and costs were possible with composite nozzle extension as compared to metallic nozzle extensions. Even greater cost and mass reductions may be possible if the regeneratively-cooled portion of the metallic nozzles can be shortened and longer C-C extensions are used see.

It should be noted that the extendible nozzle adds extra weight and requires actuators, a power supply, mechanisms for moving the extension into position during flight, and fastening and sealing devices. Hence, it requires a reliable mechanism to move the extension into position, and sealing the hot gas between the nozzle sections.

In the past two decades NASA and industry partner Carbon-Carbon Advanced Technologies (C-CAT) have been working towards advancing the technology readiness level of large-scale, domestically-fabricated, C/C nozzle extensions. These C/C extensions can reduce the overall costs of extensions relative to heritage metallic and composite extensions and to decrease weight by 50%. NASA and C-CAT have designed, fabricated and hot fire tested multiple subscale nozzle extension test articles of various C/C material systems, with the goal of assessing and advancing the manufacturability of these domestically producible materials as well as characterizing their performance when subjected to the typical environments. See Gradl and Valentine [12].

For the current NASA SLS, the upgraded RL10B-2 engine was selected to power the Interim Cryogenic Propulsion Stage (ICPS). It appears that for the SLS Artemis I mission configuration the engine was flown with the C-C nozzle extension fully deployed, as depicted in Fig. 4.5. If this is indeed the case, then the requirement for a reliable actuation mechanism was removed.

Example 4.1 Aerojet Rocketdyne's RL10 engine has an impressive legacy powering the upper stages of the Titan III-Centaur, United Launch Alliance's (ULA) Atlas V and Delta IV Heavy launch vehicles. Versions of this bipropellant engine have helped send spacecraft to explore every planet in our Solar System, including Voyager 1 and Voyager 2. More recently, a single RL10B-2 powered the Interim Cryogenic Propulsion Stage on Artemis I, an un-crewed test flight of the NASA SLS and Orion spacecraft. Four RL10 engines (Model RL10C-3) will support the more powerful Exploration Upper Stage that is being developed for future versions of SLS. Determine the exit area and the expansion ratio of the RL-10 engines shown below.

	Thrust (lbf)	Weight (lbm)	Nominal mixture ratio	I_{sp}(s)	Length (in)	Nozzle diameter (in)
RL10C-3	24,230	508	5.7 : 1	460.1	124.3	73
RL10C-X	24,120	510	5.5 : 1	460.9	130.4	73.7
RL10C-1–1	23,825	415	5.5 : 1	453.8	96.7	62

4.7.4.4 Nozzle Flow Separation

Since the engine flow rate is fixed by the chamber pressure and the nozzle throat area, (see Eq. 2.73), the nozzle increases thrust by maximizing the exhaust gas velocity, Eq. (2.71), within the practical limits of size, weight, and operational altitude. Low altitude operation limits the amount of nozzle expansion available because of higher ambient pressure. When the ambient pressure is too high (at sea level), the exhaust flow will separate from the nozzle wall, causing large side structural loads due to the random uneven flowfield of the exhaust jet. The flow will separate from the wall in the divergent part of a nozzle if the chamber pressure p_c has not yet reached its nominal value. Flow separation accompanied with lateral loads may drastically affect the performance of the rocket engine and may compromise its structural integrity.

Upper stage nozzles designed for vacuum operation must have large expansion area ratios. During ground testing, these nozzles are overexpanded, often to the extent that the exhaust gas separates from the nozzle wall, displaying an unsteady asymmetric separation that result in large lateral structural loads, especially during the startup. It was reported that during hot firing of the SSME the flow separated, and the lateral loads caused three coolant feed lines to rupture. Similarly, during ground testing of the J-2 engine, gimbal bolts failed in tension from the large nozzle side loads. On the J-2X engine, engineers added an uncooled extension of a regeneratively cooled nozzle to expand the combustion gases to a targeted exit pressure defined by the altitude for the desired maximum performance.

Separation occurs when the gas in the boundary layer is unable to adjust to the rise to ambient pressure p_a at the end of the nozzle. The exact pressure at which flow will separate from the wall cannot be predicted accurately. Many studies with overexpanded conical nozzles have yielded rules of thumb to predict separation, including the Summerfield criterion. It predicts that flow separates as soon as the wall pressure p_w at the nozzle exit is lower than about 0.35–0.4 times the ambient pressure, that is, $p_{w,sep} \approx 0.4 p_a$. The NASA Design guide (SP-8120) provides a fit of experimental data for short contoured nozzles over a broad range of nozzle area ratios, indicating that separation will occur at

$$\frac{p_{w,sep}}{p_a} = 0.583 (p_a/p_c)^{0.195} \tag{4.1}$$

where $p_{w,sep}$ is the exhaust gas static pressure on the wall at separation, p_a is the ambient pressure, and p_c is the chamber pressure.

There are many other empirical correlations to predict flow separation. Schmucker published in 1973 the status of nozzle research and separation data, which he used to develop the following separation criterium:

$$\frac{p_{w,sep}}{p_a} = \left(1.88 \cdot M_{sep} - 1\right)^{-0.64} \tag{4.2}$$

where M_{sep} denotes the separation Mach number. This means that to avoid flow separation, the nozzle exit area must be sized so that the gas exhaust pressure $p_e > \left(1.88 \cdot M_{sep} - 1\right)^{-0.64}$ for a particular nozzle exit flow Mach number.

Since then, many other correlations for flow separation have been developed. However, general agreement on the available methods has not been reached. Research continues to date, incorporating numerical models and computational tools, attempting to understand the physis of flow separation in rocket nozzles and find practical solutions to mitigate it. This task is challenging because the turbulent flow in the expanding nozzle flow causes side loads and complex shock patters. The models of separation must include boundary layer interactions with the shocks and account for shear layers, vortices, separation shocks, wall heat transfer effects, and other flow phenomena that develop by the hot gases attempting to reattach to the walls after the separation point. Ultimately, techniques for developing nozzle contours that result in the best compromise between performance and nonperformance considerations abide by optimization tools that continue to be refined as new nozzles are designed.

4.7.5   Heat Transfer

The thrust chamber of high pressure LRE's is subjected to huge heat flux, especially in the nozzle throat, and the walls may reach temperatures as high as 3889 K (7000 °R). The largest heat transferred is by convection, and about 5–35% of the total heat transferred is by radiation. In high-thrust rocket engines such as the core or booster stages of heavy lift launch vehicles, the pressure of the combustion chamber may exceed 20 MPa, and the heat flux in the throat region may reach more than 160 MW/m^2. No material could sustain the high heat flux without melting.

The melting point of conventional metals is much lower than the gas temperature in the rocket chamber. Therefore, it is essential to cool the engine to ensure that the wall material withstands the high heat load. However, cooling may affect drastically the specific impulse. And thus, the design of the rocket engine must assess carefully how it is cooled and what potential losses the chosen thermal management method may incur.

Design of the thrust chamber cooling assembly utilizes semi- or fully empirical correlations for convection heat transfer. Provided the coolant remains in the liquid phase at all points in the flow, predictions of forced convection heat transfer in the cooling ducts of rocket engines can be made with sufficient accuracy. For design optimization, engineers use correlations that take into consideration known or measured quantities that depend on design geometry, and on thermodynamics and fluid properties that affect the local heat transfer. These correlations take forms such as:

$$Nu = CRe^a Pr^b \left(\frac{T_w}{T_b}\right)^n \left(1 + \frac{2D}{L}\right)^m \tag{4.3a}$$

or

$$Nu = CRe^a Pr^b \left(\frac{\rho}{\rho_w}\right)^c \left(\frac{\mu}{\mu_w}\right)^d \left(\frac{k}{k_w}\right)^e \left(\frac{\bar{c}_p}{c_p}\right)^f \left(\frac{p}{p_{cr}}\right)^g \left(1 + \frac{2D}{L}\right)^m \tag{4.3b}$$

where the coefficients and exponents quantify the effect of (1) geometry, i.e. cooling channel dimensions and their change along the length of the chamber and its curvature); (2) chemistry, i.e. finite rate kinetics, catalytic wall effects, dissociation and recombination in the boundary layer, and pyrolysis of the coolant; (3) thermodynamics, i.e. real gas behavior, near critical behavior, and varying fluid properties; and (4) fluid dynamics, i.e. turbulence, atomization, mixing, and stratification. All the correlations are extracted from different experiments at different operating conditions. Thus, the value of the coefficients depends on the experimental facilities, measuring techniques, and operating conditions.

Cooling methods are classified in two broad areas: passive and active cooling. Passive cooling includes ablative and radiative methods. Ablative cooling means that the walls are covered with a material called *ablation*, which has a high heat capacity and absorbs heat by transforming itself chemically and/or physically. The ablation burns slowly and removes heat with the gases created in the process. This method is limited by time and by temperature: ablative materials cannot withstand very high temperatures and since they are gradually removed, ablation works only for a limited time. However, ablative cooling is simple and reliable, so it can be incorporated into large engines such as the RS-68, the most powerful LOX/LH$_2$ engine used as first stage by the Delta IV launch vehicle, and for the heavy-lift Delta IV.

Radiative cooling occurs naturally when a hot wall loses heat by radiation. This method is limited to thin surfaces with relatively moderate incident heat fluxes such as nozzle extensions that are much cooler than the main nozzle. The design of radiation cooling is based on the Stefan-Boltzmann equation,

$$q = \varepsilon \sigma T_{wg}^4 \tag{4.4}$$

where q is the heat transfer per unit area, T_{wg} is the maximum wall temperature, ε is the emission coefficient (a number between 0 and 1 that is characteristic of the surface of the radiating material), and σ is Stefan-Boltzmann radiation constant with a value 5.67×10^{-8} W/m^2-K^4.

Maneuvering bipropellants thrusters can use just radiation cooling since the heat produced during the short burns is absorbed by the thick conductive wall of the chamber, which is typically made from cooper alloy. Iridium/rhenium combustion chambers for NTO/MMH thrusters, such as Aerojet's HiPAT and others used to insert satellites into GEO, operate at temperatures up to 2200 °C (3992°F). The chamber is covered with a black rhenium coating to provide an emittance of nearly 1.00, which results in enhanced radiation cooling. Upper stage engines, such as the RL10-B and the Vinci, use ceramic matrix composite (CMC) nozzle extensions that are radiation cooled (see Fig. 4.14).

Active cooling methods include regenerative cooling and film cooling. In regenerative cooling, the flow of a cold fuel is circulated through a cooling jacket along the hot wall of the engine, and in the convection process the propellant carries away excessive heat. The heated propellant is injected in the main chamber and combusted with the oxidizer. The regenerative cooling jacket can be formed with two walls separated by a folded metal sheet, with the liquid propellant flowing along the folds. The liquid may also flow along rectangular channels machined or formed into a liner fabricated from high-conductivity material (like cooper alloys).

In a regenerative cooled chamber, heat is transferred from the hot gas to the wall by convection. Formally, transfer of heat energy includes three processes: (1) heat transfer between the hot gas and the hot-gas side wall, (2) heat conduction through the chamber

Fig. 4.14 **a** Vinci 180 kN, LOX/LH2 upper stage expander cycle engine with ceramic CMC nozzle extension. *Credit*: ESA/ Arianespace. **b** Apogee thruster (HiPAT) 445 N, with iridium-coated rhenium chamber. *Credit* Aerojet (L3Harris)

Height: 3.22 m   Height: 0.72 m

(a)   (b)

wall, and (3) convective heat transfer between the coolant and the cooled side wall. The heat flux increases if the convective heat transfer coefficient h_g increases (or if the temperature difference of recovery temperature and wall temperature becomes higher). The heat conduction within the wall is determined by the wall thickness t_w and the thermal conductivity k of the chamber wall material. On the cooling channel side, the heat is transferred to the coolant. The heat transfer to the coolant is dependent of the heat transfer coefficient h_l and the liquid coolant temperature T_l. For steady-state, heat transfer on the hot gas side, thermal conduction in the wall and convection on the cooling channel are in equilibrium. For one-dimensional heat transfer, this yields a simple correlation for the heat flux $\dot{q} = Q/A$:

$$\dot{q} = h_g\left(T_g - T_{wg}\right) = \frac{k}{t_w}\left(T_{wg} - T_{wl}\right) = h_l(T_{wl} - T_l)$$

or

$$\dot{q} = \frac{T_g - T_l}{\frac{1}{h_g} + \frac{t_w}{k} + \frac{1}{h_l}}$$

Regenerative cooling requires a cooling jacket around the thrust chamber to circulate the liquid fuel through suitable channels before the propellant is injected into the combustor. The inner liner of the rocket chamber is typically made of stainless steel or copper-based alloys as these materials offer high thermal conductivity, thus improving the cooling efficiency, having the required mechanical strength to withstand the thermal and mechanical loads that are typical in rocket engines. Bipropellant rocket engines of medium to large thrust and high chamber pressure use regenerative cooling, including the RL10 in the NASA SLS. The Merlin 1-D engine developed by SpaceX also uses a regenerative cooling approach, circulating the RP-1 kerosene fuel through channels surrounding the combustion chamber and nozzle before it is injected into the combustion chamber proper. The application limit of regenerative cooling is reached when local heat flux exceeds 80 MW/m^2 [15].

Regenerative cooling is highly efficient, but it presents some design challenges. For example, at the throat of the nozzle when the surface area of the wall is too small, it is difficult to pump enough liquid through the channel. To increase the velocity of the flux would require raising the pressure to force the propellant through the cooling jacket. Thus, film cooling may be added to compensate. With film cooling, some of the fuel is injected through ports arranged in the hottest parts of the thrust chamber wall. The liquid fuel absorbs heat by boiling and evaporating, thus forming a cool boundary film that protects the wall from the hot combustion gas. This cool film is spread along the wall by the combustion gas flow moving along the chamber. The nozzle extension of the F-1 engine was film cooled using turbine exhaust gas.

The original first stage Vulcain on Ariane 5 had regenerative cooling, while the Vulcain 2 incorporated a unique cooling system utilizing regenerative and film cooling for the combustion chamber, plus radiation and film cooling for the lower part of the nozzle. In this scheme, hydrogen dump cooling is used for the upper part of the nozzle, where the coolant flows through helical rectangular arranged Inconel (R) tubes. At a location where the nozzle expansion ratio is 32, hydrogen flowing at supersonic velocity is dumped into the nozzle in the direction of the main flow, as shown in Fig. 4.15 (see Suslov et al. [47]). Downstream of this point, hydrogen from the turbine exhaust gas is injected and a coolant film develops adding a thermal barrier to protect the lower nozzle section from heat impact of the accelerated combustion gases. To ensure equal circumferential coolant distribution, the turbine exhaust gas is guided by two pipes from the upper part of the engine alongside the upper nozzle part to a manifold ring arranged before the coolant injection, as depicted in Fig. 4.15.

Another active cooling method is transpiration cooling, a variation of film cooling in which the coolant gets into the chamber through a porous chamber wall. Regenerative and transpiration cooling were both initially considered for the SSME (RS-25 rocket). In the regenerative design, the liquid hydrogen is introduced into the pre-burner gases upstream of the injection into the thrust chamber, thus eliminating a propellant mixing process within the thrust chamber. With the transpiration coolant design, the hydrogen would be introduced through discrete apertures into the combustion chamber walls. The hydrogen coolant flow would carry away heat transferred to the wall from the combustion gases and

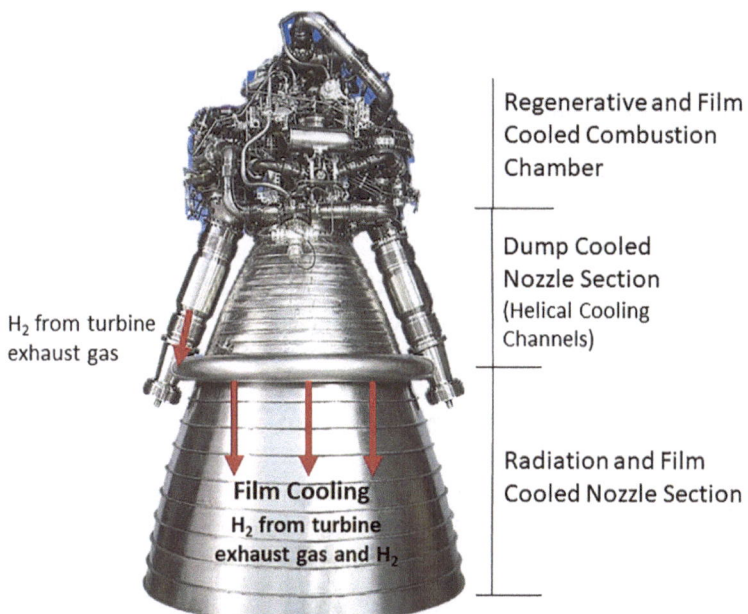

Fig. 4.15 The cooling scheme for the Snemca Vulcain 2 first stage in Adriane 5 launch vehicle. Figure adapted from cooling description by Suslov et al. [47]

subsequently mix with the boundary flow to produce a lower energy wall barrier. However, it was found that transpiration cooling caused a specific impulse loss due to (1) mixture ratio maldistribution, and (2) reduced expansion of the portion of the coolant flow injected downstream of the throat. The loss due to reduced expansion was minor and transpiration coolant specific impulse loss primarily resulted from mixture ratio maldistribution caused by incomplete mixing of the hydrogen coolant with the primary combustion gases.

The chamber of typical regenerative cooled engines includes a liner, jacket, throat ring, coolant inlet manifold, and coolant outlet manifold. The SSME nozzle (see Fig. 4.11) was cooled as follows: the hydrogen fuel entering a diffuser was split to flow to the main combustion chamber, to the three fuel transfer ducts, and through the chamber coolant valve (CCV) to the mixer. The fuel flowing through each transfer duct was split to enter the nozzle coolant inlet manifold at six points. Then the fuel goes through a single up-pass through the 1080 tubes to the outlet manifold, and then to the mixer, to join the bypass flow from the CCV.

For large bipropellant thrust chambers, the liner walls of the cooling jacket are typically made of copper (for high conductivity) alloyed with small amounts of Zr, Ag, or Si for higher strength. In some cases, this material may need a protective coating on the gas side. For the SSME, the liner was made of NARloy Z (North American Rockwell alloy Z), which is mostly copper, with silver and zirconium added. Circulating the liquid propellant through the cooling passages serves two purposes: (1) keeps the engine walls cool and, (2) increases fuel enthalpy prior to entering the combustion chamber. In some engines, such as the SSME, the coolant (liquid hydrogen) coming out of cooling channels is also used to run the turbo-pumps in the fed system.

Regenerative cooling requires large turbopump power and thus is not recommended for moderate pressure engines (GG cycle engines). For these engines, dump cooling may be used, a type of film cooling that uses a warm gas. In this method a single wall, usually made of stainless steel, is gas cooled by the warm turbine exhaust gases which are then injected or "dumped" into the diverging section of the nozzle. Dump flow cooling is used in the nozzle extension of the Vulcain 2 core stage engine.

4.8   Rocket Engine Preliminary Design

Development of rocket propulsion systems requires a precision, integrated, multi-disciplinary approach to design and manufacture. Adhering to proven design practices and process controls during design and development results in a high quality, high reliability rocket. The design process begins with the definition of technical requirements, followed by the preliminary design phase, which includes the initial conceptual design and preliminary analysis, then followed by trades studies and a design optimization step before the final design stage is reached.

The major components of a typical liquid propellant rocket engine include the combustion chamber, the exhaust nozzle, and the propellant tanks, plus the systems to deliver the propellants to the combustion chamber (see Fig. 4.2). In the combustion chamber, by means of chemical reaction of fuel and oxidizer the internal energy of the propellant is converted into heat. The main combustion chamber contains the combustion process, accelerates the gas flow to throat sonic velocity, and initiates the hot gas expansion process through its diverging section. The combustion chamber throat connects to the nozzle, requiring that it joins smoothly on its inner surface (see Fig. 4.11). Understanding of the mechanisms of heat transfer, cooling techniques, and advances in materials is crucial for optimizing the design and thermal management of reliable and efficient rocket combustion chambers.

4.8.1   Correction Parameters and Analytical Tools

In the actual design of a rocket, certain correction factors are applied to the results derived from the ideal analysis. The characteristics of the flow are derived with the ideal relations in the previous sections neglect friction, heat transfer, real gas behavior, nonuniformities of the flow, and changes in the gas composition. The performance of the real nozzle is affected by these and other losses. Therefore, we apply appropriate correction factors to the parameters derived theoretically using the ideal flow assumptions.

Internal and external flow phenomena of a rocket nozzle are rather complex. Understanding such phenomena is crucial for developing highly reliable rocket engines to power future spacecraft. Many modeling techniques are available for the thermokinetic processes within the nozzle. Due to the extremely complex flow mechanisms involved, these models are largely supported by empirical correlations which are somewhat configuration sensitive. They are successfully applied to provide initial design values. However, some losses can be reduced by experimental techniques which establish performance relationships to support the optimization of the design.

Nozzle flow fields are multi-dimensional, as the flow properties vary in both the axial and radial directions. In a conventional bell nozzle, the gas from the combustion chamber and converging through the nozzle throat must be turned away from the nozzle axis to accelerate or expand the flow. Subsequently, at the exit, the flow must be turned back parallel to the nozzle axis to maximize nozzle efficiency. There are computer programs that address such complex supersonic flow fields and are used successfully for designing and analyzing rocket nozzles.

Example 4.2 A rocket engine burning LOX/LH$_2$ produces 44.482 kN thrust operating with a chamber pressure of 6894.76 kPa, a combustion temperature of 2415.556 °C, and a mixture ratio of 3.40. For the exhaust gases assume a molecular mass $\mathcal{M} = 8.9\,\text{kg/mol}$ and a specific heat ratio of $\gamma = 1.25$. Calculate the actual specific impulse and the nozzle area ratio for optimum operation at an altitude where the ambient pressure is 10.89 kPa.

Solution: The ideal exhaust velocity is given by Eq. (2.33), using $\mathcal{R} = 8.3146\frac{\text{kJ}}{\text{kmol·K}}$

$$v_{ex} = \sqrt{\frac{2\gamma}{\gamma - 1}\frac{\mathcal{R}T_c}{\mathcal{M}}\left[1 - \left(\frac{p_e}{p_c}\right)^{\gamma - 1/\gamma}\right]} = 4266.57 \text{ m/s}$$

This gives a theoretical specific impulse of $I_{sp} = v_{ex}/g_0 = 431$ s.

The actual specific impulse is 97% of the theoretical value: $I_{sp_{actual}} = 418$ s.

Now calculate the theoretical or ideal thrust coefficient with Eq. (2.49):

$$C_F = \left\{\frac{2\gamma^2}{\gamma - 1}\left(\frac{2}{\gamma + 1}\right)^{(\gamma+1)/(\gamma-1)}\left[1 - \left(\frac{p_e}{p_c}\right)^{(\gamma-1)/\gamma}\right]\right\}^{1/2} + \left(\frac{p_e}{p_c} - \frac{p_a}{p_c}\right)\frac{A_e}{A_t} = 1.76$$

where the pressure ratio is

$$\frac{p_e}{p_c} = \frac{10.89}{6894.76} = 0.00158$$

The actual thrust coefficient is 98% of the ideal value, that is $C_{F_{actual}} = 1.72$.

Now we determine the throat area:

$$A_t = \frac{F}{C_F p_c} = \frac{0.689 \times 10^6 \text{ N}}{(1.72)(1.01325 \times 10^8 \text{ Pa})}$$

$$= \frac{0.689 \times 10^6}{1.74279 \times 10^8} = 0.0039 \text{m}^2 \rightarrow D_t = 0.070 \text{ m}$$

The area ratio given by Eq. (2.46), from which we determine the optimum nozzle area ratio:

$$\varepsilon = \frac{A_e}{A_t} = 42$$

In general, the nozzle shape is selected to maximize performance within the constraints placed on the rocket system. This goal is relatively easy to achieve with the tools presently available to propulsion engineers. Problems arise when unusual performance requirements introduce additional constraints that are not readily handled. For example, we may wish to maximize the expansion area ratio to obtain high vacuum performance from an upper-stage engine, and at the same time we want to be able to ground test the engine without the added expense of an altitude ground facility.

Full-scale experimental testing is time-consuming and costly, adding to the development cycle of a new rocket propulsion system. Advances in computational fluid dynamics (CFD) algorithms and computing speeds have made it more common to incorporate numerical simulation into the design process. Robust and accurate CFD tools play a major role in the development process of rocket engines built today. CFD codes simulate or model real physical phenomena, thereby helping us to understand and "see" what is happening inside a rocket engine. Sophisticated CFD codes continue to be developed to include more accurate physics models, including thermo-chemistry, diffusion, and turbulence. These codes provide better answers, and their use is changing from being a reference for experimental results to becoming the benchmark to preliminary design. The design and development process of future rocket engines will be greatly improved with the help of reliable and highly effective CFD modeling and simulation.

Accurate values for the rocket performance parameters can be obtained with JANNAF methodology (see JANNAF [18]), including the two-dimensional kinetics (TDK) computer program, which is combined with a boundary layer module (BLM) and accounts for nozzle losses. JANNAF refers to the Joint Army-Navy-NASA-Air Force Rocket Engine Performance Prediction and Evaluation Manual, which establishes a US national standard for the analytical and experimental evaluation the performance of LREs. Rocket engine designers have developed other tools that provide modern nozzle analytical programs that are equally applicable. However, for commonality and standardization the solutions with the JANNAF codes are normally used to assess the relative merit of the analysis as well as the design when comparing 2 or more engine concepts.

Before we leave the discussion of liquid propellant rocket propulsion, let us consider the potential of methane rockets to take us to Mars and beyond.

4.9   Methane Rockets for Interplanetary Missions

Methane is the propellant of choice for spacecraft propelled by chemical rockets considered for interplanetary missions. The possibility of using liquid methane rocket engines to power spacecraft to Mars has been studied for decades. With fuel available at the destination, a spacecraft from Earth would not have to carry so much propellant. Feasibility studies suggest that methane can be produced from the carbon dioxide in the atmosphere surrounding Mars, and therefore methane rockets could be refueled there for the return voyages.

Research and development of methalox rocket engines has accelerated, and studies suggest that conventional LOX/LH2 engines could be converted to LOX/LCH4 systems. For example, the RL10 engine powering the Interim Cryogenic Propulsion Stage of the SLS and subsequently the Exploration Upper Stage, could be adapted to run on methane, instead of hydrogen, making it more attractive for Mars exploration missions. Of course, there are technical challenges of converting traditional rocket engines to burn methane and we must pay attention especially to propellant injector redesign, ignition system, and the regenerative cooling characteristics of methane, especially for long term, reusable engines. The propellant mixing, ignition and stable combustion involve complex physical and chemical processes, which need to integrate effectively with the nozzle design and its thermal management.

Much R&D work has been devoted to methalox ignition. NASA researchers developed the Augmented Spark Impinging (ASI) igniter, a plasma-assisted design that generates a secondary torch flame to ignite a main engine (augmented ignition). It uses a spark to create the ignition kernel (spark ignition), and uses an impinging style injector to deliver and mix the propellants (an impinging igniter) (see Marshall et al. [21]). The ASI igniter provides a portion of oxygen flow upstream of the spark igniter tip, and then provides impinging oxygen and methane flows downstream of the spark igniter to create the combustible mixture. The ASI igniter was tested with methalox to help support internal and commercial engine development programs, such as those in the Lunar CATALYST. One challenge with spark exciter systems, especially at altitude conditions, is the ignition lead that transmits the high voltage pulse from the exciter to the spark igniter. The ignition lead can be prone to corona discharge, reducing the energy delivered by the spark and potentially failing to ignite the propellant mixture.

Glossary

Bipropellants A combination of rocket propellants including an oxidizer and a fuel. For example, liquid oxygen (LOX) and liquid hydrogen (LH_2) are a bipropellant combination used in many rocket engines.

Centaur A cryogenic LOX/LH_2 high-performance upper stage used by Atlas, Titan, and Delta III/IV/V launch vehicles.

Cryogenic propellants Propellants that are liquefied gases at low temperatures. Cryogenics are gases at normal temperatures but are used in propulsion systems chilled below their boiling points. Examples: oxygen (LOX), hydrogen (LH_2), methane (CH_4).

Earth storable propellants Propellants that remain stable over a range of terrestrial pressures and temperatures and can be stored in a closed vessel for long periods of time. Examples: Kerosene, Hydrazine, Monomethylhydrazine, Hydrogen peroxide (H_2O_2), Water, Nitrogen tetroxide (N_2O_4), Hydroxylammonium nitrate (HAN)-based propellants, green propellants (LMP-103S, AF-315E), Ammonium dinitramide (ADN)-based propellants, Ionic liquids. Earth storable propellants are liquids at ambient temperature and pressure and are hypergolic on contact— do not need a separate ignition system.

Earth Departure Stage The upper stage of a SLV with performance capability to send the payload out of low Earth orbit to its destination, e.g. Moon, planet. The Earth Departure Stage (EDS) is the name given to the second stage of the Block 2 Space Launch System.

Gelled Propellants Gelled and metallized fuels are a class of thixotropic (shear-thinning) fuels that improve the performance of rocket and air-breathing systems. Gelled propellants behave like a jelly or thick paint, not spilling or leaking. These gelled substances can flow under pressure, will burn, and are safer in some respects. Examples: Gelled oxygen (O_2)/hydrogen (H_2), Gelled MMH/IRFNA propellants.

Hydrazine (N_2H_4) Rocket fuel, a highly toxic, flammable liquid compound, primarily used as propellant due to its high heat of combustion, important for many spacecraft. Propellant grade per MIL-P-26536.

Hypergolic A self-igniting propellant substance. It includes fuels and oxidizers that self-ignite within about 75 ms after contact with each other.

Interim Cryogenic Propulsion Stage (ICPS) Because integration is far more complex a task than simply placing Orion on top of the NASA SLS launch vehicle, several hardware elements are required, collectively known as the Integrated Spacecraft and Payload Element (ISPE). In both Crew and Cargo configurations, ISPE connects the Core Stage with the Block 1 upper stage, the Interim Cryogenic Propulsion Stage (ICPS). Powered by Aerojet Rocketdyne RL10B-2 engine, the ICPS provides the in-space propulsion necessary to place the Orion spacecraft on a trans-lunar injection orbit trajectory.

Kerolox An acronym for a bipropellant comprised of RP-1 fuel and liquid oxygen.

LCH_4 Liquid methane fuel, a hydrocarbon with chemical formula CH_4. It is the main component of liquified natural gas (LNG); requirements for different grades of pure CH_4 are defined by MIL-PRF-32207 standard.

LNG An acronym for liquified natural gas.

LOX An acronym for liquid oxygen, the oxidant used in most bipropellant rocket engines. Propellant grade by MIL-P-25508.

Methalox An acronym for a bipropellant comprised of methane fuel and liquid oxygen.

MON An oxidizer comprised of a mixture of oxides of nitrogen. It consists of nitrogen tetroxide (N_2O_4) and nitric oxide (NO). An associated number represents the NO percentage by weight (e.g., MON-15 contains 15% NO).

Monopropellants Propellants formulated of unstable chemical substances that will decompose, under suitable conditions, releasing energy and producing hot gases that can be used in a rocket engine.

Nanopropellants Propellants with nano-sized particle additives incorporated to enhance the combustion behavior and increase specific impulse. For example, nano-sized particles of hafnium or aluminum added to a propellant can enhance its burning rate. Nanogellant for gelled cryogens has a surface area of about 1000 m2/g, resulting in the gelling of cryogenic fuels with 25%–50% less mass compared to traditional gellant material.

Nitrogen Tetroxide (N_2O_4) Hypergolic oxidizer. Propellant grade per MIL-P-26539.

RP-1 An acronym for rocket propellant 1, a hydrocarbon fuel similar to kerosene. Standard U.S. kerosene rocket fuel RP-1 is defined by Military Specification MIL-R-25576. In Russia, similar specifications for kerosene exist as T-1 and RG-1 (e.g. RD-170 engine in Table 4.4)

Space storable propellants Propellants that can be stored in a space environment for extended time without boil-off, freezing, or decomposition. These propellants are liquid in the environment of space. They are mildly cryogenic or have high vapor pressure on Earth, but because of low ambient temperatures in space they can be kept liquid under a modest vapor pressure. The storability depends on the specific tank design, thermal conditions, and tank pressure. Examples are ammonia, oxygen difluoride, diborane, and dinitrogen tetrafluoride.

Storable Propellant Type of propellants that exhibit no corrosion, decomposition, or deterioration over a specified long period, typically 5–10 years; some satellites require up to 20 years. A low freezing point is often required (30°F (272 K) for the Titan II launch vehicle). The upper temperature range can also vary, about 90°F (305 K) for an environmentally controlled system. Storable can be fuel and oxidizer.

UDMH An acronym for Unsymmetrical DiMethyl Hydrazine, a storable hyperbolic rocket fuel. Propellant grade per MIL-P-25604.

Recommended Reading and References

1. Alliot, P., Delange, J-F., De Korver, V., et al. (2015). "VINCI®, the European reference for Ariane 6 upper stage cryogenic propulsive system." 6th European Conference for Aeronautics and Space Sciences (EUCASS). https://www.eucass.eu/component/docindexer/?task=download&id=3647.
2. Ariane 6 User's Manual, Issue 2, Revision 0, Arianespace, February 2021. https://www.arianespace.com/wp-content/uploads/2021/03/Mua-6_Issue-2_Revision-0_March-2021.pdf.
3. Atlas V Launch Services User's Guide (2010), United Launch Alliance, https://www.ulalaunch.com/docs/default-source/rockets/atlasvusersguide2010.pdf.
4. Bonnani, M., Mohaddes, D., Nguyen, L., et al. (2021). "Toward Numerical Investigation of Ignition and Combustion Transition in a Subscale LOX/Methane Rocket Combustor." AIAA SciTech Forum, 11–15 and 19–21 January 2021.
5. Burkhadt, H., Sippel, M., Herbertz, A., and Klevanski, J. (2002). "Space Launcher Liquid Propulsion," in Comparative Study of Kerosene and Methane Propellant Engines for Reusable Liquid Booster Stages, December 2002.
6. CANTERA - An object-oriented software toolkit for chemical kinetics, thermodynamics, and transport processes.: http://code.google.com/p/cantera/.
7. CHEMKIN Web Overview: http://www.sandia.gov/chemkin/index.html.
8. DoD, Department of Defense, United States of America: Performance Specification, Propellant, Methane, MIL-PRF-32207, 10th October 2006.
9. Frey, H. M. and Nickerson G. R. (1989). "TDK - Two Dimensional Kinetic Reference Program", SEA Inc. NAS8 36863, 1989.
10. Fukushima, Y., Nakatsuzi, H., Nagao, R., Kishimoto,K., et al. (2002). "Development Status of LE-7A and LE-5B Engines for H-IIA Family." Acta Astronautica, Vol. 50, Issue 5, March 2002, pp. 275–284.
11. Gordon, S. and McBride, B.J. (1994). Computer Program for Calculation of Complex Chemical Equilibrium Compositions and Applications. NASA Reference Publication 1311, Oct. 1994. (http://www.grc.nasa.gov/WWW/CEAWeb/RP-1311.htm).
12. Gradl. P. R. and Valentine, P. (2017). "Carbon-Carbon Nozzle Extension Development in Support of In-space and Upper-Stage Liquid Rocket Engines." Paper AIAA 2017–5064, 53rd AIAA/SAE/ASEE J. Propulsion Conference, 10–12 July 2017, Atlanta, GA.
13. Hagemann, G., Immich, H., Nguyen, T.V, and Dumnov, G.E., "Advanced Rocket Nozzles," J. of Propulsion and Power, Vol. 14, No. 5, September –October 1998.
14. Hagemann, G., Immich, H., Nguyen, T. and Dummov, G. E. (2004). "Rocket Engine Nozzle Concepts," Chapter 12 of *Liquid Rocket Thrust Chambers: Aspects of Modeling, Analysis and Design*, V. Yang, M. Habiballah, J. Hulka, and M. Popp (Eds.), Progress in Astronautics and Aeronautics, Vol. 200, AIAA, 2004.
15. Haidn, O.J. (2008), *Advanced Rocket Engines*. In Advances on Propulsion Technology for High-Speed Aircraft (pp. 6-1–6-40). Educational Notes RTO-EN-AVT-150, Paper 6. Neuilly-sur-Seine, France: RTO. Available from: http://www.rto.nato.int.
16. Huzel D. K. and Huang D. H. (1992). "Modern Engineering for Design of Liquid Propellant Rocket Engines," Revised Edition. AIAA Progress in Astronautics and Aeronautics, Vol. 147. Washington, DC: AIAA 1992.
17. Isakovic, S.J., Hopkins Jr., J.P., Hopkins, J.B. (2004). *International Reference Guide to Space Launch Systems*, 4th Ed. AIAA, 2004.
18. JANNAF (2012). "Test and Evaluation Guidelines for Liquid Rocket Engines," JANNAF-GL-2012-01-R0, December 2012 (available from DSIAC and CPIAC).

19. Katsumi, T. and Hori, K. (2021). "Successful development of HAN based green propellant." Energetic Materials Frontiers, Vol. 2, Issue 3, September 2021, Pages 228–237.
20. Kirchberger (2014).
21. Marshall, W. M., Osborne, R. J., and Greene, S.E. (2017). "Development of Augmented Spark Impinging Igniter System for Methane Engines." Paper AIAA 2017-4665, presented at the 53rd AIAA/SAE/ASEE Joint Propulsion Conference, 10–12 July 2017, Atlanta, GA, https://doi.org/10.2514/6.2017-4665.
22. Masse, R. K., Allen, M., Driscoll, E., Spores, R., et al. (2016). "AF-M315E Propulsion System Advances and Improvements." 52nd AIAA/SAE/ASEE J. Propulsion Conference, 2016. https://doi.org/10.2514/6.2016-4577 .
23. Masse, R. K. and Glassy, B. A. (2019). "Low-Vapor-Toxicity Hydrazine Propellant Blends," J. Meeting of the 49th Combustion (CS), 37th Airbreathing Propulsion (APS), 37th Exhaust Plume and Signatures (EPSS), and 31st Propulsion Systems Hazards (PSHS) Subcomm.; and 66th JANNAF Propulsion Meeting (JPM) and meeting of the Programmatic and Industrial Base (PIB), June 3–7, 2019.
24. McBride, B.J., Reno, M., and Gordon, S. (1994). "CET93 and CETPC: An Interim updated version of the NASA Lewis computer program for calculating complex chemical equilibria with applications." NASA TM-4557. March 1994.
25. NASA Computer program CEA2 (Chemical Equilibrium with Applications 2) (http://www.grc.nasa.gov/WWW/CEAWeb).
26. NASA Facts: Propellants http://www-pao.ksc.nasa.gov/kscpao/nasafact/count2.htm.
27. NASA SP-8089, NASA Space Vehicle Design Criteria (Chemical Propulsion): Liquid Rocket Engine Injectors. March 1976.
28. NASA SP-8120, NASA Space Vehicle Design Criteria (Chemical Propulsion): Liquid Rocket Engine Nozzles. July 1976.
29. Neill, T, Judd, D., Veith, E. and Rousar, D. (2009). "Practical uses of liquid methane in rocket engine applications," Acta Astronautica, Vol. 65, Issues 5–6, September–October 2009, pp. 696–705.
30. Nickerson, G.R., Berker, D.R., Coats, D.E., and Dunn, S.S. (1993). "Two-Dimensional Kinetics (TDK) Nozzle Performance Computer Program Volume II, User's Manual." Program prepared by Software and Engineering Associates, Inc. for George C. Marshall Space Flight Center under contract NAS8-39048, March 1993.
31. Northrop Grumman Bi-Modal Thruster, https://www.northropgrumman.com/space/tr501-dual-mode-thruster.
32. Parkinson, D., VanLerbergue, W.M, and Rahman, S.A. (2017). "JANNAF Test and Evaluation Guidelines for Liquid Rocket Engines: Status and Applications." AIAA Paper presented at 2017 AIAA/SAE/ASEE J. Propulsion, Atlanta, GA, July 10–12, 2017.
33. Pichon, T. D. and Barreteau, R. (2017). "Composite Nozzle Extension Assembly for the RL10 Engine Family." Paper AIAA 2017-5066, 53rd AIAA/SAE/ASEE J. Propulsion Conference, 10–12 July 2017, Atlanta, GA.
34. Ponomarenko, A., RPA: Tool for Liquid Propellant Rocket Engine Analysis C++ Implementation, May 2010.

35. Rao G. V. R., "Approximation of Optimum Thrust Nozzle Contour." ARS Journal, Vol. 30, No. 6, p. 561, 1960.
36. Rocketdyne Propulsion and Power (1998). "Space Transportation System Training Data: Space Shuttle Main Engine Orientation," June 1998. Available: http://www.lpre.de/p_and_w/SSME/ SSME_PRESENTATION.pdf.
37. RL10 Engine specifications: https://www.rocket.com/space/liquid-engines/rl10-engine; https:// www.l3harris.com/all-capabilities/rl10-engine.
38. RS-25 Engine specifications: https://www.rocket.com/space/liquid-engines/rs-25-engine.
39. Russia's Space Rocket Fleet: http://www.russianspaceweb.com/rockets_launchers.html.
40. Schartz, W. T., Cannova, R. D., Cowley, R. T., and Evans, D. D. (1979). "Development and Flight Experience of the Voyager Propulsion System." AIAA Paper 79-1334.
41. Schmucker, R. H. (1973). Flow Processes in Overexpanded Chemical Rocket Nozzles. Part 1: Flow Separation. NASA Report TM-77396.
42. Simons, J. R. (2014). "Design and Evaluation of Dual-Expander Aerospike Nozzle Upper Stage Engine." Ph. D. Dissertation, Air Force Institute of Technology, https://apps.dtic.mil/sti/pdfs/ ADA609649.pdf.
43. Song, J. and Sun, B., "Thermal-structural analysis of regeneratively-cooled thrust chamber wall in reusable LOX/Methane rocket engines." Chinese Journal of Aeronautics, Available online 6 May 2017. cja@buaa.edu.cn; www.sciencedirect.com.
44. Space Launch System Mission Guide (2018). NASA Document No: ESD 30000. Release Date: December 19, 2018. https://ntrs.nasa.gov/api/citations/20170005323/downloads/20170005323. pdf.
45. SpaceX Falcon Users Guide (2021), https://www.spacex.com/media/falcon-users-guide-2021- 09.pdf.
46. Spores, R. A., Masse, R. Kimbrel, S. and McLean, C. (2013). "GPIM AF-M315E Propulsion System." Paper presented at 49th AIAA/ASME/SAE/ASEE J. Propulsion Conference and Exhibit, San Jose, California, USA, 15–17 July 2013.
47. Suslov, D.I., Arnold, R., and Haidn, O.J. (2010). "Convective and Film Cooled Nozzle Extension for a High Pressure Rocket Subscale Combustion Chamber." AIAA 2010-1150. 48th AIAA Aerospace Sciences Meeting Including the New Horizons Forum and Aerospace Exposition, 4 - 7 January 2010, Orlando, Florida.
48. Sutton, G.P. and Biblarz, O. (2001). *Rocket Propulsion Elements*. Seventh Edition, John Wiley & Sons, Inc. New York, 2001.
49. TDK'08(TM): http://www.seainc.com/tdk.d.html, Internet Source, retrieved on 31st March 2011.
50. Tinker, D.C., Fedotowsky, T.M., Bardsley, R.A., and Gradl, P.R. (2024). "Development of Large-Scale Thrust Chambers using Laser Powder Directed Energy Deposition." AIAA Paper, AIAA SciTech, 8–12 January 2024, Orlando, FL.
51. Togo, S., Hori, K., Shibamoto, H. (2004). "Improvement of HAN-Based Liquid Monopropellant Combustion Characteristics." Intl. Conference HEMs-2004, Belokurikha, Russia 2004.
52. Turner, M.J.L., *Rocket and Spacecraft Propulsion, Principles, Practice and New Developments*. Second Edition, Springer-Praxis Books (2006).
53. Valentian, D., Sippel, M., Gronland, T.-A., Baker, A. Adam, Van Den Muelen, J., Fratacci, G, and Caramelli, F., "Green Propellants Options for Launchers, Manned Capsules and Interplanetary Missions." ESA-02 (2004).
54. Valentine, P.G. and Gradl, P.R. (2019). "Extreme-Temperature Carbon- and Ceramic-Matrix Composite Nozzle Extensions for Liquid Rocket Engines," 70th International Astronautical Congress (IAC), Washington D.C., U.S., 21–25 October 2019, IAC-19-C2.4.9.

55. Xu, J., Zhuo, D, Xing, L., Gao, Y. and Jin, P. (2022). "Energy Estimation and Testing Verification on Ignition of a Torch Ignition System." Journal of Physics: Conf. Ser. 2235 012052.

56. Yang, V., Habiballah, M., Hulka, J. and Popp, M. (Eds.) (2004). *Liquid Rocket Thrust Chambers: Aspects of Modeling, Analysis and Design*. Progress in Astronautics and Aeronautics, Vol. 200, AIAA, 2004.

57. Yost, B. and Weston, S. (2024). "State-of-the-Art Small Spacecraft Technology." Small Spacecraft Systems Virtual Institute, Ames Research Center, Moffett Field, CA. NASA/TP-20240001462, February 2024. https://ntrs.nasa.gov/api/citations/20240001462/downloads/2023%20SOA_final.pdf.

Solid Propellant Rocket Propulsion

5

> *Delivering 16 MN or 3.6 million lbf at lift-off, the NASA SLS booster is the largest, most powerful solid propellant motor ever built. The twin SRBs provide more than 75% of the total SLS thrust at launch.*
>
> —*Dora Musielak*

A solid rocket motor is a powerful propulsion system that burns a solid propellant—a chemical compound comprising both the fuel and the oxidizer admixed. Upon ignition, the solid propellant is transformed into hot combustion gases at the grain surface. The hot gases are then expanded and accelerated in the exhaust nozzle to produce thrust. Solid rocket motor (SRM) propulsion has a wide range of applications in astronautics, from lift-off boosters for launch vehicles to in-space maneuvering of spacecraft; thus, motors are designed in many sizes, with different propellant formulations, and varying in thrust from approximately 2 N to over 16 MN (0.4 to over 3.6 million lbf). For a given application, the SRM is designed to produce a total impulse and prescribed thrust versus time profile to fulfill its mission. This is accomplished by coupling the solid propellant grain internal ballistic characteristics, such as burning rate and chamber pressure, with nozzle performance. The burning rate and solid propellant burning area directly control the motor pressure and thrust. In this chapter we wish to answer questions such as

- Are there substantial performance or operational advantages of SRM over LRE propulsion for space launch vehicles?
- What is the typical chemical composition of a solid propellant?
- Why the solid propellant is called "grain"?

© The Author(s), under exclusive license to Springer Nature Switzerland AG 2025
D. Musielak, *Introduction to Rocket Propulsion for Astronautics*, Synthesis Lectures on Engineering, Science, and Technology, https://doi.org/10.1007/978-3-031-86141-3_5

- What is the propellant burning or regression rate in the SRM combustion chamber?
- What is burn time? what is action time?
- How do we determine the chamber pressure for a SRM?
- Can we reduce the thrust of a SRM? If so, how?
- Is it possible to control the burn of SRM in flight? If so, how?

5.1   Solid Rocket Motor

A solid rocket motor (SRM) is a chemical propulsion system burning a propellant in solid state. A SRM consists of a combustion chamber or case containing the hardened propellant grain, a nozzle, insulation, and an igniter (Fig. 5.1). The motor case is a pressure vessel made of metal or composite materials and filled with the solid propellant charge, called the grain. The solid propellant grain contains all the chemical constituents needed for complete combustion, and is formed in the combustion chamber with a hole down the middle (the interior cavity). When ignited, the propellant burns rapidly and radially outward in the combustion chamber, passing through the interior cavity and expelling hot gases from the supersonic nozzle to produce thrust. Initial burning takes place at internal surfaces of cylinder perforation and slots. The internal cavity grows as the propellant is consumed through combustion. The propellant burns from the center out toward the sides of the case. The shape of the center channel determines the rate and pattern of the burn, thus providing a means to control thrust.

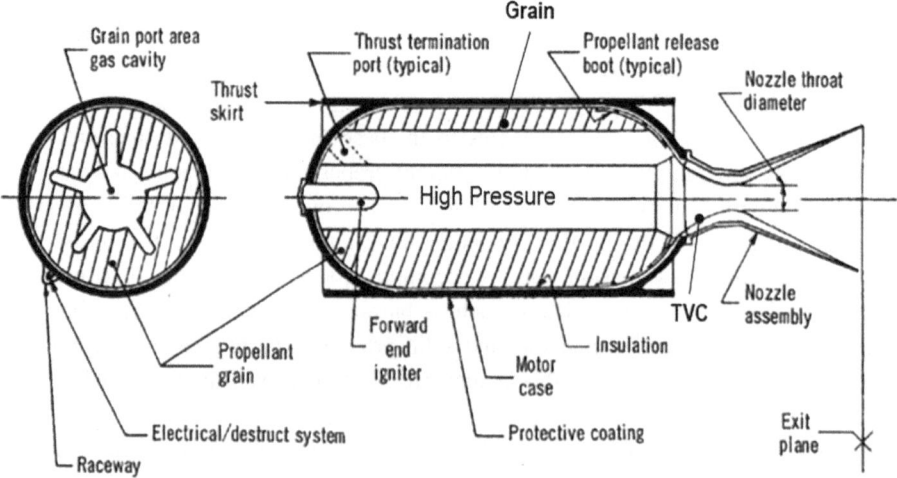

Fig. 5.1 Main components of a solid rocket motor. Image adapted from NASA SP-8025

The SRM achieves "burnout" when all its propellant is consumed and combustion is completely extinguished. At this burnout time the motor has finished its mission, having delivered the impulse for which it was designed. Figure 4.1 is a schematic of a generic SRM, showing the main components and the thermophysical process that must be carefully evaluated.

- Propellant Grain or Charge: is the solid body of hardened propellant when enclosed by the chamber; typically accounts for 82 to 94% of the total motor mass. The central opening in the grain is sometimes called the port. A typical grain consists of a rubber-like binder substance mixed with aluminum powder, ammonium perchlorate, and additives.
- Combustion Chamber or Motor Case: is a pressure vessel containing the combustion process. It is typically made of titanium, mild steel, and Kevlar.
- Igniter: provides energy to start the combustion of solid propellant. It raises the grain surface temperature to the ignition point, increasing the chamber pressure to self-sustaining levels.
- Nozzle: it provides the converging- diverging duct through which the hot gases are expanded and accelerated. Since the nozzle reaches very high temperatures, it is made of graphite epoxy or carbon-carbon (C-C) with a carbon throat insert. The nozzle closure protects the grain from exposure to hard vacuum.
- Insulation: it is added to protect the motor case from the hot combustion gas. The insulated areas are typically at the ends of the propellant grain.
- Liner: its purpose is to inhibit grain burning and insulate the combustion chamber when the flame front arrives. Added at the propellant-case joint, the liner is usually propellant binder.
- Thrust Termination: mechanism to stop or extinguish burning of a SRM before all propellant has been consumed.
- Thrust Vector Control (TVC): controls the direction of thrust, adjusting its attitude or angular velocity. TVC is achieved by redirecting the exhaust gas flow using an adjustable obstruction, such as jet vanes, jet flaps, or tabs.

The propellant compound is poured into the motor case with special tooling installed inside the chamber to provide the required propellant shape, which most commonly will have a center bore (see Sect. 5.3). The solid propellant is usually bonded to the inner wall of the case, and occupies (before ignition) the greater part of its volume. The case inner surfaces have a thermal protection insulation layer to keep the case from becoming too hot, and ensure it carries its pressure and thermo-structural loads. Large propellant grains are cast in segments, which are then stacked to form a motor. The huge solid rocket boosters (SRBs) are made in segments. For example, the SRBs for the former Space Shuttle were comprised of four segments; the SRBs for NASA's new SLS have an additional segment to increase the lift-off thrust.

Ignition of the solid propellant occurs with an igniter positioned at one end of the case. The igniter, a pyrotechnic device, or a small rocket, starts the motor operating when an electrical signal is received. The motor case is a high-pressure combustion chamber where the propellant burns producing a very high temperature gas. Once ignited the propellant burns at a very rapid rate until totally consumed, producing hot exhaust gases.

The combustion products expand through a nozzle, contoured to accelerate the hot gases to supersonic velocity to produce thrust. The burning characteristics, which determine the thrust, mass flow rate, operating pressure, and burn duration of an SRM, depend on the size and shape of the exposed burning surface.

Once ignited, a simple solid rocket motor cannot be shut off, because the solid propellant contains all the ingredients necessary for combustion, burning from the center out towards the sides of the casing. The shape of the center channel determines the rate and pattern of the burn, thus providing a means to control thrust. Shaping the propellant charge determines the amount of exhaust gas that is produced, which is a result of the propellant surface area ignited. There are some SRM designs that can be extinguished and then re-ignited by controlling the nozzle geometry or using vent ports.

Solid rocket motors must meet very stringent performance requirements including thrust imbalance and pressure/thrust oscillations. The design, operation, and performance of a SRM depend on the combustion characteristics of the solid propellant, its burning rate, burning surface, and the geometry of the grain poured into the combustion chamber. The study of these processes is known as internal ballistics, a model of internal flowfield conditions during the motor combustion time, from ignition to burnout. Internal ballistics includes the complex interactions of the solid propellant physical and chemical processes that result in SRM's behaviour and ultimately its performance and mission capability. Details of these phenomena remain as topics of research, and studies aim to developing ballistics numerical simulation models to better understand the complex interacting phenomena within SRMs, including better grain propellant formulations, to design and advance solid propulsion for space applications.

Accuracy of the thrust-time curve prediction is crucial; the propellant burning rate is the controlling factor together with its physical properties, e.g. grain size, particle size distribution and morphology, propellant viscosity. In the following sections we discuss solid grain geometry, and burning rate, and will provide examples of different SRMs currently in use.

5.2   Solid Propellant Composition

Solid propellants are pre-mixed oxidizers and fuel substances. The oxidant is usually an inorganic salt such as potassium nitrate or potassium chlorate and perchlorate. Metallic powders are added to increase the energy release, and therefore increase the combustion temperature. Solid propellants can be double-base (DB) (homogeneous, with fuel and oxidizer within the same molecule), or composite (heterogeneous, with mixture of oxidizer crystals and fuel, which is a rubber-like binder). A common binder is hydroxyl terminated polybutaliene (HTPB), while common oxidizing crystals are ammonium perchlorate (AP), ammonium nitrate (AN), nitronium perchlorate (NP), potassium perchlorate (KP), potassium nitrate (KN), cyclotrimethylene trinitramine (RDX), and cyclotetramethylene tetranitramine (HMX).

Solid propellants can also be composite, a type known as modified double-base (CMDB) propellant. Aluminum particles are commonly added to propellants, since the exhaust products will contain aluminum oxide, a compound with high molecular mass. The combustion of this heterogeneous mixture is rather complex, and thus determining the ballistic properties of such propellants is challenging.

The most common solid propellants for space applications are Polybutadiene Acrylic Acid Acrylonitrile Prepolymer (PBAN), and Hydroxyl Terminated Poly Butadiene (HTPB). PBAN is a composite powder, polybutadiene acrylonitrile copolymer (synthetic rubber) compound fuel mixed with ammonium perchlorate oxidizer (a crystallized or finely ground mineral salt, as oxidizer. The propellant is held together by a polymeric binder, usually polyurethane or polybutadienes, which is also consumed as fuel. Metal powders are added to solid propellants to increase the specific impulse I_{sp} and fuel density. Aluminum powders that constitute 12–20% of propellant mass are added. PBAN may contain 11.78% polybutadiene-acrylic acid-acrylonitrile, 70% of ammonium perchlorate, 16% of aluminum powder. This basic formulation is widely used for the strap-on boosters, including the two SRBs for NASA SLS, the Space Shuttle SRBs, the twin boosters for Ariane 5, and many upper stages. The propellant for the SLS boosters is PBAN with 86% solids (TP-H1148 VIII), while for the Launch Abort System rockets the propellant is HTPB Polymer with 6% aluminum (TP-H1264).

Did you know? Delivering approximately 16 MN or 3.6 million lbf at lift-off, the NASA SLS five segment booster is the largest, most powerful solid propellant booster ever built. The twin SRBs provide more than 75% of the total SLS thrust at launch and for the first two minutes of flight. The major physical difference between the Space Shuttle and SLS boosters is the addition of a fifth solid propellant segment to the four-segment Shuttle SRB, allowing NASA's new launcher to lift more payload. Additionally, the SLS booster is optimized for a single use, while the Shuttle's SRB was designed to be reused. The SLS booster incorporates improvements in design, process, and testing, which results in greater launch capability, and better performance, safety, and affordability.

The combustion temperature of PBAN without aluminum is about 3000 K with 90% of ammonium perchlorate. By adding the aluminum, the temperature is increased by several hundred degrees (to 3600 K for the boosters strapped to Ariane 5), reducing the oxi-dizer concentration accordingly. Additional compounds are sometimes included, such as a catalyst to help increase the burning rate, or other agents to make the powder easier to manufacture. The final product is rubberlike substance with the consistency of a hard rubber eraser. Ultimately, the objective of adding specific compounds to the basic solid propellant is to produce a formulation that has an average molecular mass of about 25, which results in the highest characteristic velocity (see Eq. 2.52).

In recent years, much research has devoted to developing nanopropellants. These are propellants with nano-sized particle additives incorporated to enhance the combustion behavior and increase specific impulse I_{sp}. For example, nano-sized particles of hafnium or aluminum added to a propellant can enhance its burning rate. In experimental studies (e.g. Yaman et al. [35], the burning rate of the double base propellant without aluminum (DB-1) was compared with other double base fuels in which aluminum was added by 2% (DB-2) and 4% (DB-3). They found increase burning rates and burning heat of new fuels manufactured by adding aluminum to the content of the standard double base fuel (DB-1).

A SRM is often designed with inhibitors, layers of a material that restrict or control burning on the propellant surface. Inhibitors are often located in and around the grain to control overall burn rate. Chunks of inhibitor or solid propellant may be released from the grain surface and react while convecting through the nozzle. These particles can drasti-cally affect the performance of the motor. At the same time, thrust vectoring capability can be added to direct the thrust in a direction other than parallel to the vehicle's longitudinal axis.

A propellant grain is characterized by its surface burning rate and geometric shape. Hence, the most critical aspect of SRM design depends on knowing the burning rate as the thrust performance over time depends on this parameter. However, burning phenom-ena are not well understood and it takes considerable amount of testing for the burning characteristics of a new propellant grain to be adequately measured, and this requires testing of full-scale SRMs. As shown below, burning rate models rely on empirical data.

5.3   Solid Grain Geometry

The geometry of the grain is designed according to ballistic requirements and mission; the design is also influenced by processing, manufacturing, and economic factors. The magnitudes of thrust, chamber pressure, and duration of burning needed to accomplish the intended mission are specified to select the best configuration. As shown in Fig. 5.2, a variety of initial grain geometry shapes of a solid propellant are commonly employed. Internal-burning grain is most widely used, least expensive and permits a variety of port shapes and sizes.

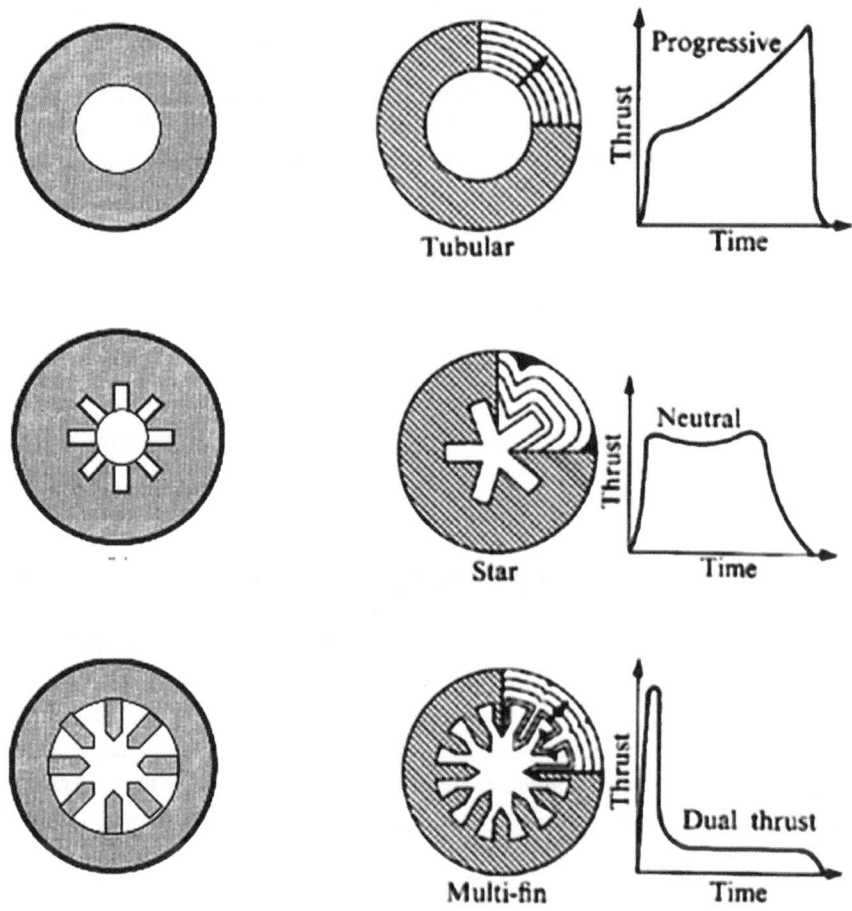

Fig. 5.2 Propellant grain geometry and variation of the thrust profile. Adapted from Ward [33]

Internal Burning Tube. This is the simplest progressive burning type, where the circumference of the circular cross section increases linearly with time, just as the area of the burning surface. With internal burning, there is a linear increase in mass flow rate and thus the thrust developed by this motor increases progressively. High thrust SRM propulsion systems requiring large burning area (A_b) are usually of this type.

Star Shape Neutral Burning: This configuration produces quasi-constant thrust because the initial burning area A_b is rather large and remains approximately constant. The large A_b results from the convolutions of the shape, since, as the star points burn away, the loss of burning area is compensated by the increasing area of the cylindrical part. Many SMR grains are shaped for neutral burning. The propellant of the SLS SRBs is shaped as a 11-point star with a double taper cylinder; the star shape propellant grain pattern was chosen to optimize burning surface area and meet the requirements of the heavy lift SLV.

Multi-fin Dual Thrust. This grain burning configuration is conceived to tailor the thrust profile, as the narrow fins of solid propellant initially produce a very high surface area, and therefore the initial thrust is very high. Once the fins burn away, a slower burning in the cylinder results in a lower thrust. This grain design can be appropriate for applications when a strong acceleration followed by sustained flight is required.

5.4   Burning Rate and Chamber Pressure

In solid propellant propulsion, precise control of the exposed grain surface area and the burning rate of the propellant mixture is required to control the gas flow exhausting through the nozzle.

5.4.1   Rate of Gaseous Propellant Mass Generated

In a motor, the solid propellant becomes combustion gas right at the grain surface, where the solid propellant regresses in the direction normal to itself in parallel layers (see Fig. 5.3). The rate of regression is known as the burning rate of the propellant. Propellant burning rate at any instant governs the mass flow rate of hot gas generated and flowing out from the motor. Therefore, in the combustion chamber of a SRM, the generation rate of gaseous propellant mass $\dot{m}$ is equal to the rate of solid propellant burned, that is

$$\dot{m} = \rho_b r A_b \qquad (5.1)$$

where ρ_b is the density of solid propellant prior to SRM starting, A_b denotes the instantaneous surface area of the burning surface area of propellant grain at the flame front, and r is the surface recession rate or burning rate (length per unit time). We could account for both the density of the solid propellant and that of the propellant gas and write the mass flow rate as $\dot{m} = r A_b (\rho_s - \rho_g)$, where ρ_s denotes the solid propellant density at ambient temperature, and ρ_g is the propellant gas density. However, here we assume $\rho_g \ll \rho_s$.

Now use Eq. (5.1) to determine by integration the total mass m of effective propellant burned, since r and A_b vary with time and pressure:

$$m = \int \dot{m} dt = \rho_b \int_0^{t_b} r A_b dt \qquad (5.2)$$

where t_b represents burning time.

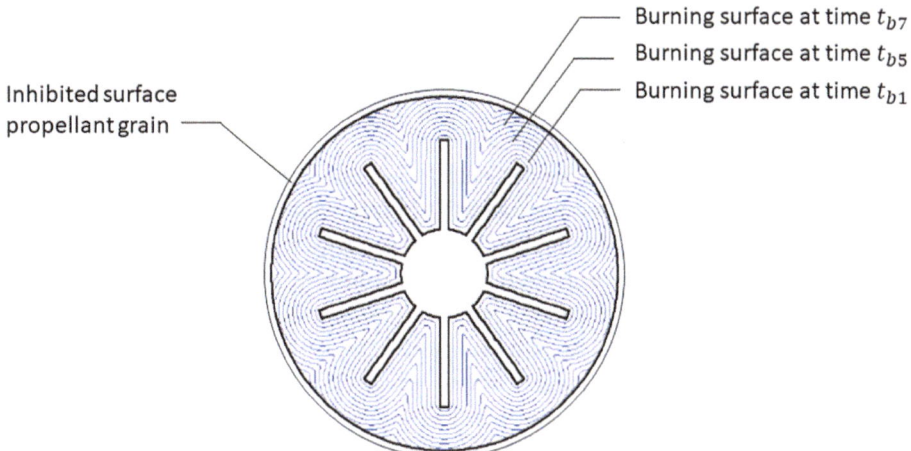

Inhibited surface
propellant grain

Burning surface at time t_{b7}
Burning surface at time t_{b5}
Burning surface at time t_{b1}

Fig. 5.3 Ten slot star-shaped grain SRM indicating the direction of the burning rate r, showing successive burning surface contours, each a fixed small time apart. Adapted from Pons [28]

Together with the burning area A_b, the burning rate r determines the combustion processes, the mass flow rate, and therefore directly controls the pressure and thrust of the motor. Burning rate is defined as the recession of solid propellant surface per unit time or rate of regression (cm/s, mm/s, or in/s). Burning rate is a function of propellant composition and other parameters. As illustrated in Fig. 5.3, the surface of a propellant grain recedes in a direction perpendicular to the surface. Thrust is extremely sensitive to the burning rate of the grain in a SRM. For example, a variation in burning rate of 1% will result in a thrust variation of 1.5 to 2%. Hence, it is important to determine burning rate as accurately as possible.

To date, there is no combustion theory that can predict accurately the burning rates of solid propellants. Therefore, rocket propulsion engineers rely on experimental methods and empirical laws to determine surface recession rates for different propellant compositions operating at different conditions, including initial propellant temperature, chamber pressure, and velocity of the gaseous combustion products over the solid surfaces. Under steady conditions and for a given initial temperature, the Vieille or Saint-Robert law is typically used to describe the linear burning rate dependence on pressure. The empirical equation (known as Saint Robert-Vieille law gives the linear burning rate r (in mm/s or inch/s) as a function of chamber pressure, p_c:

$$r = a p_c^n \tag{5.3}$$

where a is the burning rate coefficient (an empirical constant that depends on propellant grain composition and initial ambient temperature prior to ignition), and n is the burning rate pressure exponent or combustion index, sensitive to the composition of the propellant.

These ballistic parameters depend on the chemical composition and initial temperature of the propellant and are experimentally determined over some limited measurement range. The burning rate r, defined as the regressing distance (perpendicular to the burning surface) per unit of time, is very sensitive to the value of n. Clearly, high values of the exponent n give a rapid change of burning rate with pressure p_c. For stable operation, $0 < n < 1.0$. Brown (1996) provides data for several solid propellants, showing for example, for PBAN/AP/AL a burning rate of 0.55 in/s, with a pressure exponent of 0.33, while for CTPB/AP/AL, the burning rate is 0.45 in/s, with a pressure exponent of 0.40.

Equation (5.3) is accurate over the range of pressures generated during SRM testing. This empirical law assumes that burning proceeds normally to the surface of the propellant grain. The propellant burning rate, in its simplest form, is the rate of conversion of solid to gas in a direction normal to a planar burning surface. However, under certain conditions the burning rate can be affected by the crossflow velocity of the hot combustion gases past the solid burning surface. The thrust of a SRM is dependent on the combustion chamber pressure, which in turn depends on the burning rate. In general, a higher thrust is the result of a high burning rate. However, if an application requires a long burning time, this can be accommodated with a lower burning propellant.

Figure 5.3 shows the burning rate of a star-shaped grain SRM. It shows growth of the internal cavity. The lengths of contour lines are approximately the same, suggesting the burning area is roughly constant. It is clear from Eq. (5.3), burning rate becomes more sensitive to changes in pressure increases as $n \to 1$, and this can cause drop in performance. A change in the burning rate pressure exponent may promote combustion instability.

The variation of burning area (A_b) with time depends on the grain burning rate and its initial geometry. In the design of SRMs, engineers may select charge designs with geometric variations to increase thrust (progressive burning), maintain constant thrust (neutral burning), or decrease thrust (regressive burning) with time (see Fig. 5.2). Figure 5.3 shows a ten-slot star-shaped grain with several internal-burning sections and the subsequent partially burned shapes. The evolution at equally spaced time steps is shown as a representation of how grain geometry evolves during the grain burnback (evolution of the grain propellant burning surface S_b during burning time t_b).

The burning rate of a SRM depends on propellant composition and is affected by chamber conditions (p_c, T_c). Even a small change in p_c produces substantial changes in the amount of hot gas produced. In practice, as $\boldsymbol{n} \to 1$, burning rate and chamber pressure become sensitive to one another and disastrous rises in chamber pressure can occur in a few milliseconds. If n value is low $n \to 0$, burning can become unstable and the propellant grain may even extinguish itself.

An SRM has no control over oxidizer to fuel ratio once the propellant grain is ignited. The values of $\boldsymbol{a}$ and $\boldsymbol{n}$ are determined empirically for a particular propellant formulation, and cannot be theoretically predicted. Experimental data for both double base (DB) and composite modified double-base (AP-CMDB) propellants show that burning rate increases with chamber pressure p_c and with higher concentration of AP. Regression rate increases with decreasing particle size. DB propellant has a combustion index in the range $0.6 < n < 0.8$.

Equation (5.3) is valid only when the gas cross-flow velocity over the propellant surface is low enough so convection effects on the burning rate can be neglected. A different equation must be used if the combustion gases over the propellant surface move at high-velocity, as such situation leads to erosive burning. For a full background on burning rate and solid propellant combustion, readers may consult the wide literature available (see the References).

Moreover, the initial temperature of the grain affects the chamber pressure, and consequently rocket thrust. Burning time decreases with temperature. For in-space thrusters, the temperature must be carefully controlled to limit variation in required performance during its mission time, and to minimize thermal stress in the grain. Electrical heaters, thermostats, and insulation can be used to control the motor temperature.

There are numerical codes developed exclusively to perform grain burnback analysis to predict the performance of SRMs. These codes couple internal ballistic characteristics with different flow models, and have been applied successfully to calculate the grain burnback evolution of 2D and 3D grain configurations (see Pons [28]).

5.4.2 Chamber Pressure

The rocket chamber pressure p_c reaches steady state when the net mass flow of propellant gas being generated equals the mass flow rate of propellant gas leaving the nozzle. Substitute the burning rate r in Eq. (5.1) and use Eq. (2.51) to obtain this mass flow rate equality

$$\dot{m} = \frac{p_c A_t}{C^*} = \rho_b a p_c^n A_b \tag{5.4}$$

where A_t is the nozzle throat area, and C^* denotes the characteristic velocity.

Stability requires that the burning rate pressure exponent $n < 1$. If $n > 1$, increases in p_c above the operating point it will cause propellant gas generation to increase faster than the nozzle flow rate. Such result would lead to a rapid rise in chamber pressure with an accompanying explosion. Solving for p_c from Eq. (5.4), we obtain

$$p_c = \left(a \rho_b C^* \frac{A_b}{A_t} \right)^{1/(1-n)} \tag{5.5}$$

Assuming that the change in mass of the gas in the chamber is negligible, the chamber pressure can also be expressed as

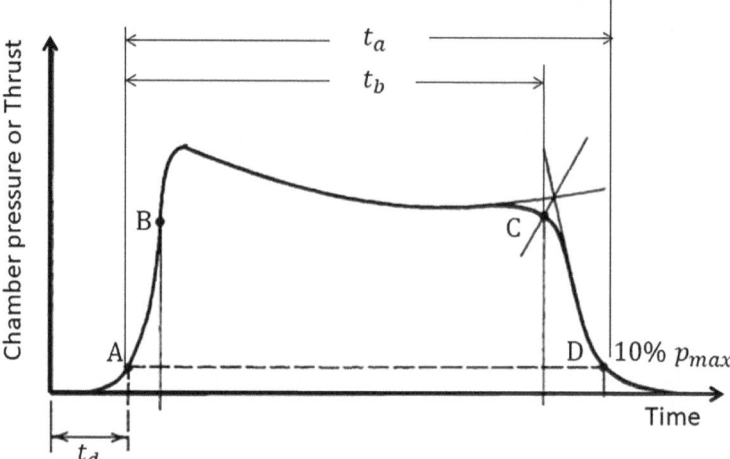

Fig. 5.4 Chamber pressure profile of a SRM and definition of burning time t_b and action time t_a

$$p_c = \left(\frac{A_b}{A_t}\right)^{1/(1-n)} \tag{5.6}$$

If the chamber pressure is known, then use Eq. (5.5) to solve for the area ratio of the SRM.

The main objective of the solid grain designer is to fulfill the requirements of the thrust-time schedule for the mission. A representative profile of experimentally measured chamber pressure is shown in Fig. 5.4, to illustrate how this pressure varies with time after ignition. Burning time t_b is measured starting from 10% thrust after ignition to 90% maximum thrust, at the approximate time when the flame front reaches the inhibitor at the chamber wall. Action time t_a is the time measured starting also at 10% thrust after ignition to 10% thrust, signaling shutdown.

In Fig. 5.4, action time is shown between ignition delay time t_d (point A) and the time corresponding to 10% peak pressure after ignition time (point D). For example, for the Star 48B motor used as upper stage for spacecraft orbit injection, delivering a "burn time average thrust" of 15,430 lbf, with a "burn time average chamber pressure" of 579 psia, the burn time was 84.1 s while the action time was 85.2 s. Burning time and action time are estimated from internal ballistics using standard burning rates, or can be measured on a test stand.

For a given propellant, the surface area versus time profile determines the shape of the thrust-time curve. An example of thrust-time curve is given in Fig. 5.5 to illustrate the performance of the upper stage Star 48B motor. This design (TE-M-711-18) first flew in 1985 from the Space Shuttle Discovery to place the Mexican *Morelos* satellite in

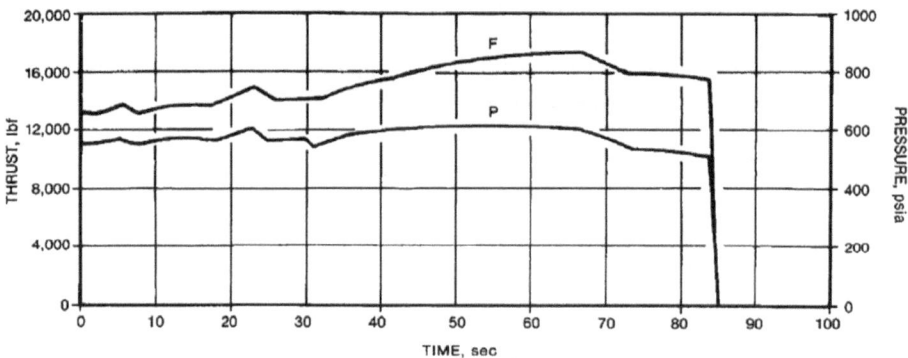

Fig. 5.5 Thrust-time curve for the long nozzle STAR 48B motor (TE-M-711-18). Motor performance: F = 17,490 lbf (max thrust); p_c = 618 psia (max chamber pressure); I = 1,303,700 lbf-sec (total impulse). *Credit* Northrop Grumman

geostationary orbit. The Star 48B uses a high-energy propellant (TP-H-3340) and high-strength titanium case. The submerged nozzle has a carbon-phenolic exit cone and a 3-D carbon-carbon throat.

Example 5.1 Consider the ballistic parameters for two solid propellants as follows: Propellant 1 has $a = 0.04(\text{mm/s})/\text{bar}^n$ and $n = 0.77$; Propellant 2 has $a = 0.055(\text{mm/s})/\text{bar}^n$ and $n = 0.87$. Determine the variation of burning rate for a range of combustion pressures from 10 to 100 atm, and compare these values with the burning rate of a AP-CMDB propellant that has $r = 0.8\text{cm/s}$ at 20 atm, and $r = 1.7\text{cm/s}$ at 100 atm,

Solution: The burning rate as a function of pressure for the two propellants yields the following results.

Pressure (psia)	Propellant 1 burning rate r (mm/s)	Propellant 2 burning rate r (mm/s)
10	0.367	0.720
20	0.685	0.95
50	0.994	1.11
100	1.28	1.26
200	1.56	1.33

Example 5.2 During a SRM test with a new propellant, its regression rate at a chamber pressure of 7 and 17 MPa is found to be 25 and 45 mm/s, respectively. Determine the chamber pressure when the burning rate is 35 mm/s.

Solution: From Eq. (5.3), the burning rate for these two conditions are $0.025 = a(7 \times 10^6)^n$.

and $0.045 = a(17 \times 10^6)^n$, respectively. Therefore,

$$\frac{0.045}{0.025} = \frac{(17 \times 10^6)^n}{(7 \times 10^6)^n} \rightarrow n = \frac{\ln(1.8)}{\ln(2.43)} = 0.662$$

The burning rate coefficient is

$$a = \frac{r}{p_c^n} = \frac{0.025}{(7 \times 10^6)^{0.662}} = 7.35 \times 10^{-7} (\text{mm/s})/\text{bar}^n$$

Therefore, the chamber pressure of the test SRM for a burning rate of 35 mm/s is

$$p_c = \left(\frac{r}{a}\right)^{1/n} = \left(\frac{0.035}{7.35 \times 10^{-7}}\right)^{1/0.662} = 11.57 \times 10^6 \text{ Pa}$$

5.5   SRM Performance

We can estimate the thrust-time curve of a SRM by assuming that specific impulse I_{sp}, grain density, and burning rate are constant, and by calculating instantaneous values of flow rate to obtain the thrust profile:

$$F = \dot{m}I_{sp} = \rho_b r A_b I_{sp} \tag{5.7}$$

This expression requires iteration as the burning surface changes with time as the port diameter of the SRM increases. The central opening in the propellant grain is called the port.

5.5.1   Total Impulse and Effective Exhaust Velocity

In Chap. 2 we defined total impulse with Eq. (2.10). For a rocket motor, we have the total burning time impulse I_b, which is equal to the thrust integral over the motor burning time t_b,

$$I_b = \int_0^{t_b} F dt = \overline{F}_b \cdot t_b \tag{5.8a}$$

where $\overline{F}_b$ denotes the burn time average thrust.

Similarly, the action time impulse I_a is the thrust integral over the action time t_a:

$$I_a = \int_0^{t_a} F dt = \overline{F}_a \cdot t_a \qquad (5.8b)$$

With these quantities known, we can calculate the average thrust based on either burning time impulse or action time impulse, although $\overline{F}_b = I_b/t_b$ is most often used.

For the Star 48 SRM, $\overline{F}_b = 15,000$ lbf. It is important to keep in mind that the maximum thrust is the most critical value for the design of the motor case. This thrust is calculated from internal ballistics for a hot grain propellant using its maximum burning rate.

With the average specific impulse $I_{sp} = I/w_p$, where w_p is the weight of propellant during a test, we can obtain the average effective exhaust velocity

$$C_{avg} = g_0 I_{sp} \qquad (5.9)$$

The average mass flow rate is

$$\dot{m}_{avg} = \frac{m_p}{t_b} = \frac{w_p/g_0}{t_b} \qquad (5.10)$$

With the thrust coefficient C_F (see Eq. 5.14), the net thrust is

$$F = C_F p_c A_t \qquad (5.11)$$

The combustion chamber pressure of an SRM is much higher compared to that in an LRE. Also, an SRM has higher thrust coefficient.

Example 5.3 Consider a SRM with combustion index $n = 0.68$ and burning a propellant of density equal to 1350 kg/m^3. The thrust coefficient is 0.95, and the characteristic velocity is 1850 m/s. When the unburnt propellant temperature is 10 °C and the chamber pressure is 4.5 MPa, the burning rate is 25 mm/s. The temperature sensitivity of the burning rate is 0.006/°C, and the burning surface area to throat area is 65. The throat area of the SRM is 450×10^{-6} m^2. Determine the chamber pressure and the thrust at the unburnt temperature of 50 °C, assuming C_F, and C^* remain constant.

Solution: **Input Data**: $n = 0.68$, $\rho_b = 1350$ kg/m^3, $C_F = 0.95$, $C^* = 1850$ m/s, $p_c = 4.5$ MPa, $r = 25$ mm/s, $A_b/A_t = 65$. The temperature sensitivity of the burning rate is 0.006/°C.

The chamber pressure p_c is given by Eq. (5.5),

$$p_c = \left(a \rho_b C^* \frac{A_b}{A_t} \right)^{1/(1-n)}$$

which requires to know the burning rate coefficient a, which we do not know for the new temperature condition given. Thus, we use the burning rate from Eq. (5.3) to find a from the condition at 4.5 MPa and 10 °C, that is

$$a = \frac{r}{p_c^n} = \frac{25 \times 10^{-3}}{\left(4.5 \times 10^6\right)^{0.68}} = 0.94 \times 10^{-6} (\text{mm/s})/\text{bar}^n$$

Given a temperature sensitivity of the propellant burning rate as 0.006 per degree Celsius, it means that the percentage change is rather small, and thus this value should not affect the remaining analysis. Now we determine the **chamber pressure for the new unburnt temperature** 50 °C condition of the SRM:

$$p_c = \left(0.94 \times 10^{-6} \times 1350 \times 1850 \times 65\right)^{1/(1-0.68)} = 6.7 \times 10^6 \text{ Pa}$$

The **thrust** at this condition is simply

$$F = C_F p_c A_t = 0.95 \times 6.7 \times 10^6 \times 450 \times 10^{-6} = 2864.25 \text{ N}$$

$$F = 2.86 \text{ kN}$$

Note: the burning rate temperature sensitivity is expressed as percent change of burning rate per degree change in propellant temperature at a particular value of chamber pressure p_c.

5.5.2 Specific Impulse as a Function of Theoretical C^* and C_F

The performance of solid propellant propulsion is based on a complex analysis as the SRM is designed to produce a total impulse and prescribed thrust versus time to fulfill its mission. This is accomplished by coupling the propellant grain characteristics with nozzle performance. Here we wish to elucidate the theoretical performance based on a specific impulse formula that can be calculated from theoretical characteristic velocity C^* and thrust coefficient C_F, as these two metrics can be obtained from propellant thermodynamic properties as we did in Chap. 2. First, we express the motor's specific impulse:

$$I_{sp} = \frac{I_a}{w_p} = \frac{C^* C_F}{g_0} \tag{5.12}$$

As we noted before, this expression can be used to optimize the specific impulse of a rocket engine, knowing the thrust coefficient, which is related to the nozzle shape and the external pressure, and the characteristic velocity, which is related to the propellant characteristics. Using a rocket model based on thermodynamic principles, we obtain an expression for the characteristic velocity C^* as a function of the specific heat ratio γ, the gas constant, R, and the chamber temperature T_c, indicating that C^* is a propellant property:

$$C^* = \frac{\sqrt{\gamma R T_c}}{\gamma \sqrt{\left(\frac{2}{\gamma+1}\right)^{(\gamma+1)/(\gamma-1)}}} = \left\{ \gamma \left(\frac{2}{\gamma+1}\right)^{(\gamma+1)/(\gamma-1)} \frac{\mathcal{M}}{\Re T_c} \right\}^{-1/2} \tag{5.13}$$

As shown, the motor's characteristic velocity is proportional to the ratio $T_c/\mathcal{M}$, suggesting that a solid propellant with a lower molecular mass $\mathcal{M}$ will yield a higher value of C^* for the same chamber temperature. For real motors, C^* is about 93% of the theoretical value.

The theoretical thrust coefficient is also obtained from thermodynamic principles, as a function of γ, nozzle area ratio, and pressure ratio:

$$C_F = \sqrt{\frac{2\gamma^2}{\gamma-1}\left(\frac{2}{\gamma+1}\right)^{(\gamma+1)/(\gamma-1)}\left[1-\left(\frac{p_e}{p_c}\right)^{(\gamma-1)/\gamma}\right]} + \left(\frac{p_e-p_a}{p_c}\right)\frac{A_e}{A_t} \tag{5.14}$$

In practice, about 98% of theoretical C_F is achieved in steady-state performance.

Table 5.1 summarizes the theoretical performance of three solid propellants. It includes specific impulse I_{sp} (effective exhaust velocity C) values based on a frozen flow model through a nozzle with optimum nozzle expansion ratio ε, expanding from a chamber pressure of 1000 psia to an exhaust pressure of 14.7 psia. Note the highest specific impulse results with the burning of AP propellant with a higher AP concentration, and lower polymer binder fuel, resulting in a higher molecular mass, and thus higher characteristic velocity C^* compared with the two other propellants.

5.6 Main Components of a Solid Rocket Motor

5.6.1 Igniters

A powerful ignitor is required for SRMs to ensure the entire inner surface of the grain reaches ignition temperature. Ignitors can be of two basic types: pyrotechnic igniters and pyrogen igniters. Pyrotechnic igniters use solid explosives or energetic chemical substances (usually small propellant pellets) to heat the propellant in the motor; the pyrotechnic charge is released directly to the propellant grain. A significant charge of pyrotechnic material is required. For example, the Igniter for the common Ariane 6 and Vega-C SRM boosters consists of a pyrotechnical igniter based on Boron/Potassium Nitrate pellets, and a main igniter with solid propellant. It is manufactured from carbon

Table 5.1 Theoretical performance of solid rocket propellant combinations—conditions for SRM performance: $p_c = 1000$ psia, $p_a = 14.7$ psia [29]. Data used with permission from John Wiley & Sons publisher

Oxidizer	Fuel	Molecular mass $\mathcal{M}$ (kg/mol)	Specific heat ratio γ	Chamber temperature T_c (K)	Characteristic velocity C^* (m/s)	Specific impulse I_{sp} (s)
Ammonium nitrate (AN)	11% binder and 7% additives	20.1	1.26	1282	1209	192
Ammonium perchlorate (AP) 78–66%	18% organic polymer binder and 4–20% aluminum	25.0	1.21	2816	1590	262
Ammonium perchlorate (AP) 84–68%	12% polymer binder and 4–20% aluminum	29.3	1.17	3371	1577	266

fiber composite and aluminum; the carbon fiber reinforced Main Case can withstand the thermal loads inside the motor, and keep its shape after the motor functioning phase.

Pyrogen igniters are small rocket motors required to ignite much larger SRMs. The exhaust gases due to rapid combustion of the fast-burning grain from the small motor initiate combustion on the surface of the main rocket engine. Pyrogen igniters are used for large SRMs such as the NASA's SLS SRBs. The pyrogen igniter consists of initiator and booster charge, with internal and external ignition grain, casing, nozzle, head end, and nozzle end. The metallic head end acts as an interface for initiator, pressure tapping adaptor, and so on. Insulation is required to protect head end and other parts of igniter casing and nozzle end that are exposed during firing. Igniter nozzles are convergent/orifice type. The pyrogen igniter exit is designed with multiple holes for ejecting the hot gas and spreading it over the large grain of the main SRM. Ignition of large SRMs is mainly controlled by heat transfer due to convection of hot gases ejected from the pyrogen igniter.

5.6.2 Cases and Nozzles

The SRM case is a pressure vessel that must be designed and manufactured with high-strength materials capable of sustaining enormous structural and thermal loads. The case is commonly made of high-strength metal alloys such as titanium, stainless steel, or composite materials. The powerful strap-on boosters developed for the Delta SLVs were the Graphite Epoxy Motor (GEM) series. GEM motors are manufactured with carbon-fibre-reinforced polymer casings to contain the HTPB-bound ammonium perchlorate composite propellant. For example, the GEM 46 strap-on booster designed for the Delta III was made of IM7/55A graphite composite, 3-D Carbon-Carbon (C-C) throat and carbon phenolic insulators (IM7/55A stands for filament-wound from high strength, high stiffness IM7 carbon fibers and HBRF 55A resin).

The reusable solid rocket motors (RSRM), SLS SRBs, and the Orion LAS ACM were designed with nozzle housing made of D6AC steel. This is a very high strength, medium carbon low alloy steel with excellent deep hardening characteristics which lend itself well to larger components. D6AC is also widely used in other aerospace components requiring a good combination of toughness and strength.

There are different types of exhaust nozzles for SRMs: (a) fixed geometry; (b) movable; (c) submerged; (d) extendible exit cone; and (e) blast-tube mounted. The movable nozzle has a flexible joint to allow controlled deflection of the thrust axis (thrust vector control TVC). The extendible nozzle allows a large expansion at high altitudes. The blast-tube is typically used in tactical missiles for balancing the center of gravity.

Submerged nozzles have the entry, the throat, and part of the expansion section submerged or cantilevered into the motor combustion chamber. Its design is more challenging, as a submerged nozzle is directly exposed to hot gases, particulate exhaust, and external pressure forces. Although submergence is predicted to yield a reduction in specific impulse (by about 1%), submerged nozzles are attractive and are utilized in many SRM applications due to their effective thrust vectoring characteristics. For example, the Northrop Grumman 8-feet long CASTOR 30XL Upper Stage motor has a submerged nozzle with a high-performance expansion ratio ε (55.9:1) and a dual density exit cone well suited for high altitude operation (Fig. 5.6). Nominally designed as an upper stage, the 119,900 lbf thrust CASTOR 30XL can operate as a second or third stage, depending on launch vehicle configuration. It features an electro-mechanical thrust vector actuation system.

The nozzle is the hottest part of a motor, especially at the throat, where the hot combustion gases cause erosion and ablation. Throat erosion affects motor performance, including variation of thrust. For submerged nozzles, the inlet part becomes enlarged. Hence, throat inserts made of refractory composite materials are incorporated. The strap-on booster GEM 46 has a carbon-carbon (C-C) throat insert.

Cooling is a challenge for SRM nozzles. Since the chamber temperature is much higher than the melting point of most metals, the combustion gas cannot be allowed to contact the walls of the motor. That is why the solid propellant (which acts as insulator) is bond to the case and other parts of the surfaces are protected with an insulating layer (case bonding). For motors used as strap-on boosters, the nozzle is made of D6AC steel alloy with a thermally insulating liner to ensure the structure remains intact until burn-out.

Fig. 5.6 CASTOR 30XL upper stage motor, 92 inches in diameter, 236 inches in length and weighs ~58,000 lbm. The submerged nozzle is eight feet long, with a high-performance expansion ratio (56:1) and a dual density exit cone well suited for high altitude operation. *Credit* Northrop Grumman

5.6.3 Thrust Vector Control and Termination

Thrust vector control (TVC) refers to changing the net thrust vector through small angles, an important in-flight operation for vehicle stability and maneuvering. TVC is crucial for the solid boosters of all launch vehicles (SLVs). This is because the boosters thrust governs the overall thrust at lift off and during the first few minutes as the SLV ascends. During the ascent trajectory, course corrections require the booster thrust to be carefully directed.

There are several TVC methods. The most common methods incorporate either movable nozzles or use fluid injection in the nozzle expansion section. The first approach is used by large SRBs, which incorporate a movable nozzle mounted on a gimballed flexible bearing, allowing it to traverse in two orthogonal directions. Large forces are required to move the nozzle quickly, contriving the motion using a hydraulic mechanism. For example, the SRB of the Space Shuttle used a nozzle that incorporated gimbal movement.

The fluid-injection TVC injects a liquid through the wall of the expansion section, and the jet causes an oblique shock in the exhaust stream, thus deflecting the thrust vector. This approach is not adequate for many SRMs as the injection jet cannot produce sufficient transverse thrust to control a large spacecraft.

It is also possible to use an auxiliary propulsion system as a means of TVC when needing to stabilize a vehicle in flight. For example, the Magellan spacecraft used its monopropellant attitude-control system to provide three-axis control with or without the solid motor firing. For orbital injection, TVC may not be needed since the burn is too short for a spacecraft to need course changing.

A SRM may require a thrust termination mechanism since the delivered impulse cannot be predicted exactly. As stated before, the motor impulse depends on the temperature of the grain, the actual propellant weight, the exact composition, and the weight of inert parts consumed. Most importantly, having thrust termination allows a spacecraft motor to measure velocity gained and shutdown when the desired velocity is reached, and the impulse uncertainty is reduced. There are several thrust termination methods. By suddenly reducing chamber pressure below a certain limit may lead to thrust termination. In some applications, thrust termination may not be required if the SRM is designed to burn out before the desired impulse is achieved. Any thrust shortfall can be made up by liquid propellant thrusters that can be readily shut-down on command (see Brown 1996).

5.7   Selected Applications of SRMs

Solid propellant rocket propulsion is used whenever high reliability, fast operational readiness, and simple storage are required. Propellant grain can be stored in the combustion chamber for a long time, up to 15 years. SRMs are used to boost SLVs, and as propulsion for ballistic missiles, sounding rockets, assisted take-off, air-launched missiles, etc. When very high thrust is required for a short period of time, SRMs are preferred over LREs. Many SLVs utilize solid propellant rockets as strap-on thrust boosters at lift off. In addition, small SRMs can also power the final stage of a launch vehicle, the crew modules, or attach to payloads to boost them to higher orbits. Medium size SRMs such as the Payload Assist Module and the Inertial Upper Stage (IUS) provide the added boost to place satellites into geosynchronous orbit or on planetary trajectories.

Figure 5.7 provides a graphical comparison of the relative sizes of Northrop Grumman's large motors series (Orion, CASTOR, GEM, and Heavy-Lift Boosters). These SRMs span a wide range of size and boost capability, with thrust ranging from approximately 2000 lbf (8896 N) up to 1.6 million lbf (7.2 MN).

The tallest and more powerful is the vectorable, 5-segment SRB specifically for the NASA SLS. It has a total mass of 1,616,123 lbm, with a burn time of 132.8 s, and delivers a total impulse of 298,000,000 lbf-sec (1325.57 N-s). The smallest SRM in Fig. 5.7 is the Orion 38, which was developed as a low-cost, high-performance third stage for the Pegasus launch vehicle. It also functions as the standard third-stage motor for other launch vehicles such as the Pegasus XL, Taurus, Taurus XL, Taurus Lite, and Minotaur-C launch vehicles, and as the fourth stage of Minotaur-I and Minotaur IV vehicles. It has a total

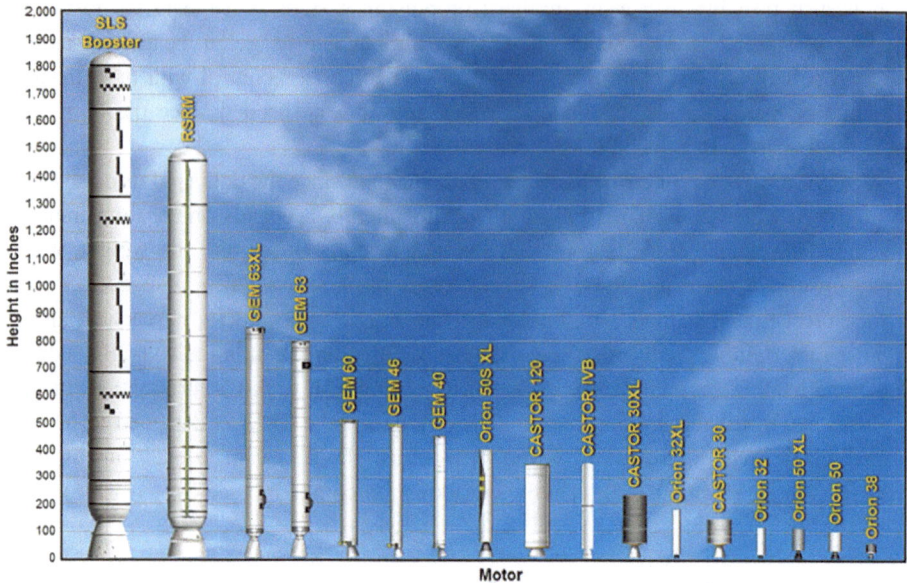

Fig. 5.7 Graphic comparison of the relative sizes of Northrop Grumman's large motors series (Orion, CASTOR, GEM, and Heavy-Lift Boosters). These SRMs span a wide range of size and boost capability, with thrust ranging from approximately 2000 lbf (8896 N) up to 1.6 million lbf (7.12 MN). *Credit* Northrop Grumman

mass of 1924 lbm, with a burn time of 66.8 s, and delivers a total impulse of 491,140 lbf-sec (2.185 MN-s).

The total impulse of 26 representative large current SRMs as a function of motor total mass is shown in Fig. 5.8. These are Northrop Grumman's large motor series (Orion, CASTOR, GEM, and Heavy-Lift Boosters). The motor with the highest impulse to weight ratio in this class is the CASTOR 30XL, a state-of-the-art upper stage motor, with $I/m = 277.8364$ lbf·sec/lbm. It has a burn time of 155 s and delivers a burn time average thrust of 104,350 lbf (464.2 kN).

How massive are SRMs? The total mass can be expressed as

$$m = \frac{I}{\eta I_{sp}} \tag{5.15}$$

where η denotes the propellant mass fraction, m_p/m_i. In the SRM community, the mass of the motor is given as weight, and it is understood to be the mass multiplied by g_0, the standard acceleration of gravity at the surface of the Earth ($g_0 = 9.80665$ m/s^2).

In the following sections we review important applications of solid propellant propulsion.

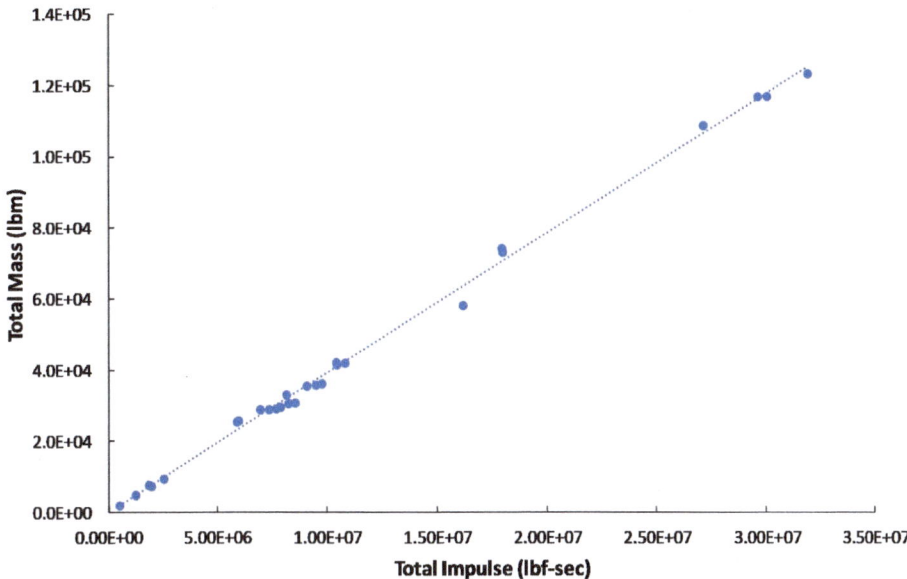

Fig. 5.8   Total impulse versus motor mass for large Orion, CASTOR, GEM, and heavy-lift boosters. Data from Northrop Grumman Catalog (2016)

5.7.1   Strap-On Booster Stage for Launch Vehicles

Large solid propellant rockets are used extensively to boost a launch vehicle, providing additional thrust during lift off. Small solid rocket motors can be used to power the final stage of a launch vehicle, or attach to payload elements to boost satellites and spacecraft to higher orbits. Solid fuel rocket motors have a typical specific impulse that is lower than that of bipropellant liquid rocket engines. However, they provide the high thrust to weight ratio required at lift-off from Earth. Launch vehicles such as the Titan, Delta, Vulcan, and Ariane 5 launchers use strap-on solid propellant rockets to provide added boost thrust. The former Space Shuttle used two solid rocket boosters (SRBs), which were recovered and reused. Table 5.2 provides technical characteristics of selected strap-on boosters for SLVs.

ESA's powerful solid rocket booster is a monolithic design, meaning that the motor is manufactured and cast as a single piece. In addition to the two liquid propulsion modules (LLPM and ULPM), ESA's Ariane 6 launch vehicle incorporates two or four P120C solid propellant rockets, known as Equipped Solid Rocket (ESR), the number depending on the configuration of the launch vehicle: Ariane 62 or Ariane 64. The P120C motor will power both Ariane 6 and Vega-C vehicles.

The P120C motor is ESA's new solid propulsion launch pillar. The P120C will burn for about 130 s using 142 t of propellant (HTPB 1912) to deliver a liftoff thrust of about 4500

Table 5.2 Performance characteristics of selected solid rocket boosters for SLVs

SLV	SRM common name	Propellant	Thrust (MN) [sl]	$I_{sp}(s)$ [sl]	Total impulse I (MN-s)	(lbf-s)	p_c(MPa)	Burn time, t_b (s)
SST	4 segment booster	PBAN	14.685	267.3	1321	297,001,731	4.27	124 (122.2)
SLS	5 segment booster	PBAN	16.01	269	1326	298,000,000	3.94	126 (132.8)
Atlas V	GEM 63 monolithic	HTPB	1.688	279.3	120.6	27,110,000	8.96	97.6
Vulcan	GEM 63XL monolithic	HTPB	463,249 lbf	280.3	131.5	29,570,000	8.96	87.3
Ariane 5	3 segment EAP	PBAN (18% Al)	7.0	274.5			6.134	130
Ariane 6	P120C monolithic	Al-HTPB 1912	4.65	278.5	380		11	132.8

kN (1 million lbf), and a burn time of 135.7 s. The solid rocket booster was developed by Europropulsion, which is owned jointly by Avio in Italy and Ariane Group in France. The P120C is 13.5 m long and 3.4 m in diameter. Avio made the carbon composite material for the 25 cm-thick monolithic casing. The Al-HTPB 1912 propellant is a blend of 69% ammonium perchlorate (AP) with 12% of hydroxyl terminated polybutadiene binder, and 19% aluminum powder.

The NASA Space Launch System (SLS) incorporates two five-segment solid rocket boosters (Fig. 5.9). Producing 16 MN thrust each, this is the largest, most powerful solid propellant booster ever built for a launch vehicle. Together, the SLS twin boosters provide more than 75% of the total SLS thrust at launch. The physical difference between the SST SRBs and SLS boosters is the addition of a fifth propellant segment, giving added thrust (from 14.6 MN to 16 MN), to allow the SLS to lift more payload than the former SST Shuttle. Having 25% more propellant, with a modified grain design, each SRB contains 625,000 kg (1.6 million lbm) of polybutadiene acrylonitrile (PBAN), designation TP-H1148 VIII. Each SRB is designed to yield a specific impulse of 267.3 s at sea level, with an average chamber pressure is 572 psia. The nozzle has an exit diameter of 149.6 in, and an expansion ratio (avg) of 7.72.

Standing 177 ft tall and burning approximately six tons of propellant every second, the twin boosters provide more than 81% of total SLS thrust at launch. The two SRBs burn up to altitude of ~ 45 km. Separation is initiated when chamber pressure p_c reaches ~ 340 kPa. After separation, SRBs continue to rise to ~ 67 km altitude and then fall. Parachutes slow down the spent structure to prevent damage on ocean impact.

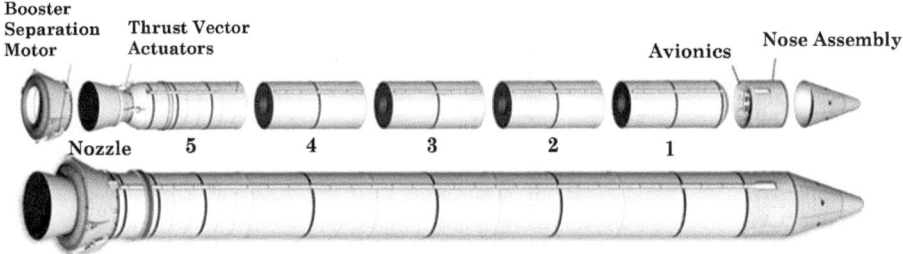

Fig. 5.9 The five-segment solid rocket booster (SRB) for NASA's Space Launch System (SLS) with thrust vector control (TVC). The SRB has new upgrades and parts, including a new case insulation-liner configuration, new exhaust nozzle design, new asbestos-free insulation and liner configuration, and new avionics systems to control flight. *Credit* NASA

The NASA SLS also includes the booster separation motors (BSMs), and the Launch Abort System's (LAS) launch abort motor (LAM) and attitude control motor. The BSMs were originally designed to thrust the spent RSRMs safely away from the Space Shuttle orbiter. These same motors are a critical part of NASA's SLS, which utilize four BSMs installed in the forward frustum of each five-segment booster, and four are installed in the aft skirt, for a total of 16 BSMs per SLS launch.

An integral part of the LAS is the launch abort motor (LAM). The LAS is designed to safely pull the Orion crew module away from the SLS launch vehicle in the event of an emergency on the launch pad or during ascent. Table 5.3 summarizes the performance characteristics for each of the SRMs designed specifically for the NASA SLS.

Another powerful monolithic SRB is the GEM 63XL developed by Northrop Grumman. It made its debut on the inaugural launch of United Launch Alliance's (ULA) Vulcan Centaur vehicle on 8 January 2024. The GEM 63XL (63-inch diameter, extended length Graphite Epoxy Motor) is a new low-cost, robust, state-of-the-art strap-on booster designed for use on ULA's heavy lift Vulcan launch vehicle. With an overall length of 865.3 inches (72 feet), the GEM 63XL is the longest monolithic SRB ever produced, a designation previously held by its predecessor, the 66-foot long, 63-inch diameter GEM 63. It holds 105,497 lbm of QDL-4 propellant (HTPB Polymer, with 19% Al), burning for 88.31 s.

Northrop Grumman states that manufacturing and casting the GEM 63XL as a single piece makes the motor more reliable and efficient by reducing joints, and overall hardware mass. This is extremely important for large SRBs that can deliver over 463,000 pounds of thrust to enhance the capabilities of rockets launching critical payloads.

Table 5.3 Solid rocket motors for NASA space launch system (SLS)

SRM	Nozzle	Diameter (m)	Overall length (m)	Propellant mass (kg)	Total mass (kg)	Total Impulse (N-s)	Burn time (s)	Chamber pressure (MPa)	Burn time average thrust (MN)
SLS booster (5-segment)	Vectorable	3.71	47.36	647,642.36	733,061	1.324×10^9	132.8	3.944	9.9996
BSM	Fixed	0.327	0.7899	34.92	75.75	81.847	0.68		0.098
LAM	Vectorable	0.932	5.6819	2154.56	3460.46	4.655×10^6	4.3		1.16

5.7.2 Upper Stage, Apogee Kick Motor and Other In-Space Propulsion

An upper stage is the launch propulsion system attached to the payload, the last piece of the SLV that serves to inject the payload into its orbit. An apogee kick motor (AKM) is a rocket motor that is regularly employed on spacecraft to provide the final impulse to change the trajectory from the transfer orbit into its final orbit (most commonly circular). The rocket firing is done at the highest point of the transfer orbit, known as the apogee. Table 5.4 summarizes the performance of representative kick motors.

The former three-stage Titan IIIE with the Centaur upper stage (also known as the Titan III-Centaur), had an option to use a Star-37E SRM as additional upper stage. Because of the energy required to achieve a Jupiter ballistic trajectory with an 825 kg (1819 lbm) payload, the twin Voyager spacecraft launched by the Titan III E/Centaur included a Star 37E rocket motor (TE-M-364-4) as the final propulsive stage required to add the final injection velocity increment of about 2 km/s (4475 mph). With a nominal total impulse capability of 2910.03 kN-s (654,200 lbf-s), the solid-rocket motor (part of the spacecraft propulsion module for launch, Fig. 5.10) was ignited 15 s after the Voyager separated from the Centaur, burning for about 43 s. Figure 5.10 depicts the stack (Voyager plus Star 37E motor) fixed to the Centaur stage (powered by two LOX/LH$_2$ RL10 engines).

Table 5.4 Representative apogee kick motors

Spacecraft	SRM (apogee or injection stage)	Max. Thrust, kN (lbf)	Max. Chamber pressure, MPa (psia)	Total impulse kN-s (lbf-sec)	Burn time/ action time (s)
Pioneer 10 and 11 (1972, 1973)	Star 37E (TE-M-364-4) 3rd stage	68.00 (15,287)	~4.482 (~650) (estimated)	2910.03 (654,200)	43.6
Voyager 1 and 2 SLV: Titan 3E/ Centaur-D1T (1977)	Star 37E (TE-M-364-4) 4th stage	68.00 (15,287)	~4.482 (~650) (estimated)	2910.03 (654,200)	43.6
IBEX SLV: Pegasus XL (2008)	Star 27H (TE-M-1157) 4th stage	23.35 (5250)	4.364 (633)	975.027 (219,195)	46.3/47.3
New Horizons SLV: Atlas V-551 (2006)	Star 48B (TE-M-711-17) 3rd stage	76.109 (17,110) (short) 77.799 (17,490) (long)	4.261 (618)	5674.77 (1,275,740) (short) 5799.14 (1,303,700) (long)	84.1/85.2

Fig. 5.10 Voyager with its
Star 37E SRM sitting atop of
the Centaur upper stage. *Credit*
Richard Kruse, HistoricSpac
ecraft.com

Voyager
Star 37E

Centaur
Upper stage

Approximately 11 min after solid-rocket burnout, the propulsion module was jettisoned, and the Voyager began their interplanetary mission.

A similar launch configuration was used for several interplanetary missions between 1974 and 1977, including the launch of the two Pioneer spacecraft. For those historical missions, Star-37E stages were considered part of the payload instead of part of the launch vehicle. The Star-37E (max thrust 12,325 lbf) belongs to the Star family of American SRMs originally developed by Thiokol (now Northrop Grumman) and used by many space propulsion and launch vehicle stages. Star motors are used almost exclusively as an upper stage, often as an apogee kick motor. For example, the Star 37XFP (9550 lbf thrust) first flew from the Space Shuttle as an apogee kick motor for SATCOM in 1985 and has also been launched from Ariane and Delta launch vehicles.

> *Did you know?* The Star 37 and Star 48 SRMs served to inject the Voyager and Pioneer spacecraft on Solar System escape trajectories, at velocities fast enough to leave the Sun's orbit and travel out into interstellar space.

While developing the Space Shuttle, NASA considered the addition of an upper stage that could be used on the Orbiter to deliver payloads from LEO to higher energy orbits (GTO or GEO) or to escape velocity for planetary spacecraft. Three stages were proposed: the LOX/LH$_2$ Centaur, the Tran-stage (fueled by hypergolic storable propellants Aerozine-50 and dinitrogen tetroxide), and the Inertial Upper Stage (IUS) burning solid propellants. The IUS was developed by Boeing for the United States Air Force beginning in 1976 for lifting payloads from LEO to higher orbits, or for injecting spacecraft into interplanetary trajectories following launch aboard a Titan 34D or Titan IV vehicles as its upper stage, and it could also be adapted to launch from the payload bay of the Space Shuttle Orbiter. The IUS had a specific impulse of 296 s.

The IUS was used on important space probes launched from Titan IV and the Space Shuttle. When the Space Shuttle Columbia carried Chandra X-ray Observatory in 1999, the IUS transferred Chandra from a 300-km altitude circular LEO to an elliptical orbit with an apogee altitude of 13,200 km, which required a velocity increment Δv of 1.7089 km/s.

NASA's New Horizons, the first and, so far, the only mission to Pluto, was launched in 2006 by the two-stage Atlas V (551) AV-010 vehicle, utilizing a Star 48B motor as 3rd stage to boost the high launch energy required to send the spacecraft to the farthest regions of our Solar System. Figure 5.11 depicts the New Horizons spacecraft and the Star 48B motor.

The Centaur second stage of Atlas V lifted New Horizons attached to the Star 48B injection stage to 800 km altitude, reaching a velocity of 12.4 km/s. The Star 48B ignited, taking the New Horizons spacecraft to its escape velocity. After the kick motor completed its burn, the spacecraft separated. According to the New Horizons press kit: "After it separates from the third stage, New Horizons will speed from Earth at about 16 km per

Fig. 5.11 **a** A depiction of New Horizons attached to its STAR 48B upper stage. *Image credit* McNutt et al. (2019). **b** The long nozzle STAR 48B with 17,490 lbf maximum thrust, and 294.2 lbf-sec/lbm propellant specific impulse. *Credit* Northrop Grumman

second, or 36,000 miles per hour – the fastest spacecraft ever launched." The spacecraft continued its long hyperbolic trajectory, from then on depending on its own maneuvering propulsion system—four small 4.4N hydrazine thrusters—which are used to adjust or correct the challenging trajectory. New Horizons mission included a flyby study of the Pluto system (in 2015), and a flyby to study Kuiper belt objects. As of July 2024, New Horizons is traveling through the Kuiper belt—it is 58.3 AU (8.72 billion km; 5.42 billion mi) from Earth, and 59.3 AU (8.87 billion km; 5.51 billion mi) from the Sun. NASA expects to extend the mission until New Horizons exits the Kuiper belt, projected to occur between 2028 and 2029.

One of the smallest kick motors is Star 27 used for orbit and satellite maneuvers. With a maximum thrust of 6340 lbf, the Star 27H was developed as the apogee kick motor for NASA's Interstellar Boundary Explorer (IBEX) mission. IBEX is a small satellite (80 kg, 176 lbm) devoted to observing the outer edge of our Solar System. In 2008, the IBEX satellite was carried into LEO by the Pegasus XL, an air-launched, three-solid propellant stage launch vehicle. The IBEX was initially placed into a highly-elliptical transfer orbit with a low perigee, and then at apogee it used the Star 27H motor as its final boost stage to raise its perigee greatly and to achieve its desired high-altitude geocentric elliptical orbit.

Small sized solid propellant rockets are also used in space applications that require tiny thrust and small impulses. For example, the Star 3 motor (Fig. 5.12) is 11.36 inches long and has a diameter of 3.18 inches. It provides 461 lbf maximum thrust and delivers a total impulse of 281.4 lbf-sec. The Star 3 was developed as the transverse impulse rocket system (TIRS) for the NASA Mars Exploration Rover (MER) program led by the Jet Propulsion Laboratory (JPL). The MER was a robotic space mission involving Spirit and Opportunity, two rovers sent to explore the Red Planet. Three TIRS motors were carried on each of the MER landers. One of the TIRS motors was fired in January 2004 to provide the impulse necessary to reduce lateral velocity of the MER Spirit lander prior to landing on the Martian surface.

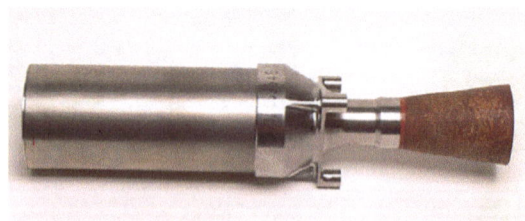

Fig. 5.12 STAR 3 (TE-M-1082-1) used in the transverse impulse rocket system (TIRS) for the Mars exploration rover (MER) program. $I_{sp} = 266.0$ s, $F_{\max} = 461$ lbf. *Credit* Northrop Grumman

5.7.3   Sounding Rockets

Sounding rockets are research space launch vehicles propelled by one or several solid rocket motor stages, carrying a payload to high altitudes over the upper layers of Earth's atmosphere. They are used for space research and for collecting weather data. A sounding rocket can have one to three stages, with the payload attached to the upper stage.

Depending on the number of stages it has, a sounding rocket can carry a science payload to altitudes between 48 and 480 km above the Earth. After launch, a sounding rocket follows a parabolic arc, flying past the Karman line, above the atmosphere where air drag does not disturb the instruments taking measurements from that region of space. After reaching its predetermined altitude (apogee), the payload follows a free-fall trajectory back to Earth.

Sounding rockets can be used to study the dynamics of Earth's ionosphere. A NASA mission known as Atmospheric Perturbations around Eclipse Path or APEP, launched three Black Brant IX rockets from Wallops Island, VA, carrying a research payload to study the perturbations of the ionosphere caused by the shadow of the Moon. In a typical APEP trajectory, the vehicle reaches an apogee of 346 km at 303.6 s after launch. At 6.2 s after launch, the first Terrier stage burnout occurs and it separates from the APEP vehicle. The second stage is ignited at 16 s, and the payload (attached to second Black Brant stage) continues its way into the Ionosphere. About 63 s after launch, at an altitude of 87.6 km the payload separates and carries out the measurements as it ascends between 180 and 325 km. After apogee, the APEP payload re-enters the atmosphere, continuing its measurements in the descending trajectory between 325 and 70 km. A parachute deploys to bring the payload safely back to Earth.

5.7.4   SRMs for Hypersonic Aircraft Flight Testing

Solid propellant propulsion is an integral component of hypersonic air-breathing propulsion flight testing. Flight is the only way to demonstrate that an air-breathing propelled hypersonic vehicle can fly on its own scramjet engine power. There are three approaches to hypersonic free flight testing: (a) air-lifted, rocket-boosted; (b) sounding rocket launch; and (c) surface-to-air missile launch.

The air-lifted, rocket-boosted approach to flight testing requires a subsonic aircraft carrier and a rocket launcher. The aircraft is required to lift a stack (the scramjet-powered vehicle attached to the SRM booster) to a pre-determined altitude and take-off speed appropriate for safe flight testing. After the stack is dropped, the booster rocket ignites and thrusts the scramjet-powered vehicle to its altitude and hypersonic take off speed. The most famous mission using this approach was the NASA X-43A Mach 10 Flight Demonstration. The X43A denotes the hypersonic cruiser that was attached to an Orion 50S SRM, which is the first stage of the Pegasus launch vehicle. The 50S booster, with

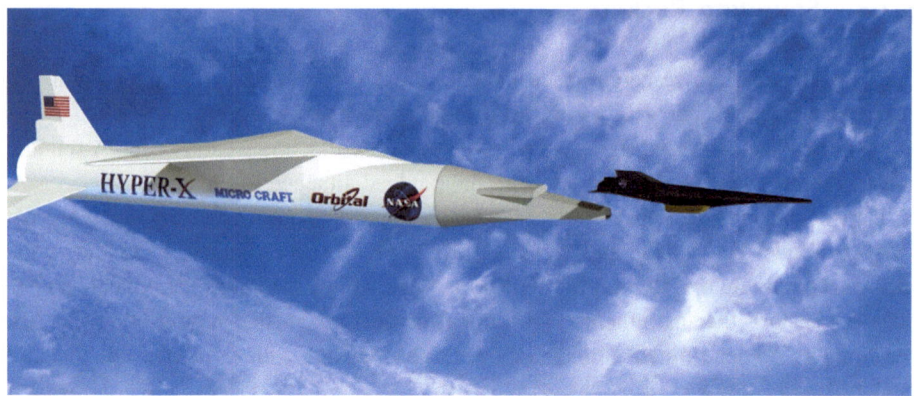

Fig. 5.13 An artist concept of NASA's X-43A vehicle stage separation. The Orion 50S rocket booster carried the research vehicle to the test altitude and speed, and the X-43A has separated from the booster prior to scramjet ignition. *Credit* NASA

some modifications to support scramjet flight-testing, is shown in Fig. 5.13 separating from the X-43A vehicle. The 50.2-inch diameter SRM has a fixed nozzle and is air ignited after a 5-s freefall drop from approximately 40,000 ft. It has a burn time of 74.9 s to 30 psia, a maximum thrust of 126,641 lbf, and delivers a total impulse of 7,873,000 lbf-sec (Ref. Northrop Grumman Product Catalog).

Figure 5.13 illustrates the X-43A HXRV separation event, which occurred at an altitude of about 28.96 km. The artist's rendition shows the Orion 50S rocket after completing its task of carrying the hypersonic X-43A to the test altitude and speed, The two vehicles separated prior to scramjet ignition to demonstrate hypersonic flight (Mach 10) with air-breathing propulsion.

Sounding rockets used to test scramjet propulsion systems such as in HIFiRE Program and Australia's HyShot program. In the second launch of the HyShot (2002), the Terrier-Orion Mk70 rocket boosted the payload (hypersonic vehicle), which remains attached to the second stage Orion motor, toward a planned apogee of 330 km. After peaking at 314 km, the trajectory was designed so that the payload was moving at Mach 7.6 between 35 and 23 km on the way down, at which point the measurements of supersonic combustion were made. This was claimed to be the first successful test of a scramjet in flight. For additional details on hypersonic air-breathing propulsion flight testing, see [23].

Example 5.4 You are required to design a solid rocket motor burning ammonium nitrate-hydrocarbon to meet these specs: average Thrust (sl): 2000 lbf, Chamber pressure: 1000 psia, Burn time: 10 s, Operating temperature: Ambient (~70°F). Determine I_{sp}, A_t, A_e, $\dot{m}_p$, w_p, I, A_b, and estimate loaded gross weight of the SRM. Properties for propellant are: $\gamma = 1.26$;

$T_c = 2700°F = 3160$ R; $r = 0.10$ in/s at 1000 psia; $C^* = 4000$ ft/s; $\rho_b = 0.056$ lbm/in^3; $\mathcal{M} = 22$ lbm/lb · mol; gas constant $R = \frac{1544}{22} = 70.2$ ft · lbf/lbm · R.

Solution: With $\gamma = 1.26$ and optimum expansion at sea level, the nozzle pressure ratio is $1000/14.7 = 68$, and $\varepsilon = A_e/A_t = 7.8$. The Ideal thrust coefficient is $C_{F_i} = 1.57$. This value must be corrected for nozzle losses. Assume a correction of 0.98 so corrected thrust coefficient is

$$C_F = 0.98 \times 1.57 = 1.54$$

$$I_{sp} = \frac{C^* C_F}{g_0} = 191 \text{ s}$$

$$A_t = \frac{F}{p_c C_F} = 1.30 \text{ in}^2$$

$$A_e = \varepsilon A_t = 7.8 \times 1.30 = 10.1 \text{ in}^2$$

With the weight flow rate of propellant

$$\dot{w}_p = \frac{F}{I_{sp}} = 10.47 \text{ lbf/s}$$

The effective propellant weight for a burn of 10 s is ~105 lbf. Allowing for residual propellant and for inefficiencies on thrust buildup, the total loaded propellant weight is assumed to be 4% larger, that is $105 \times 1.04 = 109$ lbf.

The total impulse is

$$I = F_{avg} t_b = 2000 \times 10 = 20,000 \text{ lbf} \cdot \text{s}$$

Since propellant burning surface area is $A_b = 1840$in^2, obtain $A_b/A_t = 1415$.
If A_b were not given, we use Eq. (5.5) to find it

$$p_c = \left(a \rho_b C^* \frac{A_b}{A_t} \right)^{\frac{1}{1-n}}$$

Loaded gross weight of SRM (not the vehicle) can only be estimated *after* a detailed design has been made. However, we can make an approximate guess by choosing a reasonable total impulse to weight ratio.

SRM Performance Summary:

$$F_{avg} = 2000 \text{ lbf}; \quad p_c = 1000 \text{ psia}; \quad t_b = 10 \text{ s}; \quad T_{amb} = 70°F$$

$$I_{sp} = 191 \text{ s}; \quad I = 20,000 \text{ lbf} \cdot \text{s}; \quad C_F = 1.54$$

5.8   Thrust Reduction and Differences in SRM Requirements

It should be clear by now that the performance requirements for solid propellant rocket propulsion are different, depending on whether the SRM is intended to be an initial boost stage for a launch vehicle or as an apogee post boost stage. One main difference is the thrust-to-weight requirement. The strap-on booster must produce high enough thrust to help the fully loaded launch vehicle ascend and accelerate overcoming the strong gravitational force. Of course, apogee motors do not have this requirement as they are already in orbit.

The design requirements for these two types of SRM are also different. For example, the mass/volume allocated for mounting each type of motor is different. For the large strap-on booster, the design constraint has to do with the incurred aerodynamic forces, requiring a slender design to minimize energy-loss that diminish the ideal velocity increment required to ascend to orbit. In the case of SRMs designed as post-boost injection or apogee stage, the mass/volume constraint is dictated by the size and shape of the launch vehicle payload fairing (the nose cone of the vehicle housing the spacecraft). This is because the apogee motor is attached to the spacecraft and both must fit in the allocated space. If the apogee motor is too heavy and/or large, this may reduce the mass/volume available for the spacecraft itself. There are other design and performance requirements for apogee motors, which include ignition operation in low pressure conditions, and restart capability (see Chap. 4).

Before we conclude this discussion, let us ask this important question: Is it possible to reduce the thrust in a solid rocket motor? If so, how? The answer is yes. Thrust can be reduced by changing the grain design, lowering the burning rate, or reducing the nozzle expansion ratio. In addition, thrust can be lowered by reducing the propellant burning rate r, since this reduces the internal chamber pressure (see Eq. 5.5), affecting the level of thrust achieved (Eq. 5.7). The question is then, how to reduce the burning rate of a propellant? The burning rate can easily be lowered by changing the oxidizer modal distribution to include coarser particle sizes.

It is also possible to lower the thrust by reducing the nozzle expansion ratio $\varepsilon = A_b/A_t$. This can be done by a simple truncation of the nozzle exit cone, which in fact reduces the exit area until one obtains the desired thrust level or specific impulse. As deduced from Eq. 5.5, an increase in throat area (A_t) has the same effect except that the burn time and chamber pressure are affected. Basically, by lowering the expansion ratio, the thrust coefficient C_F is reduced (see Eq. 5.14), and this lowers the specific impulse I_{sp}.

5.9 Hybrid Rocket Propulsion

Hybrid propellant rockets are propulsion systems that combine solid and liquid compounds. Typically, the fuel is solid and the oxidizer is liquid. The liquid oxidizer is injected into the solid fuel, which is shaped in such a way that it also serves as the rocket's combustion chamber (Fig. 5.14). A hybrid rocket requires three key components: (1) fast burning paraffin-based fuels shaped with a single circular fuel port geometry, (2) unique internal ballistic design to enable stable and efficient combustion, and (3) advanced carbon composite motor cases. The paraffin-based fuels are non-toxic, non-carcinogenic, non-hazardous, and environmentally friendly.

Hybrid propellant engines have demonstrated high performance at the level of solid propellant rockets. For example, a LOX/paraffin-based hybrid motor could deliver a vacuum specific impulse of 340 s (with a nozzle expansion ratio of 70:1). The hybrid propellant rocket is an attractive propulsion option since the combustion can be moderated, stopped, or even restarted by control of the liquid oxidizer. SpaceShipOne, an experimental air-launched rocket-powered aircraft with sub-orbital spaceflight capability, was powered by hybrid propellant rocket engines, with nitrous oxide as the liquid oxidizer and HTPB as the solid fuel. Hydroxyl-terminated polybutadiene (HTPB) propellant is a hard rubbery material that binds together the fuel and oxidizer. The rocket motors are made of a graphite-epoxy shell that is about 4 to 5 times lighter than metals. One hybrid rocket can generate 88 kN (20,000 lbf) of thrust, and can burn for about 87 s (1.45 min).

Despite the many advantages of hybrid solid–liquid propellant engines, as demonstrated by R&D studies and performance verification in flight, hybrid propulsion has not

Fig. 5.14 Schematic of a hybrid rocket system

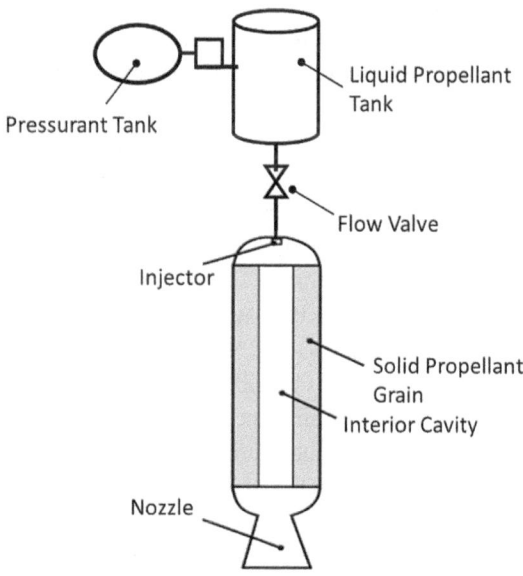

achieved its full potential in space applications to date. For a comprehensive review of hybrid rocket propulsion, the reader may consult [13]. They discuss the technical challenges that limit the use of hybrid propulsion for space applications and evaluate the technologies and approaches necessary to bridge the gaps in hybrid rocket development.

5.10   R&D Related to Solid Propellant Rocket Propulsion

Advances in composite materials and controls have made possible to design and manufacture more powerful strap-on boosters and upper stages for SLVs and more efficient solid rocket motors for spacecraft control. However, improper utilization of new technologies may result in failures (e.g. cracking or complete rupture of the motor case in service), or cause unnecessary weight penalties or high costs. For example, the rupture of one of the nine GEM boosters led to the failure of the Delta II launch on 17 January 1997, which was destroyed thirteen seconds after liftoff. According to reports (ADA 2000), the SRM catastrophically failed because its outer composite layers were damaged through some type of impact to the outside of the case after it was proof tested. Hence, the case was fully pressurized during launch, but it split longitudinally, leading to failure of the Delta II mission.

Efforts to reduce the overall weight of SRMs include new advanced light-weight strong materials (e.g. composites) and improvements in design and manufacturing techniques. In the U.S., Northrop Grumman is leading such efforts. The company is pushing the boundaries of technologies, including improvements in motor design using virtual reality to build the next generation of boosters. Northrop Grumman's GEM 63XL solid rocket booster benefits from flight-proven designs and components, advancing the GEM motors legacy of increasing capabilities. For example, their engineers use virtual reality to immerse into 360-degree virtual environments to observe and test life-size boosters. In minutes, engineers create 3-D computer-aided design (CAD) renderings, and with virtual reality headsets they can thoroughly examine a full-scale model to validate the GEM motor design before manufacturing begins.

To address booster obsolescence issues in design and manufacturing, in 2019 Northrop Gruman began the Booster Obsolescence and Life Extension (BOLE) program to develop new solid rocket boosters for NASA SLS Block 2. These boosters are derived from the composite-casing SRBs that were in development for the canceled OmegA launch vehicle. The goal of BOLE is to increase the payload of the SLS to 130 ton (290,000 lbm) to LEO, and at least 46 ton (101,000 lbm) to trans-lunar injection TLI (Tobias, et al. 2021). In 2024, the company announced it had completed the first BOLE motor segment, expecting to conduct a full-scale static test with all five segments integrated and horizontally fired in a test bay. The BOLE motors add nearly five metric tons of payload capacity for SLS Block 2 configured for Moon and Mars missions above the enhancements already in work for the SLS Block 1B slated to fly Artemis IV. According to Northrop Gruman, the new

SRBs will be used on SLS Block 2 beginning with Artemis IX when all the recovered and refurbished shuttle-era steel cases have been expended.

Research and development efforts continue to increase the current level of understanding of structural strength degradation of SRMs caused by local damage. Efforts to characterize failure modes using nondestructive examination, experimental determination of the failure modes of SRMs, and analytical predictions of strength degradation in SRMs due to damage are ongoing.

Glossary

Action Time Time comprised of the burning time of a solid rocket grain plus the time to burn the sliver; usually taken as the time from 10% thrust at ignition to 10% thrust at burnout.

Attitude control system (ACS) A thruster system used to maintain spacecraft positioning and orientation. Also known as reaction control system (RCS).

Apogee kick motor (AKM) A motor used to circularize the orbit of a spacecraft, often to geosynchronous Earth orbit (GEO). In general, a kick motor is a propulsion system used to accelerate a payload already in space, and is therefore integrated into a mission at the expense of a launch vehicles existing payload capacity.

Binders Complex hydrocarbons that serve as solid fuels and in addition they provide mechanical strength, binding all the grain ingredients.

Burnback A term used in the SRM internal ballistic analysis to model to the evolution of the grain propellant burning surface during burning time.

Burning Rate The linear velocity at which a strand of solid propellant will burn as measured by the velocity of the flame front, usually in inches per second.

Burning Rate Exponent Burning rate, in a solid propellant, increases proportionately with the absolute pressure raised to the burning rate exponent.

Burning Time Time required for vigorous combustion of a solid grain; taken as the time from 10% thrust during ignition to 90% thrust during burnout (also called burn time).

Carboxyl-terminated polybutadiene (TPB) A polymer used as a propellant binder.

Chamber Pressure The total pressure in a solid rocket motor measured in the combustion chamber before the gas enters the converging nozzle.

Characteristic Exhaust Velocity Indicative of performance for a SRM with a given propellant combination.

Composite Propellants A solid rocket propellant, a mixture of oxidizers, fuels, and additives.

Double-Based Propellants A solid propellant composed of more than one substance, each of which acts as oxidizer and fuel.

Effective Exhaust Velocity The rocket gas exhaust velocity that has been adjusted to allow the small pressure area term of a vacuum engine to be neglected.

Effective Propellant Weight The loaded weight less the burnout weight of a solid rocket motor (may be more or less than the propellant weight).

Erosive Burning Burning that erodes the surface of the grain as well as consuming it chemically.

Ethylene propylene diene monomer (EPDM) A class of elastomeric rubber insulation materials typically used to insulate motor cases.

Expansion Ratio Nozzle area ratio, $\varepsilon = A_e/A_t$.

Grain A charge of fuel and oxidizer in a solid-propellant rocket motor, mixed and molded to some desired shape.

Graphite epoxy motor (GEM) Orbital ATK developed GEM designs for the Delta II launch vehicle. Designed to take advantage of proven, off-the shelf technologies, the GEM system provides increased performance and heavier lift capability.

Geosynchronous Earth orbit (GEO) At an altitude of 22,600 miles (36,371 km) over sea level, GEO is an orbital location where satellites remain over a fixed point over the Earth.

Hybrid Propellants Propellant combinations consisting of separate liquid and solid propellants.

Hydroxyl terminated polybutadiene (HTPB) A type of polymer used as a propellant binder that handles higher strain levels inherent to composite motor cases.

Impulse The product of thrust and time. A parameter defined by the area under the thrust-time curve of a rocket motor. Impulse is also the product of total propellant weight and specific impulse.

Kick Motor A solid propellant propulsion system used to accelerate a payload already in space, and is thus integrated into a mission at the expense of an SLV existing payload capacity. The manufacturer Northrop Grumman defines the apogee kick motor (AKM) as "a motor used to circularize the orbit of a spacecraft, often to geosynchronous earth orbit (GEO)." Examples of kick motors include the Star series of solid rockets.

Launch Abort System (LAS) Propulsion system designed to pull the Crew Module (CM) off an exploding or out of control launch vehicle while still on the launch pad and get the CM far enough out to achieve a safe landing under parachutes The LAS has three SRMs: the large abort motor responsible for full abort, the attitude control motor (ACM) that uses multiple pintle values to rapidly actuate and direct steering forces near the nose of the rocket, and the jettison motor that fires just enough to pull the LAS safely away from the CM.

Liner Insulating layer for the motor case serving also to extinguish the flame.

Low Earth orbit (LEO) A position reached by many launch vehicle systems prior to orbital adjustments that are typically made using perigee kick motor (PKM) and apogee kick motor (AKM) propulsion.

Neutral Burning A solid motor grain that produces constant, or quasi-constant, thrust.

Oriole A 22-inch-diameter, high-performance, low-cost rocket motor used as a first, second, or upper stage for sounding rockets, medium-fidelity target vehicles, and other trans atmospheric booster and sled test applications.

Perigee kick motor (PKM) A motor typically used to raise a satellite into elliptical orbit.

Polybutadiene acrylic acid acrylonitrile polymer (PBAN) A binder formulation widely used on large rocket boosters such as the Titan III and the NASA SLS.

Progressive Burning A solid grain that burns and produces thrust increasing with time.

Rocket-assisted deceleration (RAD) Designation for motors used to decelerate payloads such as the Mars RAD motors.

RApid VEctoring Nozzle (RAVEN) A rocket nozzle that incorporates control mechanism for high-angle thrust vector control (TVC) and high nozzle slew rate.

Reaction control system (RCS) An auxiliary rocket propulsion system used to provide trajectory corrections (small Δv additions), and for correcting rotational or attitude positions in almost all spacecraft and all major launch vehicles.

Reusable solid rocket motor (RSRM) Designation used for the SLS solid boosters.

Solid Propellant A rocket propellant combination in which both propellants are solids at normal temperatures.

Space Launch System (SLS) NASA new super heavy-lift rocket launch vehicle designed to transport astronauts to the Moon as part of the Artemis missions. Based on NASA's bold vision for future human spaceflight, the SLS is conceived as an evolvable architecture that will grow more capable through upgrades to the rocket engines, boosters, and upper stage, providing a flexible platform for a variety of human and robotic deep space missions, including trips to Mars.

Solid rocket motor (SRM) A rocket engine where the oxidizer and fuel matter are stored in the combustion chamber as a mechanical mixture in solid form.

Solid strap-on booster (SSB) Solid rocket first stages, which are characterized by high thrust and short burn time, used by space launch vehicles to boost their thrust at lift-off. SSBs ensure that the thrust level far exceeds the weight of the entire launch vehicle stack, including payload.

Trans-Lunar Injection (TLI) Designation for a motor system used to inject a satellite into a lunar orbit. This specific designation applies to the STAR 37FM-based TLI stage used for the Lunar Prospector spacecraft.

Transverse impulse rocket system (TIRS) Designation for motors used to stabilize the lander during descent as part of the Mars Exploration Rover mission

Thrust vector actuation (TVA) Refers to the system used to actuate a TVC nozzle.

Thrust vector control (TVC) Refers to a type of rocket nozzle designed to move the net thrust vector through small angles for vehicle stability and maneuvering control.

Volumetric Loading Fraction In a solid motor, the ratio of the grain volume to the case volume, excluding the nozzle.

Web The maximum radial thickness of a solid propellant grain.

Recommended Reading and References

1. Ariane 5 User's Manual (2020). Arianespace. www.arianespace.com.
2. Atlas V Launch Services User's Guide (2010), United Launch Alliance, www.ulalaunch.com/docs/default-source/rockets/atlasvusersguide2010.pdf.
3. Beckstead MW. Overview of combustion mechanisms and flame structures for advanced solid propellants. In: Yang V, Brill TB, Ren WZ, editors. Solid propellant chemistry, combustion, and motor interior ballistics, vol. 185. Progress in Astronautics and Aeronautics, AIAA; 2000. pp. 267–85.
4. Beckstead, M.W., Puduppakkam, K., Thakre, P., and Yang, V. (2007). "Modeling of combustion and ignition of solid-propellant ingredients." Prog. Energy Combust. Sci., 33 (6) (2007), pp. 497–551.
5. Cavallini, E. (2009). "Modeling and Numerical Simulation of Solid Rocket Motors Internal Ballistics." Ph.D. Dissertation, Dipartimiento di Meccanica e Aeronautica, Sapienza Universita di Roma.
6. Chaturvedi, S., and Dave, P. N. (2019). "Solid propellants: AP/HTPB composite propellants." Arabian Journal of Chemistry, Vol. 12, Issue 8, December 2019, pp. 2061–2068.
7. Dawson, V. P. and Bowles, M. D. (2004). "Taming Liquid Hydrogen: The Centaur upper Stage Rocket (1958–2002)." NASA SP-4230. https://www.nasa.gov/wp-content/uploads/2023/04/sp-4230.pdf.
8. DeLuca, L.T. and Annovazzi, A. (2023). "Survey of burning rate measurements in small solid rocket motors." FirePhysChem, Available online 7 December 2023. https://www.sciencedirect.com/science/article/pii/S2667134423000573.
9. Derek, et al. (2021). "Experimental and numerical investigation of high-pressure nitromethane combustion." Proceedings of the Combustion Institute, Volume 38, Issue 2, 2021, pp. 3325–3332. https://www.sciencedirect.com/science/article/abs/pii/S1540748920302923.
10. French, J. (2000). "Analytical Evaluation of a Tangential Mode Instability in a Solid Rocket Motor." AIAA 2000-3698. 36th AIAA/ASME/SAE/ASEE J. Propulsion Conference, Huntsville, AL (July 2000).
11. Fry, R. S. (2002). "Solid Propellant Subscale Burning Rate Analysis Methods for U.S. and Selected NATO Facilities." CPTR 75, Technical Report, Oct 1997-Mar 2001.
12. Germani, T., Bandelier, E., Cloutet, Ph., et al. (2023). "P120C Solid Rocket Motor Synthesis of the Development of the Common Propulsive SRM for Ariane 6 and Vega-C and P160C Way Forward." Aerospace Europe Conference 2023 – 10TH EUCASS – 9TH CEAS. https://doi.org/10.13009/EUCASS2023-096.
13. Glaser, C., Hijlkema, J., and Anthoine, J. (2023). "Bridging the Technology Gap: Strategies for Hybrid Rocket Engines." Aerospace 2023, 10(10), 901; https://doi.org/10.3390/aerospace10100901.
14. Glaser, et al (2022). "Evaluation of Regression Rate Enhancing Concepts and Techniques for Hybrid Rocket Engines." Aerotecnica Missili & Spazio, Vol. 101, pp. 267–292 (2022). https://doi.org/10.1007/s42496-022-00119-4.
15. Hasue, K., Miura, R., and Yoshitake, K. (2013). "Equation for Burning Rate as a Function of Pressure and Temperature." Sci. Tech. Energetic Materials, Vol. 74, No. 5, 2013. https://www.jes.or.jp/mag/stem/Vol.74/documents/Vol.74,No.5,p.113-117.pdf.
16. Heister, S. D., Anderson, W. E., Pourpoint, T. L., and R. Joseph Cassady, R. J. (2019). Rocket Propulsion. Cambridge University Press. p. 250. ISBN 978-1-108-42227-7.
17. Isakovic, S.J., Hopkins Jr., J.P., Hopkins, J.B., *International Reference Guide to Space Launch Systems*, 4th Ed. AIAA, 2004.

18. Kamps, L., Hirai, S., and Nagata, H. (2021). "Hybrid Rockets as Post-Boost Stages and Kick Motors." Aerospace 2021, 8(9), 253; https://doi.org/10.3390/aerospace8090253, https://www.mdpi.com/2226-4310/8/9/253.

19. Kubota, N. (2002). "Variation of regression/burning rate with chamber pressure." Propellants and Explosives, Thermochemical Aspects of Combustion, Wiley-VCH, Weinheim, Germany, 2002.

20. Kubota, N. (1984). "Survey of Rocket Propellants and their Combustion Characteristics." in K. K. Kuo and M. Summerfield (Eds.), Fundamentals of Solid Propellant Combustion, Volume 90 in series on Progress in Astronautics and Aeronautics, American Institute of Aeronautics and Astronautics, New York, 1984, 891 pages.

21. Lafranconi, R. and Lopez, M. (2007). "The European Small Launcher." Technical Report Vega Programme Dept., ESA Directorate of Launchers – European Space Agency BR-257, ESA-ESTEC, ESTEC, PO Box 2200 AG Noordwijk, The Netherlands (April 2007).

22. McDowel, J. (1997). "Kick in the apogee - 40 years of upper stage applications for solid rocket motors, 1957–1997." 33rd Joint Propulsion Conference and Exhibit 06 July 1997 - 09 July 1997, Seattle, WA, U.S.A. https://doi.org/10.2514/6.1997-3133.

23. Musielak, D. (2022). *Scramjet Propulsion: A Practical Introduction.* A Wiley Aeronautic & Aerospace Engineering book. ISBN: 978-1-119-64063-9. https://www.wiley.com/en-us/Scramjet+Propulsion:+A+Practical+Introduction-p-9781119640639.

24. NASA Space Launch System Reference Guide, Space Launch System Reference Guide – NASA.

25. NASA SP-8025, NASA Space Vehicle Design Criteria (Chemical Propulsion): Solid Rocket Motor Metal Cases. April 1970. Guidelines and practices for design of solid rocket motor cases.

26. New Horizons Press Kit (2006). https://www.nasa.gov/wp-content/uploads/2015/03/139889 main_presskit12_05.pdf.

27. Northrup Grumman Propulsion Product Catalog, April 2016. https://cdn.prd.ngc.agencyq.site/-/media/wp-content/uploads/NG-Propulsion-Products-Catalog.pdf.

28. Pons, A. (2013). "Study of grain burnback and performance of solid rocket motors." M.S. Thesis, ETSEIAT- Universitat Politècnica de Catalunya.

29. Sutton, G.P. and Biblarz, O. (2001). *Rocket Propulsion Elements*, Seventh Edition, John Wiley & Sons, Inc. New York, 2001.

30. Thompson, B. E., Bouchery, O., and Lowney, K. D. (1995) "Flow through a Submerged Nozzle." J. of Spacecraft and Rockets, Volume 32, Number 6, November 1995.

31. Tobias, M.E., Griffin, D.R., McMillin, J.E., Haws, T.D., and Fuller, M.E. (2020). "Booster Obsolescence and Life Extension (BOLE) for Space Launch System (SLS)." 2020 IEEE Aerospace Conference, Big Sky, MT, USA, 2020, pp. 1–9.

32. Turner, M. J. L. (2006). *Rocket and Spacecraft Propulsion, Principles, Practice and New Developments.* Second Edition, Springer-Praxis Books.

33. Ward, T. (2010). *Aerospace Propulsion Systems.* 1st ed., John Wiley & Sons (Asia) Pte. Ltd. Singapore, 2010.

34. Willcox, M. L. and Brewster, M. Q (2007). "Solid Rocket Motor Internal Ballistics Simulation Using Three-Dimensional Grain Burnback." J. of Propulsion and Power, Vol. 23, No. 3, May-June 2007. https://doi.org/10.2514/1.22971.

35. Yaman, H., Celik, V. and Degirmenci, E. (2014). "Experimental investigation of the factors affecting the burning rate of solid rocket propellants." Fuel, Volume 115, January 2014, pp. 794–803. https://doi.org/10.1016/j.fuel.2013.05.033.

Electric Propulsion

6

For a space mission requiring large total impulse, the
thrust can be achieved with a rocket that expels ions.

—Dora Musielak

Electric propulsion is widely applicable for maneuvering spacecraft and cruising through interplanetary space. Electric propulsion (EP) refers to devices in which the acceleration of a propellant mass is accomplished by electrical heating and/or by electric and magnetic body forces, making possible to exhaust mass at high speeds (>10 km/s). The EP capability to achieve high exhaust speed translates into higher specific impulse (>1000 s) compared with chemical rockets, and this in turn, enables higher propellant efficiency, particularly important for deep space missions.

Electric propulsion is an integral part of the technologies for advancing astronautics and space exploration. Starting with classification of electric thrusters, we will learn the important performance parameters that define the application of each EP system. High-power, low-thrust propulsion systems are ideally suited to perform transfers between cis-lunar orbits. The high efficiency of low thrust propulsion to perform velocity changes (Δv) enables cislunar transfers for low propellant costs on relatively massive cis-lunar vehicles where propellant mass must be used sparingly. While these transfers could be performed with chemical propulsion systems, the amount of propellant required is greater than six times that used by a 40 kW Solar Electric Propulsion (SEP) system. Among the different electric propulsion systems conceived, the ion and Hall thrusters using heavy inert gas Xenon propellant are the leading concepts in terms of performance (thrust, specific impulse, and efficiency). Hence, details of these systems are given, due to their importance as applied to probes for space exploration missions.

The material presented in the next sections will help us answer questions such as

- How does electric thruster performance compare with that of a chemical rocket?
- How long can an electric rocket operate on full thrust mode?
- What is the difference between Hall effect and grided ion thrusters?
- Do electric thrusters utilize the same propellants as chemical thrusters?
- What are the advantages of electric propulsion for an interplanetary mission?

6.1   Introduction

Deep Space 1 was the first interplanetary spacecraft propelled by an electric propulsion system, developed as part of the NASA Solar Electric Propulsion Technology Application Readiness (NSTAR) effort. Launched in 1998 and propelled by the NSTAR ion thruster, Deep Space 1 achieved a speed of 16,200 km/h (10,066 mph) on its way to a flyby of the asteroid Braille. The 2.3 kW NSTAR ion thruster is a form of electric propulsion that creates thrust by accelerating ions. The highly successful Deep Space mission brought electric propulsion to the forefront of astronautics, and established it as a viable component of interplanetary flight.

In 2007, NASA launched Dawn, its second ion-propelled spacecraft on a mission to study Vesta and Ceres, two of the largest asteroids in the Solar System. After being deployed by the Delta rocket launcher, Dawn started its NSTAR ion rockets and monopropellant reaction control system to get the additional velocity needed to reach Vesta. Dawn used ion propulsion for the entire spaceflight: to change orbit to lower altitudes around Vesta, to leave Vesta, and to cruise to Ceres and to intersect its orbit.

The ion rockets in Dawn are a class of electric propulsion capable of achieving what chemical rockets cannot—travel farther, faster, and cheaper by efficient use of propellant. This is a most desirable benefit for long interplanetary trips beyond the Moon. The operating principle of ion rockets is quite simple. Heavy atoms such as Cesium or Xenon are ionized, accelerated through a high-voltage grid, and ejected out the back of the rocket. The momentum of the ejected atoms produces a steady, constant thrust that can be maintained for years at a time. And since the speed of the atom beam is thousands of times higher than the hot gas exhaust from chemical rocket engines, very little mass is needed to generate thrust over a long period of time. For the 1200 kg Dawn spacecraft, the propellant mass was only 425 kg, ejected steadily for 8 years, and thus it could reach a speed of over 10 km/s (36,000 km/hr or 22,300 miles/hour). This is equal to 315 million km/year or the distance from the Earth to the Sun and back.

In this chapter, we use the terms rocket or thruster interchangeably as they both denote a propulsion device that accelerates and ejects a propellant mass or stream of particles, thereby imparting thrust to a spacecraft on which it is mounted. In common practice, *thruster* refers to a propulsion system characterized by low thrust. We will soon discover that electric rockets powered by an external energy source are low thrust (< 1 N) propulsion systems. However, electric propulsion makes very efficient use of propellant by accelerating it to a velocity ten times that of chemical rockets.

6.1.1 High Impulse Spaceflight

According to the rocket equation, if the exhaust velocity v_{ex} is constant over a given period of thrust, a spacecraft achieves an increment in its velocity Δv, which depends linearly on v_{ex} and logarithmically on the amount of propellant it expends,

$$\Delta v = v_{ex} \ln \frac{m_0}{m} \tag{6.1}$$

where m_0 and m are the total spacecraft mass at the start and completion of the acceleration period. Another way of stating this is solving for the fraction of the rocket mass m/m_0 which can be accelerated through a given velocity increment Δv as a negative exponential in the ratio of that increment to the exhaust velocity, Eq. (2.15). The rocket equation implies that if the spacecraft is to deliver a significant fraction of its initial mass to its destination, the speed of the rocket exhaust v_{ex} must be comparable to this characteristic velocity increment Δv.

For long interplanetary missions requiring large Δv the propulsion system must produce very high relative exhaust velocities. This cannot be achieved by chemical rockets because, as we noted in Chap. 4, they are fundamentally limited by their available combustion reaction energies and heat transfer tolerances. The thrust of a chemical rocket is derived from the thermal expansion of the hot propellant gases exhausting through a convergent-divergent nozzle. And since the hot gases are produced by a chemical reaction of propellant, the exhaust velocity that can be obtained is limited by three factors: (1) the intrinsic energy available in the chemical reaction and convertible to enthalpy of the gas in the chamber; (2) the heat transfer that can be tolerated by the rocket walls; and (3) the frozen flow losses, that is, the unrecoverable energy deposition in the internal modes of the gas and radiation losses from the exhaust.

In electric propulsion the energy required to produce thrust is not stored in the propellant, rather it is supplied by an external power source, e.g. nuclear, solar, or batteries. Electric propulsion can heat the propellant to much higher temperatures than is possible with chemical reactions. Provided that the walls of the rocket chamber and nozzle are protected from excessive heat transfer, an electric propulsion system can apply suitable body forces such as electric and magnetic fields to accelerate the hot propellant stream and achieve a specific impulse much higher than it possible with chemical rockets. Thus, electric power is of crucial interest for astronautics.

Long-range missions and complex missions such as interplanetary and lunar flights, or long-time missions, such as station-keeping or maintenance of satellite position and orientation for several years, are characterized by large velocity Δv budgets. A round-trip mission to Mars from Earth requires a minimum Δv of approximately 14 km/s. In addition, these complex interplanetary missions require thrust of variable magnitude and direction, and specialized maneuvers by the spacecraft to maintain its trajectory.

In Chap. 2 we showed that exhaust velocity v_{ex} is directly related to the specific impulse I_{sp}, an important performance parameter of the propulsion system that defines the thrust per rate of propellant weight used $(F/\dot{m}g_0)$, Eq. (2.13):

$$I_{sp} = \frac{F}{\dot{m}g_0} = \frac{\dot{m}v_{ex}}{\dot{m}g_0} = \frac{v_{ex}}{g_0} \tag{6.2}$$

And we rewrite Eq. (6.1) as

$$\Delta v = v_{ex} \ln \frac{m_0}{m} = I_{sp}g_0 \ln\left(\frac{m + m_p}{m}\right) \tag{6.3}$$

where m_p the propellant mass, and m is the thruster dry mass (mass of the structure, including power supply systems, and the payload).

Equation (6.4) shows that for a given mission with a specified Δv and final delivered mass m, the initial mass of the spacecraft or wet mass $(m_0 = m + m_p)$ can be reduced by increasing the I_{sp} of the propulsion system. A propulsion system that provides a large propellant exhaust velocity compared to the mission Δv has a propellant mass that is only a small fraction of the initial spacecraft wet mass (see Chap. 2). That is why high Δv missions are best accomplished with electric propulsion because it offers much higher exhaust velocities and I_{sp} than do conventional chemical propulsion systems.

The thrust delivered by any rocket is directly proportional to the product of propellant mass ejected and its exhaust velocity (see Chap. 2). If the increase of thrust is due to ejecting a large amount of mass, this would mean an unacceptable large initial propellant mass fraction. Thus, for mission requiring a large total impulse, we want an engine that delivers a high exhaust velocity. Modern chemical rockets can achieve exhaust speeds up to 4 km/s, and so it is not effective for long-term or long-duration missions that require higher exhaust velocities. Electric propulsion can achieve I_{sp} orders of magnitude higher than those possible with chemical rockets because the exhaust speed of some electric thrusters can reach 10^3 km/s, depending on design and propellant used.

6.1.2 Electric Propulsion

Electric propulsion (EP) is a form of low-thrust propulsion where charged particles or plasmas are accelerated using electrostatic or electromagnetic forces and ejected at very high exhaust velocities, thus resulting in high specific impulse (I_{sp}) propulsion, which can be as high as 10 times greater than the I_{sp} of chemical propulsion systems. According to the rocket equation, having a high I_{sp} significantly reduces the propellant mass for a given velocity change (Δv). However, since the mass flow rate in EP devices is rather small, the produced thrust is also very low ($F < 1$N). This means is that an EP thruster must operate continuously for its low thrust acceleration to yield the required Δv when integrated over a considerably long time. For example, NASA's Dawn spacecraft performed its mission with a propellant mass of only 425 kg, but its ion rocket ejected it steadily for 8 years.

The electric propulsion system consists of a power source (could be solar or nuclear), power conversion and conditioning systems, a propellant source, and the thruster itself. The propellant is heated or affected by an electric current. Thrust is produced by expansion of the hot gas and acceleration of charged particles in electric or magnetic fields to high exhaust velocities. While the exhaust velocity of chemical rockets is limited by the energy contained in the chemical bonds of the propellant used, electric thrusters separate the propellant from the energy source, which is now a power supply, and thus are not subject to the same limitations.

There are two types of EP depending on their power source: solar and nuclear (Fig. 6.1). Nuclear-electric propulsion (NEP) is an electric propulsion in which the electrical energy used to accelerate a propellant comes from a nuclear power source, such as a space-based nuclear reactor. Solar Electric Propulsion (SEP) uses solar energy to heat the propellant. As the name implies, SEP utilizes solar cell panels, consisting of an extensive array of photovoltaic cells, to turn sunlight into electricity to power low-thrust, high specific-impulse ion engines or Hall thrusters. For continuous sunlight operation, current solar-cell systems have proven to be the most lightweight and reliable and the longest duration space power source available. Many communications satellites use solar energy to power satellites in geostationary orbit.

Did you know? Each of the three ion rockets that propelled the Dawn spacecraft had a specific impulse of 3100 s and could generate a thrust of 90 mN. While a chemical rocket can produce a thrust force several orders of magnitude higher, the much smaller ion rocket (30 cm diameter) can achieve an equivalent trajectory velocity change Δv by firing over much longer time. Dawn did it over 8 years.

Example 6.1 A 1000 kg spacecraft will perform a maneuver requiring a Δv of 5 km/s. Estimate the propellant mass that could be saved on this mission using an electric thruster

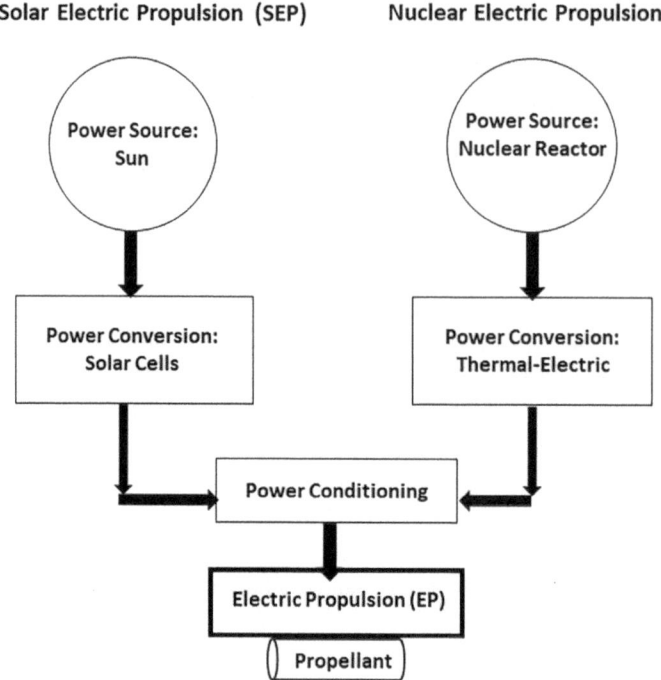

Fig. 6.1 Conceptual illustration of electric propulsion systems

with an exhaust velocity of 30-km/s by comparing with the propellant mass required by a typical chemical rocket that has an exhaust velocity of 4.41 km/s.

Solution: From Eq. (6.3), with $m = 1000$ kg, the mass of propellant required by an electric thruster with exhaust velocity $v_{ex} = I_{sp}g_0 = 30$ km/s

$$m_{pE} = m\left[e^{\Delta v/(I_{sp}g_0)} - 1\right] = 1000\left[e^{5/30} - 1\right] = 181.36 \text{ kg}$$

The propellant mass needed with a chemical rocket with $v_{ex} = I_{sp}g_0 = 4.41$ km/s

$$m_{pC} = m\left[e^{\Delta v/(I_{sp}g_0)} - 1\right] = 1000\left[e^{5/4.41} - 1\right] = 2107.40 \text{ kg}$$

The electric thruster would accomplish the mission using one order of magnitude less propellant than the chemical rocket. This highlights the attractiveness of space-worthy EP systems that lies in the conservation of propellant mass for missions of large velocity increments Δv.

6.2   Electric Propulsion Basic Concepts

6.2.1   Kinetic Energy and Power of Exhaust Jet

Since the ejected propellant mass has kinetic energy $E_{jet} = \frac{1}{2}mv_{ex}^2$, we express the **power** of the exhaust jet, P_{jet} as the kinetic energy per unit time:

$$P_{jet} = \frac{dE_{jet}}{dt} = \frac{1}{2}\dot{m}v_{ex}^2 = \frac{\dot{m}(g_0 I_{sp})^2}{2} \tag{6.4}$$

where $\dot{m}$ is the mass flow rate of propellant exhausted.

Equation (6.4) gives the energy rate that must be supplied to a propulsion system by a power source, which we can also write in the form,

$$P_{jet} = \frac{F^2}{2\dot{m}} \tag{6.5}$$

where F is the thrust. This formula shows that for a propulsion system to deliver a higher thrust without increasing the propellant flow rate, it requires an increase in its jet power.

The power of the jet is diminished by losses in the energy conversion process, including losses associated with conversion of electric energy into propulsive jet kinetic energy. The kinetic power of the propulsive jet P_{jet} per unit thrust F can be expressed by a simple relation, assuming no significant pressure thrust, i.e. ideal expansion:

$$\frac{P_{jet}}{F} = \frac{\frac{1}{2}\dot{m}v_{ex}^2}{\dot{m}v_{ex}} = \frac{1}{2}v_{ex} = \frac{1}{2}g_0 I_{sp} \tag{6.6}$$

This means that, since the power-to-thrust ratio of the propulsive jet is proportional to the exhaust velocity, electrical thrusters capable of high I_{sp} require more power per unit of thrust.

Now we define the **thruster efficiency** as the ratio of the thrust-producing kinetic energy (axial component) rate of the exhaust to the total electrical power P_e supplied to the thruster (power input), including any power required to evaporate or ionize the propellant:

$$\eta_t = \frac{\text{power of the jet}}{\text{electrical power input}} = \frac{P_{jet}}{P_e} \tag{6.7}$$

where the power input to the thruster P_e results from the product of electric current I and all associated voltages V, that is, $P_e = \sum IV$.

Use Eq. (6.6), and with electric power input P_e to the thruster (in watts), express the efficiency as

$$\eta_t = \frac{\frac{1}{2}\dot{m}v_{ex}^2}{P_e} = \frac{FI_{sp}g_0}{2P_e} = \frac{Fv_{ex}}{2P_e} \tag{6.8}$$

This thruster efficiency is a measure of how effectively electric power and propellant are used in the production of thrust. It accounts for all the energy losses that are not converted to kinetic energy. The losses include wasted electrical power, heat losses, and other process losses.

Power is the major constrain for electric thrusters. This is a reason why the thrust achieved with electric propulsion is much lower than that of chemical rockets. For example, if the power input is 1 kW and the exhaust velocity is 30,000 m/s, assuming 80 percent power conversion efficiency, the thrust produced is, from Eq. (6.8):

$$F = P_e \frac{2\eta_t}{v_{ex}} = 1000 \text{ W} \frac{2(0.8)}{30,000 \text{ m/s}} = 0.053 \text{ N}$$

Example 6.2 A spacecraft must deliver 50 N thrust for an orbital maneuver. Estimate the power input required for an electric propulsion system with a specific impulse of 2100s, and compare it with that of a chemical rocket ($I_{sp} \sim 300$ s). Assume a propulsion efficiency $\eta_t = 0.60$ for both propulsion systems.

Solution: First, we calculate the mass flow rate from the thrust equation, Eq. (6.2)

$$\dot{m} = \frac{F}{I_{sp}g_0}$$

and the power input from Eq. (6.7):

$$P_e = \frac{P_{jet}}{\eta_t} = \frac{\frac{1}{2}\dot{m}v_{ex}^2}{\eta_t} = \frac{FI_{sp}g_0}{2\eta_t}$$

For the electric propulsion system: $\dot{m}_{EP} = \frac{50\text{N}}{(2100\text{s})(9.8 \text{ m/s}^2)} = 2.4 \times 10^{-3}$ kg/s

$$P_{EP} = \frac{FI_{sp}g_0}{2\eta_t} = \frac{(50\text{N})(2100 \text{ s})(9.8 \text{ m/s}^2)}{2(0.6)} = 857.5 \text{ kW}$$

For the chemical rocket: $\dot{m}_{CR} = \frac{50 \text{ N}}{(300 \text{ s})(9.8 \text{ m/s}^2)} = 1.7 \times 10^{-2}$ kg/s

$$P_{CR} = \frac{FI_{sp}g_0}{2\eta_t} = \frac{(50 \text{ N})(300 \text{ s})(9.8 \text{ m/s}^2)}{2(0.6)} = 122.5 \text{ kW}$$

These values are theoretical, of course, and will vary due to the specific design, but they are indicative of trends we can expect from these two types of propulsion. They show how much more energy input is required for an electric propulsion system than for a chemical rocket, but the propellant flow is much smaller for the EP system. Hence, it could be more economical.

6.2.2   Ideal Performance

Propulsion systems that can deliver high exhaust velocities, or high I_{sp}, are required to maximize a mission payload mass. The payload fraction is the ratio of payload mass to the total initial mass (m_L/m_0). For a chemical rocket, payload fractions are relatively small and thus we seek other propulsion systems that allow an increase in the mass of payload that can be transported to carry out a mission. Let us then explore the performance of an electrical rocket in terms of its electric power and the payload fraction.

In Chap. 2, we defined the rocket mass ratio as the ratio of initial to final mass, where the initial mass is comprised of propellant, structure, and payload, $m_0 = m_p + m_s + m_L$. Now, we must include the mass of the electric power supply, which can be considered part of the spacecraft structure. The mass of the dry propulsion system m_{SE} consists of the thruster, propellant storage and feed system, the energy source with its conversion system and auxiliaries, and the associated structure. Then the total initial system mass is

$$m_0 = m_p + m_{SE} + m_L \tag{6.9}$$

and we use it to express the payload mass ratio,

$$\frac{m_L}{m_0} = 1 - \frac{m_p}{m_0} - \frac{m_{SE}}{m_0}$$

The power conversion efficiency relates the energy source input to the power supply, which must be larger than its electrical power output for converting the raw energy into electrical power at the desired voltages, frequencies, and power levels. This converted electrical output P_e is then supplied to the propulsion system. The *specific power* α of the propulsion system is the ratio of the engine electrical power output P_e to the mass of the dry propulsion system m_{SE}:

$$\alpha = \frac{P_e}{m_{SE}} \tag{6.10}$$

The *power-to-mass ratio* in Eq. (6.10) has typical values between 0.1 and 0.2 kW/kg. Values of 0.5 to 2.0 kW/kg are expected in the future with technological advances in power conditioning equipment and other design improvements. The specific power has a direct effect on the velocity of the vehicle for a given payload to propellant mass ratio. That is, to attain a high spacecraft velocity v, a very high specific power is required. Therefore, a higher specific power is crucial for obtaining the best performance of an electric thruster.

Now, since the electrical power is converted by the thruster into kinetic energy of the exhaust, we rewrite the electric power input P_e using Eq. (6.8) to make allowance for losses:

$$P_e = \alpha m_{SE} = \frac{\frac{1}{2}\dot{m}v_{ex}^2}{\eta_t} = \frac{m_p v_{ex}^2}{2t_p \eta_t} \tag{6.11}$$

where m_p is the propellant mass, v_{ex} is the effective exhaust velocity, and t_p the propulsive time during which the propellant is being ejected at a uniform constant rate.

Therefore, the mass of the propulsion system is

$$m_{SE} = \frac{m_p v_{ex}^2}{2\alpha t_p \eta_t}$$

This expression stresses that the high exhaust velocity of electric propulsion comes at the expense of power and mass.

Using Eqs. (6.9), (6.11), and the rocket equation, we obtain a relation for the reciprocal payload mass ratio:

$$\frac{m_0}{m_L} = \frac{e^{\Delta v/v_{ex}}}{1 - \frac{(e^{\Delta v/v_{ex}} - 1)v_{ex}^2}{2\alpha t_p \eta_t}}, \tag{6.12}$$

which assumes a gravity-free and drag-free flight.

With appropriate substitutions, we can write Eq. (6.12) in the form of the familiar rocket equation, which now shows the additional effects of the electric power supply on the overall vehicle velocity. The rocket equation for EP can be expressed as

$$\Delta v = \sqrt{\frac{2\alpha \eta_t m_{SE}}{m}} \ln\left(1 + \frac{m_p}{m_{SE}}\right) \tag{6.13}$$

where α is the power-to-mass ratio or specific power, η_t is the thruster efficiency, m_p the propellant mass, and m_{SE} the mass of the combined structure and electric power supply mass.

For a given mission, there is an optimum range of specific impulse and thus an optimum propulsion system design. To see how this is true for EP, let us define a parameter that combines the specific power α, the propulsion efficiency η_t, and the propulsive time t_p as follows:

$$v_c = \sqrt{2\alpha t_p \eta_t} \tag{6.14}$$

Because this expression has units of speed, the parameter v_c is called the "characteristic speed." It is not a physical speed but v_c serves to represent the speed the power plant would have if its full power output were converted into the form of kinetic energy of its own inert mass m_{SE} during the propulsive time t_p, which is the actual mission time. We then use the characteristic speed to normalize the change of vehicle velocity Δv which results from the propellant being exhausted at a speed v_{ex}.

The results of such analysis indicate that, for a given payload fraction m_L/m_0 and characteristic speed v_c, there is an optimum value of v_{ex} corresponding to the peak vehicle velocity increment Δv; this means that there exists a specific set of desirable operating conditions for a given mission.

The inert mass of the dry propulsion system (power plant) m_{SE} increases with I_{sp}, and the propellant mass m_p decreases with I_{sp}. The optimum values can be found by differentiating Eq. (6.12) to relate Δv, v_{ex}, and v_c for maximum payload fraction:

$$\left(\frac{v_{ex}}{\Delta v}\right)\left(e^{\Delta v/v_{ex}} - 1\right) - \frac{1}{2}\left(\frac{v_c}{v_{ex}}\right)^2 - \frac{1}{2} = 0 \tag{6.15}$$

The maximum values are found in the ranges $\Delta v/v_{ex} \leq 0.805$, and $0.505 \leq v_{ex}/v_c \leq 1.0$. This means that, for a given electric thruster, its optimum operating time t_p will be proportional to the square of the total required change in vehicle velocity, Δv. Therefore, a mission requiring high values of Δv would take very long time to accomplish.

Note that all previous equations depend on the overall propulsion efficiency. Therefore, they apply to all fundamental types of electric rocket systems. This simple analysis does not take into consideration specific design and thus only provides the overall performance trends. For well-designed electric propulsion systems, the overall efficiency ranges from 0.4 to 0.8.

In actual design, the mass budget for an electric propulsion system may be given as $m_0 = m_L + m_p + m_{pt} + m_{sa} + m_t$, where m_L is payload mass, m_p is propellant mass, m_{pt} is propellant tank mass (which is typically assumed to be 0.1 m_p), $m_{sa} = \alpha_{sa}\max(P)$ is the mass of solar arrays with α_{sa} denoting the mass-to-power ratio for the solar arrays in kg/kW, and $m_t = \alpha_t P$ is the mass of the thruster with α_t denoting the mass-to-power ratio for the system, including all power handling equipment.

The above analysis is highly idealized. Because of inherent losses, low thrust-to-weight propulsion systems experience a performance loss equivalent to increasing the effective mission Δv. Nonetheless, electric propulsion can be used to obtain additional increases in payload mass by using a solar energy device to perform part of an orbit transfer for a mission.

Example 6.3 Assess the performance of an electrical propulsion rocket for raising a spacecraft from a low to a higher orbit in 4 weeks, with a payload mass of 100 kg. Calculate the velocity increase of the spacecraft and its average acceleration during the duration of the mission. The typical Δv value for constant low thrust (acceleration < 0.001 m/s^2) orbit transfer from LEO (200 km altitude) to GEO (no plane change) is approximately 4.71 km/s. The performance of the thruster is given as follows:

I_{sp} (s)	F (N)	α (W/kg)	η_t
2000	0.20	100	0.5

Solution: The propellant flow rate is, from the ideal thrust Eq. (6.2):

$$\dot{m} = \frac{F}{I_{sp} g_0} = \frac{0.20}{2000 \times 9.81} = 1.02 \times 10^{-5} \text{ kg/s}$$

The total required propellant for the duration of the mission, $t = 4$ weeks $= 2.42 \times 10^6$ s, is:

$$m_p = \dot{m} t = 1.02 \times 10^{-5} \times 2.42 \times 10^6 = 24.69 \text{ kg}$$

The required electrical power is, from Eq. (6.11),

$$P_e = \frac{\frac{1}{2} \dot{m} v_{ex}^2}{\eta_t} = \frac{1}{2} \frac{\dot{m} (I_{sp} g_0)^2}{\eta_t} = \frac{1.02 \times 10^{-5} (2000 \times 9.81)^2}{2(0.5)} = 3.92 \text{ kW}$$

The mass of the dry propulsion system is, from Eq. (6.10),

$$m_{SE} = \frac{P_e}{\alpha} = \frac{3.92}{0.1} = 39.2 \text{ kg}$$

The total initial mass (before engine operation) is, from Eq. (6.9),

$$m_0 = 100 + 24.7 + 39.2 = 163.9 \text{ kg}$$

and the final mass (after engine operation) is,

$$m = m_0 - m_p = 139.2 \text{ kg}$$

The velocity increase of the spacecraft under ideal vacuum conditions is, from Eq. (6.1),

$$\Delta v = v_{ex} \ln \left[\frac{m_0}{(m_0 - m_p)} \right] = (2000 \times 9.81) \ln \left[\frac{163.9}{139.2} \right] = 3200 \text{ m/s}$$

The average acceleration of the vehicle is

$$a = \frac{\Delta v}{t} = \frac{3200}{2.42 \times 10^6} = 1.32 \times 10^{-3} \text{ m/s}^2 \rightarrow a = 1.35 \times 10^{-4} g_0$$

Based on the value obtained for Δv, this propulsion system would not be adequate to raise the spacecraft from LEO to GEO in 28 days, which requires $\Delta v = 4.71$ km/s. To optimize this thruster and satisfy Eq. (6.16) we must either increase the operating time or increase the thrust, or both.

6.3   Classification of Electric Propulsion

Electric thrusters are usually described in terms of the acceleration method used to produce the thrust. The many and varied electric propulsion concepts are classified in three fundamental types: electrothermal, electromagnetic, and electrostatic systems. As shown in Table 6.1, a variant of each propulsion type is possible, and their performance vary accordingly. The most advanced EP systems to date are the ion thrusters and Hall thrusters. In the next sections we highlight design and performance characteristics of each type.

6.3.1   Electrothermal Thrusters

In electrothermal thrusters, the propellant is heated by electrical means and expanded thermodynamically to produce thrust; i.e. the heated gas is accelerated to supersonic speeds through a nozzle, just as in the chemical rocket.

The overall performance of an electrothermal rocket can be assessed with the one-dimensional energy equation. As we found in Chap. 2, the exhaust speed is limited to that of a fully expanded nozzle flow, $v_{ex} \propto \sqrt{T_c/\mathcal{M}}$, where T_c is the maximum tolerable chamber temperature, and $\mathcal{M}$ is the molecular mass of the gas. This indicates that, just

Table 6.1 Electric propulsion classification

Electrothermal Propellant mass is heated by electrical means and expanded thermodynamically to produce thrust		Electrostatic Propellant mass is accelerated by direct application of electric body forces to ionized gas and then ions are accelerated		Electromagnetic Propellant mass is first ionized and then plasma is accelerated by interaction of external and internal magnetic fields with electric currents driven through stream	
Name	$I_{sp}(s)$	Name	$I_{sp}(s)$	Name	$I_{sp}(s)$
Resistojet	≤ 300	Ion engine	2500–6000	Pulsed plasma thruster (PPT)	650–1400
Arcjet	500–600	Hall thruster	1500–2000	Magnetoplasmadynamic (MPD) thruster	
Microwave electrothermal thruster (MET)	6000	Field emission thruster	>2000	Pulsed inductive thruster (PIT)	
Pulsed electrothermal thruster (PET)	850–1200	Field emission colloid thruster		Variable specific impulse magnetoplasma rocket (VASIMR)	>2000

as a chemical rocket, the electrothermal rocket operates more effectively with propellants with the lowest molecular mass. Hydrogen, however, is not practical for this application due to its frozen flow tendencies and difficulty of storage. Frozen flow losses are due to non-recovered energy "frozen" in the internal modes and dissociation of the molecules. Instead, more complex molecular gases such as ammonia and hydrazine, which dissociate into low effective molecular masses and high specific heat gas mixtures in the chamber, are better suited for electrothermal propulsion.

Several techniques are used to heat the propellant in a chamber using electrical means. We classify electrothermal propulsion in terms of the physical approach to propellant heating:

- Resistojets: the heat is transferred directly to the propellant from a solid surface, such as the chamber wall or a heater coil.
- Arcjets: the propellant is heated by an electric arc driven through it.
- Inductively and radiatively heated thruster: in this concept some form of electrodeless discharge or high-frequency radiation is used to heat the propellant.

In a typical resistojet, heating of a propellant such as helium, ammonia, nitrogen, or hydrazine is usually achieved by ohmic heating, sending electricity through a resistor consisting of a hot incandescent filament. The hot gas is then expanded and ejected through a conventional rocket nozzle. Resistojets require low operational voltage and simple power processing, making them very attractive for small spacecraft.

An important design issue to consider for a resistojet is the materials used. The material properties limit the operational temperatures, and thus the achievable performance. Typical materials used for the electrical resistance of high performance resistojets include rhenium, tungsten, tantalum, molybdenum, platinum, or their alloys.

Resistojets have been used in space since 1965 but became commercially available in 1980, after use in the first INTELSAT-V satellites. This type of propulsion is best used in situations where energy is much more plentiful than mass, and where propulsion efficiency needs to be reasonably high and where low-thrust is acceptable. Resistojet propulsion is now used for orbit insertion, attitude control, and deorbit of LEO satellites, including satellites in the Iridium satellite constellation.

The electrothermal hydrazine thruster (EHT), used for satellite station-keeping maneuvers, is a type of resistojet where the hydrazine fuel is catalytically decomposed, and the decomposition products are heated to an even higher temperature using a separate heater element, achieving an exhaust velocity of ~ 3500 m/s, delivering a thrust of 0.3 N with an efficiency of 80% when operating at a power level of 750 W. The EHT yields a nominal specific impulse of 300 s (a conventional hydrazine chemical rocket has $I_{sp} \sim 225$ s). The HIgh Performance Electrothermal Hydrazine Thruster (HIPEHT) depicted in Fig. 6.2a delivers vortex propellant flow directly over the heating element, thus allowing the propellant temperature to approach the heating element temperature. In the Augmented Catalytic

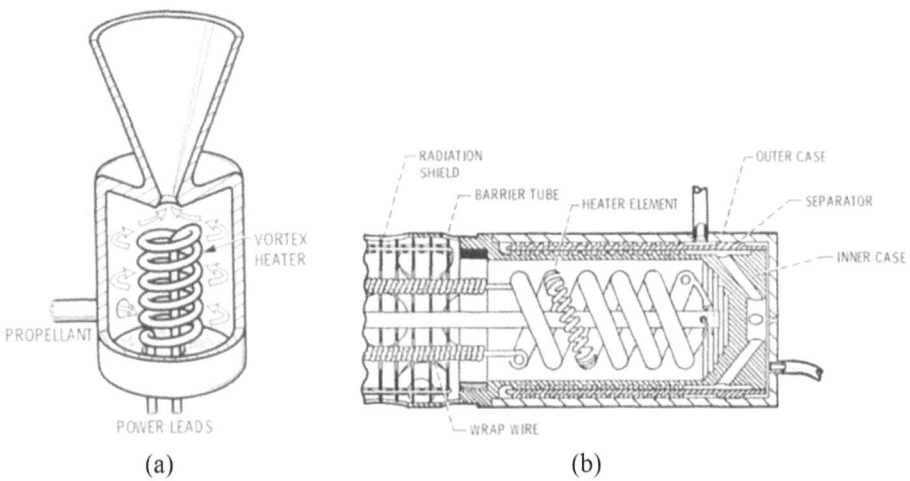

(a)   (b)

Fig. 6.2 Schematic of **a** HIPEHT resistojet; **b** Augmented catalytic thruster (ACT). *Credit* Stone [22]

Thruster (ACT) shown in Fig. 6.2b, the heating element does not contact the propellant directly; instead, it radiates to a heat exchanger through which the propellant flows.

Example 6.4 A 1 kW resistojet heats hydrogen gas to 2200 K and expands it through a nozzle. Determine the ideal exhaust velocity and thrust it could produce.

Solution: For hydrogen gas, $R = \frac{\Re}{\mathcal{M}} = \frac{8314.3 \text{ J/kg-mole-K}}{2 \text{ mole}} = 4157.15 \text{ J/kg} \cdot \text{K}$, $\gamma = 1.2$. From Eq. (2.33), the maximum exhaust velocity (perfect expansion) is

$$(v_{ex})_{max} = \sqrt{\frac{2\gamma}{\gamma - 1} \frac{\Re T_c}{\mathcal{M}}} = \sqrt{\frac{2(1.2)}{0.2}(4157.15 \text{ J/kg} \cdot \text{K})(2200 \text{ K})} = 10{,}476 \text{ m/s}$$

From Eq. (6.4) we solve for the mass flow rate, with 1 kW jet power

$$\dot{m} = \frac{2P_{jet}}{v_{ex}^2} = \frac{2(1000 \text{ W})}{(10{,}476 \text{ m/s})^2} = 1.82 \times 10^{-5} \text{ kg/s}$$

and therefore, the ideal thrust is

$$F = \dot{m}v_{ex} = \left(1.82 \times 10^{-5} \text{ kg/s}\right)(10{,}476 \text{ m/s}) = 0.19 \text{ N}$$

Although the thrust level of the resistojet is rather small, the exhaust velocity is much higher than it is possible with a chemical rocket with such a small amount of propellant.

In the microwave electrothermal thruster (MET), also known as plasma rocket, a propellant gas is heated by means of a free-floating microwave-generated plasma within a microwave resonant cavity. A microwave resonant cavity, or radio frequency (RF) cavity, is a metal-enclosed structure that contains electromagnetic fields in the microwave region of the spectrum. The microwaves bounce back and forth between the walls of the cavity, and at the cavity's resonant frequencies, they form standing waves. A plasma is created by AC induced breakdown of the propellant in the region of maximum electric power density. The plasma acts as an electrodeless resistive load, which efficiently absorbs applied continuous wave (CW) microwave power and converts it into thermal, internal, and radiant energy of the propellant gas. The heated propellant plasma expands through a converging-diverging nozzle to produce thrust.

The MET (Fig. 6.3) has the capability of high thrust and thrust density (thrust per unit cross-sectional area of the nozzle) common to thermal rockets operating at high pressure. Heating the propellant with microwave energy makes it possible to achieve high temperatures on the order of 20,000 K. The specific impulse is mainly constrained by the throat and nozzle wall material temperatures. If the hot plasma is kept away from the surfaces of the thruster to reduce heat transfer and avoid material damage, using a magnetic nozzle, the resulting specific impulse could be as high as 6000 s.

A magnetic nozzle can be created by a short, solenoid superconducting magnet applied in the throat-nozzle region of the MET. The pinch effect and high magnetic pressures generated by the magnetic field can, in principle, reduce wall losses from the hot plasma species; promote recovery of ionization, dissociation, and excitation energy from the hot species; and stabilize the plasma discharge in the cavity. The stabilizing effect of the magnetic field on the plasma can be augmented fluid-dynamically using a swirl component in

Fig. 6.3 Concept of a microwave electrothermal thruster

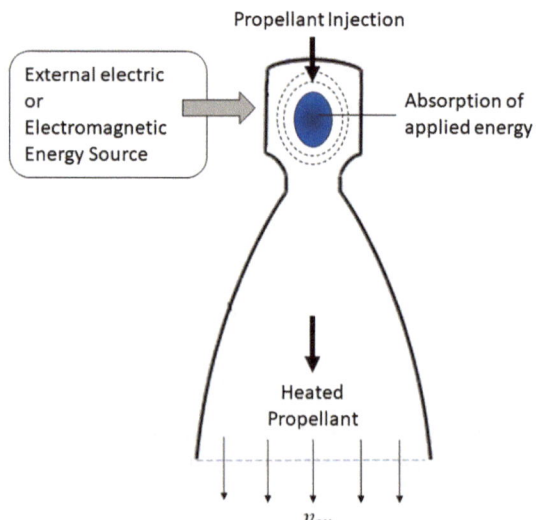

the propellant flow. And since the MET has no electrodes, this may allow for significant reductions in thruster erosion and tremendous improvements in overall lifetime, compared with other EP systems.

6.3.2 Electromagnetic Thrusters

In this concept, the propellant is first ionized and then the resulting plasma is accelerated by interaction of external and internal magnetic fields with electric currents driven through the stream. Moderately dense plasmas are high temperature or non-equilibrium gases, electrically neutral and reasonably good conductors of electricity.

The general principle of acceleration in all electromagnetic thrusters is essentially the same as that in ordinary electric motors. As illustrated in Fig. 6.4, when current flows in an electrically conducting plasma in the presence of a magnetic field $\vec{\mathbf{B}}$, it produces an electromagnetic body force acting on the plasma. The body force $\vec{\mathbf{F}}$ is equal to the vector or cross product of the current density $\vec{\mathbf{j}}$ and the magnetic field. The direction of the force is perpendicular to both $\vec{\mathbf{j}}$ and $\vec{\mathbf{B}}$. In electromagnetic thrusters, the $\vec{\mathbf{j}} \times \vec{\mathbf{B}}$ body force, known as the *Lorentz force*, acts on the plasma propellant to accelerate it and produce thrust.

Electromagnetic propulsion concepts include

- Magnetoplasmadynamic (MPD) Thruster
- Pulsed Plasma Thruster
- Helicon Plasma Engine
- Inductive Pulsed Plasma Rocket
- Variable Specific Impulse MagnetoPlasma Rocket (VASIMR).

The Magnetoplasmadynamic (MPD) thruster uses a very high current arc to ionize a significant fraction of the propellant mass, and then electromagnetic $\mathbf{j} \times \mathbf{B}$ Lorentz forces in the plasma discharge to accelerate the charged propellant plasma. Since both the current and the magnetic field are usually generated by the plasma discharge, MPD thrusters

Fig. 6.4 Electromagnetic body force on a plasma element

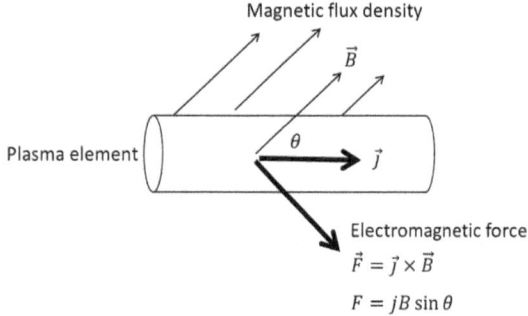

Fig. 6.5

Magnetoplasmadynamic
(MPD) thruster

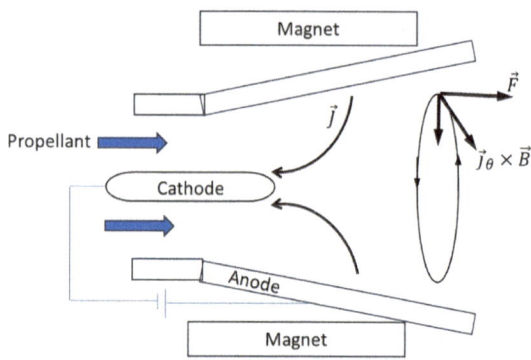

tend to operate at very high powers to generate sufficient force for high specific impulse operation. Figure 6.5 shows the model of the MPD propulsion concept.

The MPD thruster is the most powerful form of electromagnetic propulsion. Its ability to efficiently convert megawatts of electric power into thrust makes the MPD thruster an excellent candidate for economical delivery of lunar and Mars cargo, outer planet rendezvous, and sample return, and for enabling other deep space robotic and piloted planetary exploration missions. MPDs can process more power and create more thrust than any other type of electric propulsion currently available, while maintaining the high exhaust velocities associated with ion propulsion.

A pulsed plasma thruster (PPT) utilizes a pulsed discharge to ionize a fraction of a solid propellant such as Teflon ablated into a plasma arc, and electromagnetic effects in the pulse to accelerate the ions to high exit velocity. The pulse repetition rate is used to determine the thrust level. The PPT was demonstrated as a precision attitude control actuator for the Earth Observing-1 (EO-1) spacecraft. This is a small (total mass of 4.95 kg), self-contained electromagnetic propulsion system that can deliver high specific impulses (650–1400 s), very fine impulse bits (90–860 μN-sec) at low power levels (12–70 W), and an estimated total impulse of 460 N-s.

Another important concept in this category of plasma rockets is the Variable Specific Impulse MagnetoPlasma Rocket (VASIMR). This is a type of electric thruster capable of delivering very high specific impulses. A propellant gas such as Argon, Xenon, or Hydrogen is injected into a tube surrounded by a magnet and a series of two radio wave (RF) antennas. The antenna couplers turn the cold gas into superheated plasma and the expanding magnetic field at the rear end of the rocket, known as magnetic nozzle, converts the thermal energy of the plasma particles into directed flow and produce the thrust force.

The first RF coupler converts the propellant gas into plasma by ionizing it (knocking an electron loose from each gas atom). It is called the helicon section because its coupler is shaped in such a way that it can ionize a gas by sending helical waves through the gas. Downstream from the first helicon section, the cold plasma is readier for acceleration in the second stage. There the charged particles interact with magnetic fields.

The VASIMR has a second RF coupler called the Ion Cyclotron Heating (ICH) section. ICH is a technique used in fusion experiments to heat a plasma to temperatures on the order of those in the core of the Sun. The radio waves hit ions and electrons along their orbits around field lines at resonance, resulting in accelerated motion and raising the temperature of the plasma to about 10^6 K, or two hundred times the temperature of the Sun's surface.

Because the thermal motion of the ions around field lines is mostly perpendicular to the rocket's direction of travel, a magnetic nozzle is used to convert the ions orbital momentum into useful linear momentum. As the magnetic field lines expand, the spiral paths of the ions around their field lines elongate, resulting in ion speeds on the order of 50 km/s.

The name of the VASIMR represents its ability to vary thrust and specific impulse. This is a required feature of a propulsion system that allows it to optimally match mission requirements, resulting in the lowest trip time with the highest delivered payload for a given fuel load. VASIMR uses electromagnetic (RF) waves to create and energize the plasma within its core, it does not require electrodes, giving greater reliability and longer life and enables a much higher power density than ion thruster and other plasma rocket concepts. Also, since every part of a VASIMR engine is magnetically shielded and does not come into direct contact with plasma, it is expected to have longer life expectancy than other ion/plasma engine designs.

In principle, VASIMR can produce high thrust and thus it can be useful for moving large payloads in low Earth orbit, transferring payloads to the Moon, and transferring payloads from Earth orbit to interplanetary orbits. However, it requires high power. For example, for missions within the Earth's gravitational sphere of influence (SOI), VASIMR requires power levels ranging from 100 – 500 kW. The nominal parameters for these missions are 5000 s specific impulse with a total mass-to-power ratio of 10 kg/kW.

One of the key challenges in developing VASIMR for performance-demanding missions is the power supply. A high-power electric thruster requires a high amount of electricity, and generating that in space require some engineering innovations. Solar power can be efficiently used for near-Earth missions, such as drag compensation for space stations, lunar cargo transport and in-space refueling. With the new advances in solar array technology, it is expected an increase in solar power utilization by an order of magnitude.

6.3.3   Electrostatic Thrusters

In an electrostatic thruster the propellant is accelerated by direct application of electric body forces to ionized the gas. These propulsion devices can operate only in a near vacuum.

An electrostatic thruster, regardless of type, consists of the same basic components: (a) the propellant supply; (b) electric power source; (c) ionizing chamber; (d) accelerator region; and (e) means of neutralizing the exhaust beam. Neutralization is achieved by the injection of electrons downstream; this process is important to avoid a space-charge buildup outside of the spacecraft which could easily nullify the operation of the thruster.

Electrostatic propulsion is typically classified in terms of its source of charged particles:

a. *Electron bombardment thrusters.* In this concept, positive ions from a monatomic gas such as Xenon or Mercury are produced by bombarding the gas with electrons emitted from a heated cathode. Ionization can be obtained by either direct current DC or radio frequency RF.
b. *Ion contact thrusters.* Positive ions are produced by passing the propellant gas through a hot (about 1100 °C) porous tungsten contact ionizer. Cesium vapor was used extensively in the original ion engines. Today Xenon is the more common propellant gas. It is a safe inert gas that stores as a dense gas (1.1052 kg/liter) under a moderate pressure of 58.4 bar at room temperature.
c. *Field emission or colloid thrusters.* This concept charges the tiny droplets of propellant either positively or negatively, and the droplets pass through an intense electric field discharge. To date the stability of large, charged particles remains a challenge.

The simplest electrostatic propulsion concept is the *ion thruster*, a device that accelerates propellant ions by an electrostatic field. Ion thrusters create very small levels of thrust compared to conventional chemical rockets but achieve very high specific impulse by accelerating their exhausts to very high speed. There are two ways to accelerate the ions: *electrostatic ion thrusters* use the Coulomb force and accelerate the ions in the direction of the electric field, while *electromagnetic ion thrusters* use the Lorentz force.

The scalar form of Coulomb's Law relates the magnitude and sign of the electrostatic force F, acting simultaneously on two-point charges q_1 and q_2: $|F| = k_e \frac{|q_1 q_2|}{r^2}$, where r is the separation distance and k_e is Coulomb's constant. If the product $q_1 q_2$ is positive, the force between them is repulsive; if $q_1 q_2$ is negative, the force between them is attractive.

The Lorentz force is the force on a point charge due to electromagnetic fields. If a particle of charge q moves with velocity $\mathbf{v}$ in the presence of an electric field $\mathbf{E}$ and a magnetic field $\mathbf{B}$, then it will experience a force $\mathbf{F} = q(\mathbf{E} + \mathbf{v} \times \mathbf{B})$.

Another form of electrostatic propulsion is the *Hall-effect plasma thruster*. Known simply as Hall thruster, this device accelerates ions with the use of an electric potential maintained between a cylindrical anode and a negatively charged plasma that forms the cathode. The bulk of the propellant is introduced near the anode, where it becomes ionized, and the ions are attracted towards the cathode, they accelerate towards and through it, picking up electrons as they leave to neutralize the beam and leave the thruster at high velocity.

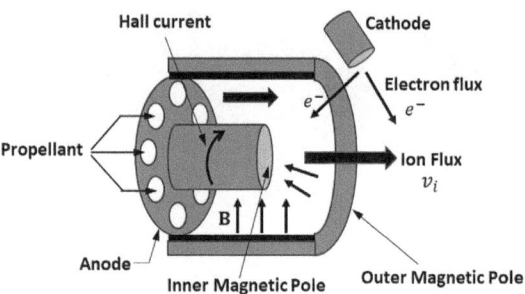

Fig. 6.6 Sketch representation of basic Hall effect thruster operation with the propellant distribution, anode, cathode, two magnetic poles, and resultant ion flow direction shown. The Hall effect means that the ions and electrons swerve in opposite directions in the magnetic field, creating an electric field. This expels the Xenon ions as a propulsive jet

The principle of Hall thrusters (Fig. 6.6) is based on the Hall effect, e.g. the creation of a voltage across a current-carrying conductor by a magnetic field. The propellant is passed through the anode, and a cross-field discharge is used to generate the plasma. An electric field is established perpendicular to the applied magnetic field, which accelerates the ions electrostatically so they reach high exhaust velocities, while the transverse magnetic field inhibits or constraints any electron motion that would tend to short-out the electric field. As illustrated in Fig. 6.6, an external cathode generates electrons to neutralize the ion beam to prevent the spacecraft from becoming electrically charged.

Hall-effect thrusters operate with a noble gas such as Xenon. After being ionized, the Xenon plasma is accelerated to speeds up to 1.7×10^4 m/s. Figure 6.7 is a photograph of a Hall thruster, showing the Xenon plasma emitting a blue glow as it exhausts the device. Xenon is a chemically inert, colorless gas. This is the propellant of choice for most electrostatic engines because of its large molecular mass, high thrust-to-current ratio, and the relative ease by which it is ionized.

Hall thrusters have been used on Soviet and Russian spacecraft since the mid-1970s. Due to its performance characteristics, the Hall thruster is now the propulsion system of choice for station-keeping and orbit raising of large commercial space vehicles. In addition, NASA's Psyche spacecraft (designed to study asteroid 16 Psyche) uses the first Hall thrusters operating beyond lunar orbit. If the mission is successful, it will demonstrate that Hall thrusters can play an important role in future deep space missions.

Although the efficiency and specific impulse of Hall thrusters are less compared with ion thrusters, the thrust at a given power level is higher. Because the design and operation are much simpler, Hall thrusters are also considered for human missions to near Earth objects; their high-power processing capabilities and their efficient operation at moderate specific impulses would reduce the trip times.

Electric power for propulsion can be obtained from either sunlight or from a nuclear reactor. In the case of SEP, solar photons are converted into electricity by solar cells. In

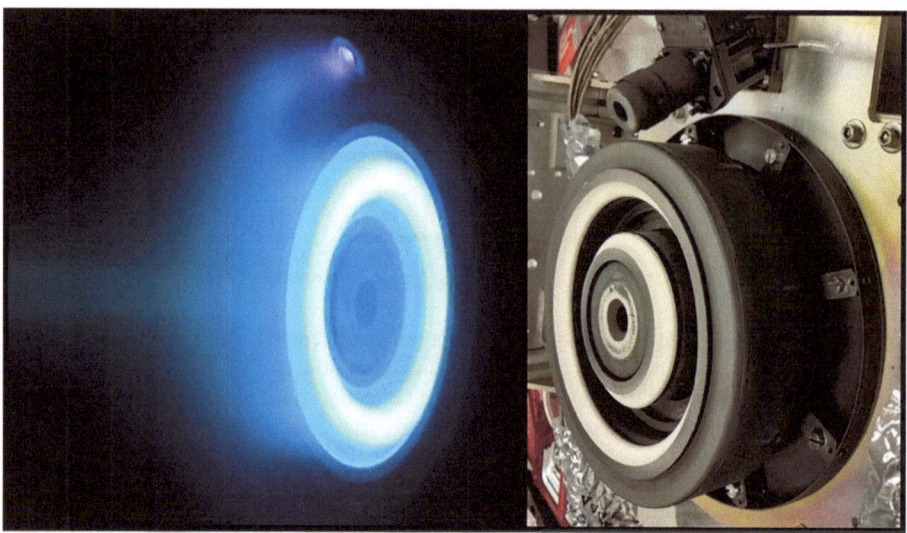

Fig. 6.7 Right image shows a NASA Hall thruster similar to one for the Psyche spacecraft. Left image captures the Xenon glow of an operating Hall thruster. $I_{sp} = 1800$ s. *Credit* NASA/JPL-Caltech

a Xenon ion SEP system, solar panels generate the power to ionize the Xenon propellant gas. Solar panels generate about 5 to 10 kW of power for a moderate planetary mission at 1 AU. Such solar array is comparable to that on a typical communication spacecraft.

A MegaFlex array is being designed for missions utilizing solar-electric propulsion requiring 350 kW or more of power. The array will measure 10 m across when fully deployed and in addition to being lightweight features a low stowage volume.

In NEP, thermal energy from a nuclear reactor is converted into electricity by either a static or dynamic thermal-to-electric power conversion system. Static systems have the advantage of no moving parts for high reliability, but they have low efficiency; dynamic systems have moving parts (e.g., turbines, generators, etc.) and do not scale well for small systems, but they are expected to operate with higher efficiency.

It is not possible to describe in detail the many electric propulsion concepts in development. However, due to the interest in ion propulsion and its recent success, the following section provides a general overview of ion thrusters.

6.4   Ion Propulsion

The *ion thruster* consists of basically three components: the plasma generator, the accelerator grids, and the neutralizer cathode (Fig. 6.8). The collisionless stream of positive atomic ions, liberated from a source, is accelerated by an electrostatic field established

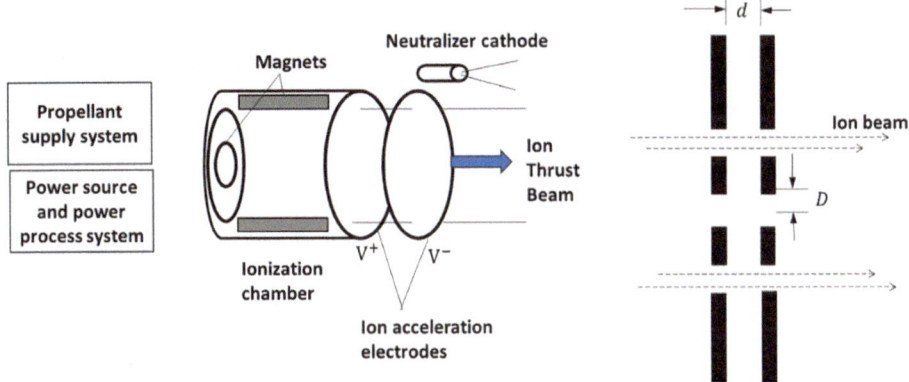

Fig. 6.8 Basic components of the grided Ion Thruster. The ion acceleration electrodes electrostatically accelerate the positive ions flowing through the grid holes. The right sketch shows the enlarged section of a dual grid with lined holes of diameter D and grid separation d

between the source surface and the accelerating electrode made up of a permeable grid. The accelerator multiple-aperture grids are used to electrostatically extract ions from the plasma and accelerate them to high velocity at voltages up to and exceeding 10 kV.

The ion thruster requires several subsystems, including the discharge chamber, the discharge cathode assembly (DCA), the grids (also called the ion optics), and the neutralizer (Fig. 6.8). During normal, steady engine operation, the neutral Xenon gas propellant is injected both through the DCA and through a ring of injectors. The DCA emits electrons that are accelerated by the electric field established between the positively biased discharge chamber walls and the negatively biased DCA. These electrons ionize the propellant by striking the gas atoms, removing one or more electrons. The "ring-cusp" magnetic field created by the magnets that surround the discharge chamber is used to improve the ionization efficiency by increasing the residence time of the electrons in the discharge chamber; the longer an electron remains in the discharge chamber, the more opportunity there is to ionize the propellant atoms.

Downstream of the grid, electrons from another source join the ion beam to produce a stream with a zero-net charge. The ion beam then exits the device at high velocity. The speed of the neutral exhaust beam is determined by the net potential drop between the ion source and the plane of effective neutralization, and by the charge-to-mass ratio of the ion species.

6.4.1   Ionization Process

Ionization is the process of electrically charging an atom or molecule by adding or removing electrons. A propellant gas is considered ionized when some or all the atoms or molecules contained in it are converted into ions. An ion is simply an atom or molecule that is electrically charged. It can be positive, when it loses one or more electrons, or negative, when it gains one or more electrons.

Plasma is an electrically neutral gas in which all positive and negative charges—from neutral atoms, negatively charged electrons, and positively charged ions—add up to zero. Plasma has some of the properties of a gas but is affected by electric and magnetic fields and is a good conductor of electricity. Plasma is an essential component in electric propulsion, where electric and/or magnetic fields are used to accelerate the electrically charged ions and electrons to provide thrust.

The propellant ionization and acceleration processes in the ion thruster are physically separated. Ionization takes place in the discharge chamber (see Fig. 6.8) utilizing different approaches such as by electron bombardment, by application of radiofrequency wave energy, or by electron cyclotron resonance. The ion acceleration process extracts ions from the discharge chamber by means of a multi-aperture grid assembly called ion optics.

The electric force depends only on the charge, and all charged particles must be of the same sign to move in the same direction. Electrostatic thrusters use charged heavy-molecular-mass atoms as positive ions. A proton is 1840 times heavier than an electron, and a typical ion of interest contains hundreds of protons. Charged colloid which can be 10,000 times heavier than atomic particles are also a possibility.

6.4.2   Ion Grids and Electrostatic Acceleration

As illustrated in Fig. 6.8, ions are electrostatically accelerated as they pass through the ion optics, which consist of two or three multi-aperture electrode grids charged to high voltage. Located at the rear of the discharge chamber, the grid in contact with the propellant plasma in the discharge chamber is called screen grid, the second one is called accelerator (accel) grid, and the third one, used in some ion engines, is called decelerator grid.

In the NSTAR ion rocket, the screen grid is maintained at DCA potential of about 1090 V above spacecraft ground (reference) at full thruster power. The second or acceleration grid is biased some 225 V below ground. The thickness of the grids is less than 1 mm and are placed about 0.7 mm apart from each other. Ions created in the discharge chamber enter through the holes in the screen grid and are accelerated by the 300 V drop in potential established by the two grids. Ions emerge from the acceleration grid at speeds greater than 39 km/s. Downstream of the acceleration grid there is an electron-emitting neutralizer to keep the spacecraft electrically neutral with respect to its environment by emitting one electron for every positively charged ion that leaves the thruster.

The thruster ion optics assembly serves three main purposes: (A) extract ions from the discharge chamber, (B) accelerate ions to generate thrust and (C) prevent the backstreaming of the electrons from the neutralizer hollow cathode.

The erosion of the acceleration grids is the main life-limiting effect of any ion thruster. Since the ion optics play a key role in the thruster performance, much effort is devoted to improve the design. The design of accelerating grids is crucial for optimum performance of ion engines.

6.4.3 Performance Relationships for Ion Thrusters

The thrust delivered by an ion thruster depends on the ion flux and is expressed by

$$F = \dot{m}_i v_i \tag{6.16}$$

where $\dot{m}_i$ is the ion mass flow rate and v_i is the ion velocity.

The energy that charged particles gain from passing through a potential difference is given by $E = eV_{acc}$, where e is the electrical charge (in coulombs), and V_{acc} is the voltage imposed across the accelerating chamber (across the grids). Conservation of energy requires that the kinetic energy of charged particles be equal to the electrical energy gained in the accelerating field. The conservation equation is, assuming no collisional losses:

$$\frac{1}{2}m_i v_i^2 = eV_{acc} \tag{6.17}$$

where m_i is the mass of the charged particles, and v_i is the speed the ions gained in the accelerator. Thus, with m_i in kilograms and V_{acc} in volts, the speed of the ions v_i (in m/s) is

$$v_i = \sqrt{\frac{2eV_{acc}}{m_i}} \tag{6.18}$$

The mass flow rate of ions is related to the ion beam current I_b by

$$\dot{m}_i = \frac{I_b m_i}{e} \tag{6.19}$$

In an ideal ion thruster, the current across the accelerator represents the sum of all the propellant mass (100% singly ionized) carried per second by the particles accelerated:

$$I_b = \dot{m}_i \frac{e}{m} \tag{6.20}$$

Substituting Eqs. (6.18) and (6.19) into Eq. (6.16), we obtain the ideal thrust from the beam of accelerated ion particles:

$$F = \dot{m}_i v_i = I_b \sqrt{\frac{2mV_{acc}}{e}} \tag{6.21}$$

For a given beam current and accelerator voltage, the thrust is proportional to the square root of the mass-to-charge ratio (m/e) of the charged particles. Here we neglected the thrust and power absorbed by the neutralizing electrons since both are rather small (about 1%). If the propellant is Xenon, $\sqrt{2m/e} = 1.65 \times 10^{-3}$, the thrust is given by

$$F = 1.65 I_b \sqrt{V_{acc}} \tag{6.22}$$

where F is in mN, I_b in amperes, and V_{acc} in volts.

The basic thrust equation given by Eq. (6.21) applies for a unidirectional, singly ionized, monoenergetic beam of ions. It must be modified to account for the divergence of the ion beam and the presence of multiply charged ions that is usually observed in electric thrusters. The total corrected thrust is given by

$$F = \tau \dot{m}_i v_i = \tau I_b \sqrt{\frac{2mV_{acc}}{e}} \tag{6.23}$$

where τ is a thrust correction factor (product of the divergence and multiply charged species terms. For example, if a 2A Xenon ion thruster has a 10-deg half-angle beam divergence and the thrust correction is $\tau = 0.958$, at 1500 V it yields a total thrust of 122.4 mN.

In practice, micronewton (μN) thrust levels are sufficient for E-W station-keeping and attitude control. Space maneuvers such as N-S station-keeping, orbit changes, drag correction, and vector positioning require millinewton (mN) thrust levels. For orbit raising and interplanetary travel the thrust required will vary from 0.2 to 10 N.

> **Did you know?** For the NSTAR 30-cm ion rocket, the thrust and corresponding total mass flow rate range from about 94 mN and 3 mg/s at full power to 20 mN and 1 mg/s at minimum power, with cathode mass flow rates as low as 0.24 mg/s.

The specific impulse relates the thrust to the rate of propellant weight. And since for ion propulsion the thrust is due mainly to the ions, we write

$$I_{sp} = \frac{v_{ex}}{g_0} = \frac{v_i}{g_0} \frac{\dot{m}_i}{\dot{m}_p} \tag{6.24}$$

where v_i is the exhaust velocity for unidirectional, mono-energetic ion exhaust.

To account for the ionized versus nonionized propellant, a *thruster mass utilization efficiency* is defined for singly charged ions as

$$\eta_m = \frac{\dot{m}_i}{\dot{m}_p} = \frac{I_b m}{e \dot{m}_p} \tag{6.25}$$

Substitute Eq. (6.23) for the thrust and Eq. (6.25) for the propellant utilization efficiency into Eq. (6.24) to obtain an expression for the specific impulse I_{sp}:

$$I_{sp} = \frac{\tau \eta_m}{g_0} \sqrt{\frac{2eV_{acc}}{m}} \tag{6.26}$$

where we used propellant utilization efficiency for singly charged ions.

The I_{sp} for any propellant is then written as

$$I_{sp} = 1.417 \times 10^3 \tau \eta_m \sqrt{\frac{V_{acc}}{m_i}} \tag{6.27}$$

where V_{acc} is the beam voltage, and M_i is the ion mass in atomic mass units (AMU); 1 AMU $= 1.6605 \times 10^{-27}$ kg. For Xenon, with $m_i = 131.29$, the I_{sp} is given by

$$I_{sp} = 123.6 \tau \eta_m \sqrt{V_{acc}} \tag{6.28}$$

For an ion rocket with thrust correction factor $\tau = 0.958$ (a 10-deg half-angle beam divergence) and a 90% propellant utilization of Xenon at 1500 V, the I_{sp} would be 4127 s.

Another parameter of importance for the ion thruster is the current density j, a measure of the density of flow of a conserved charge, or the flux of the charge. The current density that can be obtained with a charged particle beam has a saturation value depending on the geometry and the electrical field. This fundamental limit is caused by the internal electric field associated with the ion cloud opposing the electric field from the accelerator when too many charges of the same sign try to pass simultaneously through the accelerator.

The saturation current can be derived for a plane-geometry electrode configuration from basic principles. The current density in terms of the space charge density is:

$$j = \rho_e v_i \tag{6.29}$$

The voltage in a one-dimensional space-charge region is found from Poisson's equation,

$$\frac{d^2 V}{dx^2} = \frac{\rho_e}{\varepsilon_0} \tag{6.30}$$

where x represents distance, and ε_0 is the *permittivity* of free space: $\varepsilon_0 = 8.854 \times 10^{-12}$ F/m (farads per meter).

Solving Eqs. (6.18), (6.29), and (6.30) simultaneously and applying the proper boundary conditions, the result yields the following relation for the current density, known as the *Child-Langmuir law*:

$$j = \frac{4\varepsilon_0}{9}\sqrt{\frac{2e}{m}}\frac{(V_{acc})^{3/2}}{d^2} \qquad (6.31)$$

where d is the accelerator inter-electrode distance or grid separation.

The equation for the saturation current density for atomic or molecular ions is,

$$j = 5.44 \times 10^{-8}\frac{V_{acc}^{3/2}}{\mathcal{M}^{1/2}d^2} \qquad (6.32)$$

where the current density is in A/m², V_{acc} in volts, and the distance d in meters.

The current density and the area are very sensitive to the accelerator voltage as well as to the electrode configuration and spacing. For Xenon with electron bombardment, values of j vary from 2 to about 10 mA/cm².

Let the cross section be circular so that $I = (\pi D^2/4)j$, where D is exhaust beam emitter diameter. Then we use Eqs. (6.21) and (6.32) to obtain a relation for the thrust in terms of grid size and the voltage,

$$F = \frac{2}{9}\frac{\pi\varepsilon_0 D^2 V_{acc}^2}{d^2} \qquad (6.33)$$

and substituting the values of the constants we obtain a thrust formula

$$F = 6.18 \times 10^{-12}V_{acc}^2\left(\frac{D}{d}\right)^2 \qquad (6.34)$$

where the ratio of the beam emitter diameter D to the accelerator-electrode grid spacing d is the aspect ratio of the ion accelerator region (see Fig. 6.8).

For multiple grids with many holes, the diameter D is that of the individual perforation hole and the distance d is the mean spacing between grids. Due to space-charge limitations, D/d can have values no higher than about one for simple, single-ion beams. This implies a rather stubby engine design with many perforations.

In actual design, engineers minimize the diameter of each acceleration grid aperture to retain nonionized neutral gas in the plasma generator, and attempt to maximize the screen grid transparency so that that the grids extract the maximum possible number of ions from the plasma. Also, the grid separation determines the acceleration of the particles. Thus, the electrode diameters and spacing are optimized to eliminate direct interception of the beam ions on the acceleration grid, which would cause rapid erosion due to the high ion energy.

Now combine Eqs. (6.9), (6.10), and (6.20), and assume a conversion efficiency of potential energy to kinetic energy η_t, to obtain the power of the electrostatic accelerator:

Fig. 6.9 Photograph of NEXIS ion thruster, showing the 57-cm-diameter multi-aperture grids and plasma screen enclosing the thruster body. *Credit* Polk et al. [17]

$$P_e = IV_{acc} = \frac{\dot{m}v_i^2}{2\eta_t} \qquad (6.35)$$

This relation clearly shows that a high exhaust ion velocity requires a high accelerator power, which in turn can be minimized by a higher energy efficiency.

The Nuclear Electric Xenon Ion System (NEXIS), shown in Fig. 6.9, has a 65 cm diameter discharge chamber with six rings of magnets in a ring-cusp configuration. NEXIS is the result of research and development activity within NASA's Project Prometheus in 2003. The goal was to develop high specific impulse nuclear electric propulsion systems that would enable more robust and ambitious science exploration missions to the outer Solar System. NEXIS provided a huge improvement for ion propulsion, designed to achieve efficiencies $\geq 78\%$ while increasing the thruster power to 120 kW and yielding specific impulse greater than 6000 s.

In general, the overall efficiency of an electrostatic thruster is a function of the thruster efficiency as well as of other loss factors. One loss of energy is the energy expended in ionization (charging the propellant). This process resembles the dissociation energy in electrothermal devices. Ionization is necessary to make the propellant respond to the electrostatic force and is non-recoverable.

The ionization energy is found from the ionization potential of the atom or molecule times the current flow. The amount of ionization to be expected in as gas in thermal equilibrium is given by the Saha equation:

$$\frac{n_i}{n_n} \approx 2.4 \times 10^{21} \frac{T^{3/2}}{n_i} e^{-U_i/kT}$$

where n_i, n_n represent the number density per m^3 of ions and neutral species, respectively, T is the gas temperature in kelvin, k the Boltzmann's constant, and U_i the ionization potential.

Table 6.2 Ionization potential of propellant gases

Gas	Atomic mass (kg/kg-mol)	Ionization potential (eV)	Remarks
Argon (Ar)	39.948	15.76	
Krypton (Kr)	83.80	13.99	
Xenon (Xe)	131.29	12.13	Most effective
Mercury Vapor (Hg)	200.59	10.44	Difficult to store
Cesium Vapor (Ce)	132.905	3.89	Highly reactive
Neon (Ne)	20.183	21.6	
Hydrogen (H)	2.014	15.4	

The Saha equation assumes complete thermodynamic equilibrium, single U_i, and is not applicable to extremely hot or dense plasmas. The ionization potential represents the energy needed to ionize a gas (remove electrons), and it varies significantly from one gas to another. Xenon is the most used propellant due to its high atomic mass and low ionization potential. Ionization potential of some gases is given in Table 6.2 together with the atomic mass. In actual practice, to operate the ionization chamber of an electrostatic thruster higher voltages are required than the ionization potential shown in this table. The unit of ionization potential is eV (electron volts), where $1 \text{ eV} = 1.60 \times 10^{-19}$ J.

Cesium was initially regarded as a propellant for electrostatic thrusters because of its high vapor pressure and ease of ionization, but soon it was discovered that cesium is highly reactive and very difficult to isolate. Mercury vapor was also considered for its well-known ionization behavior from fluorescent lamps, but it was found too difficult to store and utilize. Eventually Xenon was selected as it was easy to handle and is relatively easy to ionize.

Plasmas in electric propulsion devices can span orders of magnitude in plasma density, temperature, and ionization fraction. Therefore, models used to describe the plasma behavior and characteristics in a thruster must be formed with assumptions that are valid in the regime being studied.

6.4.4   Power Required to Ionize a Propellant

The energy required to ionize a propellant must be kept at a minimum. Thus, the selection of propellant is crucial to ensure the thruster performance is maximized since ionization energy represents a source of inefficiency since it takes away from the kinetic energy available for thrust. Ionization energy is expressed as the energy required to ionize a neutral atom. However, for propulsion performance, ionization energy is expressed as the amount of energy required per-kilogram of mass flow, using a factor, E_{ion}, to account for the difference in ionization potentials in the propellant and the difference in atomic masses of the candidate species present.

The power required to completely ionize the neutral mass flow of propellant entering the thruster, denoted P_{ion}, is expressed as

$$P_{ion} = \dot{m}E_{ion} \tag{6.36}$$

Since the kinetic power of the exhaust beam is given by Eq. (6.7), we use it with P_{ion} to calculate the maximum theoretical efficiency of the acceleration process, assuming that the only energy loss is propellant ionization:

$$\frac{P_{ion}}{P_{jet}} = \frac{2E_{ion}}{\left(g_0 I_{sp}\right)^2} \tag{6.37}$$

This ratio gives the theoretical minimum efficiency penalty required to singly ionize the propellant as a function of I_{sp}. Kieckhafer and King [11] plotted Eq. (6.37) for the propellants they investigated. As expected, propellants with lower ionization energies were shown to require a smaller fraction of the total thruster power to ionize.

The performance of the ion thruster performance has also been assessed with different propellants. It is found that with argon, for example, a potential drop $e > 10^4$ V yields exhaust velocities greater than 2×10^5 m/s. With Xe/Cs the exhaust velocity is lower, yet it is higher than the optimum values obtained for a generic EP with a high specific power α.

Modern ion thrusters can propel a spacecraft up to 90 km/s (200,000 mph) and can deliver up to 0.5 newtons (0.1 pounds) of thrust. To compensate for low thrust, the ion thruster must be operated for a long time for the spacecraft to reach its top speed.

Example 6.5 For an electron-bombardment ion rocket the following data is available:

Working fluid   Xenon (131.3 kg/kg-mol)

Net accelerator voltage   700 V

Distance d between grids   2.5 mm

Diameter D of each grid opening   2.0 mm

Number of holes in the grid   2200

Ionization potential for Xenon   12.08 eV

Determine the thrust, exhaust velocity, specific impulse, mass flow rate, propellant needed for 91 days operation, the power of the exhaust jets, and the thruster efficiency.

Solution: The ideal thrust is obtained from Eq. (6.34):

$$F = 6.18 \times 10^{-12} V_{acc}^2 \left(\frac{D}{d}\right)^2 = 1.94 \times 10^{-6} \text{N per grid hole}$$

The total ideal thrust is then obtained by multiplying by the number of holes

$$F = 4.26 \text{ mN}$$

The exhaust velocity is obtained from Eq. (6.18):

$$v_i = \sqrt{\frac{2eV_{acc}}{\mu}} = 31,860 \text{ m/s}$$

Therefore, the specific impulse is

$$I_{sp} = 3248 \text{ s}$$

The mass flow rate, obtained from Eq. (6.19), is

$$\dot{m} = \frac{F}{v_i} = 1.34 \times 10^{-7} \text{ kg/s}$$

For a cumulative period of 91 days of operation, the amount of Xenon propellant needed (assuming no losses) is

$$m = \dot{m}t = 1.05 \text{ kg}$$

The ideal kinetic energy rate in the jet is

$$\text{K.E.} = \frac{1}{2}\dot{m}v^2 = 67.9 \text{ W}$$

In practice, this quantity will be reduced by inherent losses in the system, including ionization losses.

6.4.5 The NSTAR and NEXT Programs

The NASA Solar Technology Application Readiness (NSTAR) project was a program to validate ion propulsion technology for use on NASA deep space missions. The first NSTAR flight Xenon Ion Thruster, Power Processor Unit (PPU) and Digital Control and Interface Unit (DCIU) were used as the primary propulsion on the Deep Space 1 comet and asteroid rendezvous probe, which was launched on 24 October 1998. The highly successful mission lasted over three years. On 18 December 2001, NASA engineers shut

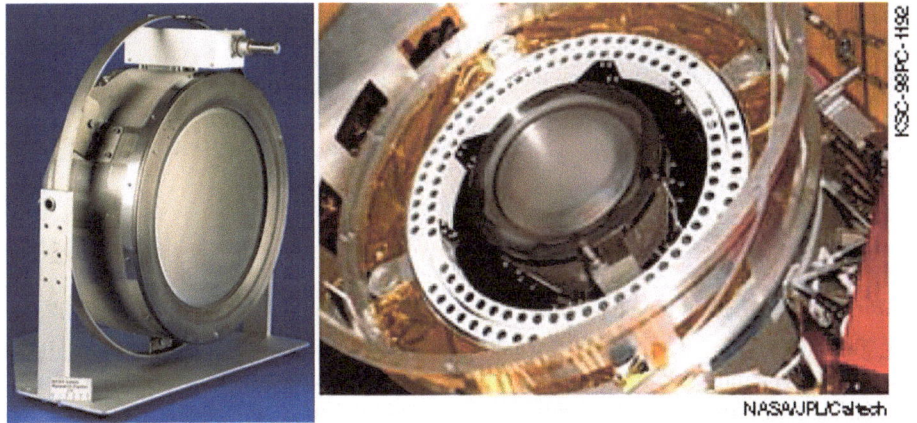

Fig. 6.10 Close-up view of the ion engine on Deep Space 1. NSTAR engine: $L = 30$ cm, $m = 8$ kg (17.6 lbm), $I_{sp} = 3100$ s, $F = 20$ to 92 mN. *Credit* NASA/JPL

down Deep Space 1, terminating the first U.S. space mission utilizing Xenon ion thrusters as its primary mode of propulsion.

At full throttle, NSTAR consumed about 2.3 kW of electric power and produced a thrust of about 90 mN (0.02 lbf). A typical chemical on-board rocket produces larger thrust (450 to 2250 N) but for much shorter times. The NSTAR ion thruster operated for 16,246 h. A close-up view of Deep Space 1 ion engine is shown Fig. 6.10.

In 2007, NASA launched the Dawn spacecraft, the second interplanetary mission using a NSTAR engine. Dawn had three redundant ion thrusters of 30 cm diameter. Dawn became the first exploratory mission to use ion propulsion to enter and leave more than one orbit. The mission featured extended stays at two very different extraterrestrial bodies: giant asteroid Vesta and dwarf planet Ceres, both found in the main asteroid belt between Mars and Jupiter. Dawn carried 425 kg (937 lbm) of on-board Xenon propellant, and was able to perform a velocity change of 25,700 mph (11.49 km/s) over the mission.

Deriving from the successful NSTAR ion thrusters that propelled Dawn, the 6.9 kW NASA Evolutionary Xenon Thruster (NEXT) was conceived for the Double Asteroid Redirection Test (DART) mission, a grided ion thruster several times more powerful than its predecessors. Launched in 2021, DART was a NASA space mission aimed at testing a method of planetary defense against near-Earth objects (NEOs). Using electricity generated by the spacecraft's solar panel, the NEXT ion thruster accelerated the Xenon propellant to speeds of up to 40 km/s, yielding a specific impulse of 4170 s, an improvement of over 1000 s compared with NSTAR thruster. The DART mission ended in September 2022 when the spacecraft successfully impacted its asteroid target.

Did you know? The Dawn spacecraft used its ion rockets for more than 2000 days, with interruptions of only a few hours each week to turn the antenna toward Earth. This thrusting time exceeds the 678 days of Deep Space 1 operating on ion propulsion. Dawn carried more propellant, 425 kg (937 lbm), Deep Space 1 carried just 82 kg (181 lbm).

Example 6.6 A Xenon ion engine operates with charged particles mass of 2.2×10^{-25} kg and charge $e = 1.6 \times 10^{-19}$ coulombs. What is the theoretical speed of the ions if the voltage across the acceleration chamber grids is 1.3 kV? Determine also the average acceleration of the ions as they cross the grids having a separation of 0.7 mm, and calculate the force they experience by this acceleration.

Solution: The theoretical speed of the ions is, from Eq. (6.21),

$$v_i = \sqrt{\frac{2eV_{acc}}{m_i}} = 4.348 \times 10^4 \text{ m/s}$$

The NSTAR ion thruster accelerates the ions to 30 km/s. Thus, if we assume an average speed of 3.0×10^4 m/s, the charged particles would move across the grids in $\sim 2.3 \times 10^{-8}$ s. Thus, the average acceleration is

$$a = \frac{\Delta v}{\Delta t} = 1.28 \times 10^{12} \text{ m/s}^2$$

The force experienced by the particles due to acceleration is, from Newton's second law,

$$F = m \cdot a = \left(2.2 \times 10^{-25} \text{ kg}\right)\left(1.28 \times 10^{12} \text{ m/s}^2\right) = 2.816 \times 10^{-13} \text{ N/particle}$$

6.5   Applications of Electric Propulsion

Electric propulsion thrusters are extensively used for attitude control, orbit maintenance, and/or primary propulsion to provide orbital changes. Mission planners and spacecraft operators now recognize electric propulsion as capable of providing substantial gains in mission performance and/or cost reductions. Science missions could use nuclear electric propulsion (NEP) to reach distant regions of our Solar System, while SEP can propel tugs used to position, service, resupply, repositioning, and salvage space assets.

Mission analyses suggest a need for higher power electric propulsion systems for deep space and cislunar missions, including both geocentric and lunar orbital station-keeping maneuvering. These diverse applications require propulsion systems with different optimal characteristics in terms of thrusting time, specific impulse, and payload capability. For example, Earth orbital applications such as space tugs, and spacecraft orbit insertion require high thrust electric propulsion systems for timely space transfers. However, missions with large delta-v such as interplanetary space missions require high specific impulse propulsion.

6.5.1 Earth Orbit Missions

Solar electric propulsion (SEP) can be used to efficiently transport heavy payloads from LEO to higher orbits. For example, a payload can be launched from the ground via conventional chemical rockets. Then, powered by electric thrusters the payload could spiral out to higher energy orbits, including the Euler-Lagrange points. Some advanced concepts are even considered as cargo tugs between LEO and the Moon. Other SEP applications include orbit insertion, orbit transfers, attitude control, orbit maintenance, and de-orbit, i.e. assist in descent.

Ion thrusters are routinely used for station-keeping on commercial and military communication satellites in geosynchronous orbit. For example, SpaceX's Starlink satellite constellation uses Hall thrusters to raise orbit, perform maneuvers, and de-orbit at the end of their use. The Gravity Field and Steady-State Ocean Circulation Explorer (GOCE) was launched by ESA on 16 March 2009. It used ion propulsion throughout its twenty-month mission to overcome the atmospheric drag while in its low orbit (altitude of 255 km) before intentionally deorbiting on 11 November 2013. China's Tiangong space station also uses, in addition to chemical thrusters, four Hall-effect thrusters to adjust and maintain its orbit.

The benefits of using SEP for Earth-orbit missions include increased operational life, higher payload mass and/or reduced launch vehicle costs. SEP is also ideal for large communication satellites in GEO that require reliable and economical propulsion systems for long-term station-keeping. For example, for north-south station-keeping (NSSK), a geostationary satellite needs a total velocity increment Δv of up to 55 m/s per year of service life. This corresponds to a requirement of 110 kN·s total impulse for a 2-ton class satellite per year. The gross propulsion system mass for a 10-year orbit life would be 250 kg for an ion thruster as compared with 600 kg for a chemical rocket.

6.5.2 Lunar Orbit Propulsion

According to NASA, Hall thrusters are considered ideal for human missions to near Earth objects because of their high-power processing capabilities and their efficient operation at moderate specific impulses, which could reduce the trip times for such missions. For the Power and Propulsion Element (PPE) of the Lunar Gateway, the space station currently developed under NASA's Artemis program, the primary propulsion will consist of a combination of 6 kW Hall thrusters provided by Busek and NASA Advanced Electric Propulsion System (AEPS) Hall thrusters. The high specific impulse of Hall thrusters will allow for efficient orbit raising and station-keeping for the Lunar Gateway's polar near-rectilinear halo orbit. The PPE is a high-power, 60-kilowatt solar electric propulsion spacecraft built by Maxar Technologies and operated by NASA. It will provide power, high-rate communications, attitude control, and orbital transfer capabilities for the Lunar Gateway and will be the first U.S. electric propulsion system on a human-rated mission.

The SMART-1, a spacecraft launched in 2003 by the European Space Agency (ESA) was designed to use a Snecma PPS-1350-G Hall thruster to transfer from GTO to lunar orbit. The 1.5 kW Xenon Hall thruster had a specific impulse of 1660 s. The SMART-1 satellite completed its mission on 3 September 2006, impacting on the Moon's surface in a controlled manner.

6.5.3 Inner Planets Propulsion

In 2018, ESA launched BepiColombo, a mission to explore Mercury. The spacecraft uses ion thrusters in its trajectory to Mercury, then a chemical rocket will complete orbit insertion. The Mercury Transfer Module (MTM), which supplies power to the spacecraft, is equipped with four QinetiQ-T6 ion thrusters, which operate singly or in pairs for a maximum combined thrust of 290 mN. Solar energy is supplied by two 14-m-long (46 ft) solar panels. Depending on its distance to the Sun, the generated power can range between 7 and 14 kW. Each T6 ion thruster requires between 2.5 and 4.5 kW, depending to the desired thrust level for the mission.

6.5.4 Deep Space Propulsion

The same concept considered for near Earth missions could facilitate missions to near Earth asteroids and other destinations in deep space. In other words, a payload or spacecraft could first be launched to LEO via conventional chemical rockets, and then use electric propulsion to propel spacecraft on low-thrust, high-specific impulse interplanetary trajectories.

Using EP for deep-space missions would reduce considerable the amount of propellant, and thus the size of the spacecraft. To date, ion engines have demonstrated high performance for interplanetary spacecraft. To achieve a velocity increment $\Delta v = 4.5$ km/s, the 486 kg Deep Space 1 spacecraft operated a single 30-cm ion thruster during its mission, consuming less than 81 kg of Xenon propellant. This translates into a lifetime at full power of 7500 h.

One of the most successful probes propelled by grided ion thrusters was designed to study a near-Earth asteroid (NEA) was Hayabusa, a Japanese sample return mission that brought a sample of material from a small NEA. Launched in December 2014, the Hayabusa 2 spacecraft operated with ion engines during the multi-year mission, the outbound journey from Earth to the asteroid and the flight back to Earth after collecting samples. The three-grid ion engines that propelled the JAXA's Hayabusa spacecraft used 66 kg of Xenon propellant, and yielded a specific impulse between 3000 and 3200 s. Each ion engine in Hayabusa 2 operated for about 6400 h on the outward journey, and for about 3000 h on the return journey.

NASA's Psyche spacecraft (launched in 2023) is propelled by a solar-powered Xenon Hall-effect ion thruster (SPT-140) on a mission scheduled to reach asteroid 16 Psyche in August 2029. The giant metal rich asteroid Psyche orbits the Sun between Mars and Jupiter at a distance ranging from 2.5 to 3.3 AU (378 million to 497 million km) from the Sun. The small (8.5 kg) 4.5 kW SPT-140 thruster can deliver a specific impulse of 1800s.

Of course, for deep-space exploration missions we must be careful in comparing the performance of an electric thruster with that of a chemical rocket. One-way minimum-energy transfers produce unacceptable long travel times to far away planets. This may result on damage to instruments due to an extended exposure to the severity of the interplanetary medium and the long exposure to radiation.

Moreover, orbital trajectories usually involve one or more gravity assists that can provide a Δv much greater than that provided by an on-board electric thruster. For example, the Voyager spacecraft incorporated a Solid Rocket Motor (SRM) to provide the impulse for the final increment of injection velocity after launch. The total launch mass was 2016 kg, including the 1046 kg SRM, to deliver a total impulse of 2897 kN·s. The gravity assist provided by Jupiter saved both Voyager spacecraft over 1600 tons of in-flight propulsive mass [20].

An interesting Solar Electric Earth Gravity Assist (SEEGA) trajectory has been considered with ion propulsion for a Neptune orbiter mission. This mission would deliver a payload (an orbiter) of 310 kg to Neptune after a 10-year trip time. Voyager 2 reached Neptune in 12 years with a 115-kg science payload.

Continuing progress in the development of both solar-cell power systems and ion thrusters will make it possible to carry out mission applications of electric propulsion for deep space interplanetary spacecraft. Many studies have concluded that electric propulsion technology can enable deep-space missions by reducing the launch mass or enhance exploration goals by reducing trip times and allowing greater payloads to the outer planets and even for interstellar missions. The reader is urged to consult the literature for details on these and other studies.

6.6   Technology Goals for Electric Propulsion

After decades of experimental research and laboratory testing, ion engine technology has now reached a level of maturity that allows EP to be considered for a variety of Earth orbit or deep space missions. Advancing new technologies for better use of electric propulsion in missions to the outer planets continues. NASA is developing a 7-kW Xenon ion thruster for near-term solar powered spacecraft, and a 25-kW ion engine for nuclear-electric spacecraft. The 7-kW ion thruster and power processor can be throttled down to 1 kW and are applicable to 25-kW flagship missions to the outer planets, asteroids, and comets.

Table 6.3 summarizes the range of operating parameters for thrusters with flight history. As shown, power levels are rather small. To achieve significant improvements in performance and cost savings, system power levels of 100's of kW are necessary for future EP systems along with the capability to store a much larger mass of propellant. With a specific impulse of 6000 s and producing a thrust of 90 mN, the NEXIS Xenon ion engines represent the SOA in ion thrusters. Ion engines utilize the least amount of propellant, e.g. the Dawn interplanetary mission used 425 kg of Xenon: 247 kg for its Vesta approach, and another 112 kg to reach Ceres.

Alternative propellants must also be considered. Although Xenon has been the propellant of choice for ion and Hall thrusters, mainly due to its low ionization energy, high atomic mass and easy storage and flow metering, there are disadvantages that preclude the use of Xenon in future high-power thrusters. One disadvantage is high cost, which can adversely affect the expense of a long duration mission. For example, studies have shown that for high-Isp missions, Kypton is a good choice over Xenon as at such high exhaust energies the large ionization cost is minimized (see Kieckhafer and King [11]).

6.6.1   Limitations of Electric Propulsion

Electric propulsion systems are limited by the need for technically sophisticated external power sources, and very low to modest thrust density capabilities. Thus, the high-specific impulse capabilities of electric propulsion come at the expense of power and mass. This

Table 6.3 Typical operating parameters for EP thrusters with flight history

Thruster	Specific impulse, I_{sp} (s)	Input power (kW)	Efficiency range (%)	Propellant
Resistojet	300	0.5–1	65–90	N_2H_4
Arcjet	500–600	0.9–2.2	25–45	N_2H_4
Ion thruster	2500–4170	0.4–6.9	40–80	Xenon
Hall thruster	1500–2000	1.5–4.5	35–60	Xenon

is referred to as the "power supply penalty" that places a premium on maximizing the specific power (in W/kg). This means that the highest attainable specific impulse of a propulsion system may not always be the optimum choice for a given mission.

The mass of the power plant is directly proportional to the exhaust velocity v_{ex}, and inversely proportional to the product of specific power and propulsion efficiency. The power supply penalty limited the application of electric propulsion on-board spacecraft until the 1990s, as spacecraft power finally began to increase to meet the growing needs of communication satellites. Nonetheless, in addition to other design constraints, such as launch vehicle selection, mission time, and payload mass, power plant mass must also be considered in the trade-offs needed to optimize a given mission profile.

Ion thrusters need to operate for extended periods of time—typically thousands to tens of thousands of hours—to impart the required Δv to a spacecraft. To maintain this continuous and long operation, crucial design challenges include propellant management necessary to control mass flow rates, and engine lifetime.

Ion engines are limited in total deliverable impulse by the maximum propellant throughput due to engine damage. A small fraction of the accelerated ion current impinges on the grids, causing some power loss and sputtering or damage caused by ions collisions with the solid surfaces. Sputtering removes material from exposed surfaces upon impact and reduces the life of the grids. Other spacecraft surfaces can be contaminated by the deposition of the material that has been sputtered away by the impacting high-energy ions. Sputtering is a huge problem for Earth-orbiting satellites where the plume ions cannot always be directed away from all the important components of the spacecraft such as solar arrays and antennas. The erosion of the discharge chamber, which causes the contamination of spacecraft surfaces, is directly related to the thruster lifetime.

Conceptually, the fixed grid gap in ion thrusters constrains the power level, specific impulse and thrust delivered. NASA is currently developing the VIPER (Variable IsP Electric Rocket), an ion thruster concept in which the grid gap of the ion optics assembly can be adjusted during flight. The variable grid gap approach may improve the performance of ion propulsion.

A disadvantage of solar electric thrusters is the reduction in I_{sp} as the power decreases, which occurs in deep space missions where the power available decreases as the spacecraft moves away from the Sun, i.e. the beam voltage decreases, resulting in a proportional reduction of I_{sp}.

Electric propulsion requires a reliable electric power supply system of low specific mass, interfaced with suitable power processing equipment, and impeccable operation in the space environment over long periods of time. This is especially an issue for Earth-orbiting spacecraft that may be damage due to an extended exposure to the severity of the Van Allen radiation belt.

One of the most significant challenges for electric propulsion in general is the need for substantial amounts of electrical power. Combinations of energy sources and conversion methods are typically utilized and some have reached sufficient technology maturity, but only solar cells (photovoltaic), isotope thermoelectric generation units (nuclear), and fuel cells (chemical) have advanced to the point of routine space-flight operation. To date, power output capacity has increased from the low one-kW range to the medium tens of kWs required for some missions. Engineers continue seeking technological breakthroughs to develop the hundred kilowatts or more needed by some applications.

In addition to achieving higher specific impulse, EP offers other operational benefits, including precision and variability of thrust levels and impulse increments, shutdown and restart capabilities, and the use of chemically passive propellants. Their major limitations are the need for sophisticated external power sources, and low to modest thrust density capabilities.

In some cases, accelerations tend to be very low ($10^{-4} - 10^{-6} g_0$), but thrusting times are typically long (several months). Thus, spacecraft propelled by electric propulsion are designed to move in multiple spiral trajectories to reach their destination. Continuous thrusting at a fixed inertial attitude lowers the apogee and raises the perigee in each orbit until it reaches the final high circular orbit. Because of the long transfer orbit durations with an electric thruster, we can also consider using two different types of propulsion for missions with trajectories other than spiral, for example using chemical rockets to arrive at a very eccentric, super-synchronous elliptical orbit and then switch to electric thrusters to attain a higher orbit.

Due to the limited thrust densities, electric propulsion is not appropriate for rapid maneuvers in strong gravitational fields. They cannot be used for launch or ascent-descent near planetary surfaces, and even outer orbit transfer maneuvers can only be performed very slowly over gentle spiral trajectories. Therefore, the near-planet application of electric propulsion is limited to attitude-control, station-keeping, drag-reduction, and modest orbit-changing functions (such as orbit phase changes in LEO constellations) where the minuteness and precision of thrust, propellant conservation, and long lifetime give them superiority over chemical rockets. For interplanetary cruise, electric propulsion offers much more substantial advantages over chemical propulsion if it delivers the levels of thrust required for heavy cargo and even crewed missions to Mars. The potential of EP to propel robotic probes to the outer planets and beyond the Solar System is now seriously considered.

Due to the limitations of the propulsion system, designers must carry out optimization studies involving multidimensional trade-offs among mission objectives, propellant and mass of the energy generation system, trip time, internal and external environmental factors, and overall system reliability.

The increasing life trend in Earth-orbit satellites from a minimum of 8 years to at least 15 years significantly increases the total impulse and durability requirements of the propulsion system. For many missions, thrusters must have the capability to operate efficiently over a range of specific impulses from below 1000 to over 40,000 s. For example, the north-south station-keeping (NSSK) function of a typical geosynchronous satellite requires 40 to 45 kN-sec (9000 to 10,000 lbf-sec) of impulse per year.

Additional increases in payload mass or decreases in launch vehicle costs are sought by using SEP for orbit transfer. EP research and development is also focused on scaling down both physical size and power level (< 100 W), for applications on micro-spacecraft.

Critical areas of R&D incorporate nanotechnology and advanced materials to develop new components and electric-power generation systems with low weight per unit electric power produced, and with increased lifetime. Research continues to develop better components that will boost the efficiency of electric propulsion and improve efficiency in conversion of electric power into thrust.

Higher power thrusters are also being developed. A flight demonstration mission is now planned on a representative trajectory through the Van Allen radiation belts to test and validate key capabilities and technologies required for future exploration elements such as a 300 kW solar electric transfer vehicle.

6.7   The Possibilities for the Future

There are huge benefits by using electric propulsion for many space missions. High-Δv missions require a propulsion system that will either significantly reduce the amount of required propellant, or that can increase the payload or spacecraft dry mass for a given wet mass associated with the mission requirement.

However, the decision to select a propulsion system to carry out a given mission is based in a complex and comprehensive analytical optimization process that considers many variable-thrust trajectories such as aerobraking, swing-by, and in-flight course corrections, availability of on-board power sources, fraction of payload to be returned, secondary mission goals in flight, and other design and operational considerations.

Electric propulsion, while highly efficient, can only produce a small amount of thrust. This requires that the thrusters operate for a long time, which corresponds to a significant fraction of the trajectory. This performance characteristic makes it more difficult to find optimal trajectories. Direct and indirect methods are used for optimizing low-thrust trajectories, each giving varying results. Direct methods parameterize the problem and use nonlinear programming techniques to optimize an objective function by adjusting a set of variables. Several direct methods are available, each subject to the limitations of the nonlinear programming techniques that are used. Indirect methods are based on calculus of variations, an analytic procedure that requires to solve a two-point boundary value problem and which solution requires to satisfy terminal constraints and targeting conditions. These indirect methods are subject to extreme sensitivity to the initial guess of the variables—some of which are not physically intuitive or known a priori. If the trajectory includes maneuvers such as gravity assist, the solution becomes much more complex and the level of uncertainty increases.

In the case of a piloted mission, for example, the optimization would also consider the length of time to accomplish the mission, internal and external environmental hazards, and other human factors that affect the crew. Today, due to the lack of experience with sufficient EP in crewed spacecraft, variables involving human factors lack fundamental databases, and theoretical representations to carry out a suitable mission assessment are merely projections.

The technology of electric propulsion has matured significantly in the last decades. Electric thrusters are limited by the power-generation technology available. High-voltage electric fields can accelerate charged particles to high exhaust velocities. However, the acceleration of electrically charged particles requires a large quantity of electric power. In terms of propellant flow rate, the amount of electric power required is given by Eq. (6.12), considering the efficiency of energy conversion, η_t.

Hence, the electric power requirements increase as the exhaust velocity is increased. Suppose we need an electric rocket capable of delivering 4.45 N (1 lbf) thrust, with $I_{sp} = 5099$ s. This would require more than 100 kW of electric power to accelerate enough charged particles to the required exhaust velocity (assuming perfect energy conversion efficiency):

$$P = \frac{1}{2}Fv_{ex} = \frac{1}{2}(4.45 \text{ N})(5 \times 10^4 \text{ m/s}) = 1.1125 \times 10^5 \text{J/s} \approx 100 \text{ kW}.$$

This is achievable. However, the required power plants to produce higher electric power could be too heavy for space applications. The development of lightweight power systems for space propulsion power is one of today's most challenging problems.

Research efforts for EP in general concentrate on extending it to new operating regimes to support deep-space missions of interest. Grided ion propulsion is a rapidly evolving research field where new areas are being pursued, including propellants, and study of new promising concepts, such as the annular ion engine.

For the Hall thruster, research focuses on answering several outstanding physics-based questions such as what is the nature and impact of large-scale ionization-driven plasma oscillations, what causes micro-instabilities and what is their impact on particle transport. Research continues to evaluate the processes that govern thruster lifetime, and study their operation on alternative propellants. Recent results of such research and development efforts are contained in a special issue of Journal of Applied Physics. The reader is urged to consult Jorns et al. [9] and the literature cited therein for complete details on the state of the art of these technologies.

Electric propulsion continues advancing and evolving. For missions with large cargo and piloted spacecraft to the planets require high-power levels (> 100 kW) and such systems may become available in the next few decades. Advanced solar power generation technology will enable low cost, modular power growth from 30 kW to MW class spacecraft and promise a revolutionary capability.

Glossary

e Electrical charge

E_{ion} Ionization potential factor

E_{jet} Kinetic energy of ejected propellant jet

I Electrical current

I_b Ion beam current

I_{sp} Specific impulse

j Charge current density

m Mass of charged particles

m_i Ion mass in atomic mass units (AMU)

$\dot{m}_i$ Ion mass flow rate

$\dot{m}_p$ Propellant mass flow rate

m_{pp} Mass of power plant

η_m Thruster mass utilization efficiency

η_t Thruster efficiency

α Specific power

P_e Total electrical power supplied

P_{ion} Power required to ionize a propellant

P_{jet} Power of the jet

P_{KE} Kinetic power in exhaust beam

τ Thrust correction factor

v_c Characteristic speed

v_i Ion velocity

V_{acc} Voltage across ion accelerator

V Voltage

Recommended Reading and References

1. Brewer, G. R. (1970). *Ion Propulsion Technology and Applications*, New York: Gordon and Breach, 1970.
2. Byers, D. C., "An experimental investigation of a high-voltage electron-bombardment ion thruster," Journal of the Electrochemical Society, Vol. 116, No. 1, pp. 9–17, 1969.
3. Curran, F. M., Sovey, J. S., and Myers, R. M., "Electric propulsion: An evolutionary technology," IAF-91-241, 42nd Congress of the International Astronautical Federation, Montreal, CA, Oct. 5–11, 1991.
4. Goebel, D. M., Katz, I., *Fundamentals of Electric Propulsion: Ion and Hall Thrusters*, John Wiley & Sons, 2008. http://descanso.jpl.nasa.gov/SciTechBook/st_series1_chapter.cfm.
5. Hargus, W. A. and Nakles, M. R. (2009). "Hall Effect Thruster Ground Testing Challenges." AFRL-RZ-ED-TP-2009-316.

6. Hutchins, M., Simpson, H. and Palencia Jimenez, J. (2015). "QinetiQ's T6 and T5 Ion Thruster Electric Propulsion System Architectures and Performances." IEPC-2015-131/ISTS-2015-b-131.

7. Jahn, R. G. (1968), *The Physics of Electric Propulsion*, McGraw-Hill, New York.

8. Jahn, R.G. and Choueiri, E.Y. (2002). *Encyclopedia of Physical Science and Technology*, Third Edition, Volume 5.

9. Jorns, B., Mikellides, I. G., et al. (2022). "Physics of electric propulsion." J. Appl. Phys. 132, 110401 (2022).

10. Kaufman, H. R. (1974). "Technology of Electron-Bombardment Ion Thrusters," in Advances in Electronics and Electron Physics, vol. 36, L. Marton (ed), Academic Press, 1974.

11. Kieckhafer, A. and King, L. B. (2005). "Energetics of Propellant Options for High-Power Hall Thrusters." J. of Propulsion and Power, Vol. 23, No. 1, January-February 2007.

12. Martinez-Sanchez, M. and Pollard, J. E. (1998). "Spacecraft electric propulsion - an overview," Journal of Propulsion and Power, Vol. 14, No. 5, pp. 688–693, 1998.

13. Mercer, C. R., McGuire, M.L., Oleson, S. R., and Barrett, M. J. (2015). "Solar Electric Propulsion Concepts for Human Space Exploration." NASA/TM-2016–218921, AIAA-2015-4521.

14. Mikellides, I. G. and Lopez Ortega, A. (2021). "Growth of the lower hybrid drift instability in the plume of a magnetically shielded Hall thruster," J. Appl. Phys. 129(19), 193301 (2021). https://doi.org/10.1063/5.0048706.

15. Nazareno, F., Gabriel, S. B. and Golosnoy, I. O. (2018). Alternative Propellants for Gridded Ion Engines. SP2018_00102.

16. Oleson, S. R., Myers, R. M., Kluever, C. A., Riehl, J. P., et al., "Advanced propulsion for geostationary orbit insertion and north-south station keeping," Journal of Spacecraft and Rockets, Vol. 34, No. 1, pp. 22–28, 1997.

17. Polk, J. E., Goebel, D. M., Katz, I., Snyder, J. S. et al. (2005). "Performance and Wear Test Results for a 20-kW Class Ion Engine with Carbon-Carbon Grids," AIAA-2005-4393, 41st Joint Propulsion Conference, Tucson, Arizona, July 10–13, 2005.

18. Pollard, J. E., Jackson, D. E., Marvin, D. C., Jenkin, A. B., et al., "Electric propulsion flight experience and technology readiness," AIAA Paper 93-2221, 1993.

19. Sangregorio, M., Xie, K., Wang, N, and Zhang, Z. (2018). "Ion engine grids: Function, main parameters, issues, configurations, geometries, materials and fabrication methods." Chinese Journal of Aeronautics, Vol. 31, Issue 8, August 2018, Pages 1635–1649.

20. Schatz, W. J., Cannova, R.D., Cowley, R.T., and Evans, D.D. (1979). Development and Flight Experience of the Voyager Propulsion System. AIAA 79-1334.

21. Soulas, G. C., Haag, T. W., Herman, D. A., et al. (2012). Performance Test Results of the NASA-457M v2 Hall Thruster. NASA/TM—2012-217711

22. Stone, J. R. (1986). "NASA Electrothermal Auxiliary Propulsion Technology." NASA TM-87281.

23. Sutton, G. P. and Biblarz, O. (2001). *Rocket propulsion elements*, 7th ed., New York, John Wiley & Sons, 2001.

24. Turchi, P. J. (1995). "Electric Rocket Propulsion Systems," Chapter 9 in *Space Propulsion Analysis and Design*, edited by R. W. Humble, G.N. Henry, and W. J. Larson, New York: McGraw-Hill, pp. 509–598, 1995.

25. Wertz, J. R. and Larson, W. J., eds. (1999). *Space Mission Analysis and Design*, third edition, New York: Springer Publishing Co., 1999.

26. Yost, B. and Weston, S. (2024). "State-of-the-Art Small Spacecraft Technology." Small Spacecraft Systems Virtual Institute, Ames Research Center, Moffett Field, California. NASA/TP-20240001462, February 2024. https://ntrs.nasa.gov/api/citations/20240001462/downloads/2023%20SOA_final.pdf.

Advanced Propulsion: Beyond Chemical Rockets

7

The journey will be long. We will voyage through an inky dark and lonely elliptical path, but the new world we seek in the cosmos will amaze us with unforeseen treasures.

—*Dora Musielak*

Human exploration missions beyond Earth orbit require propulsion systems with high specific power, outstanding performance, and reliable power supply systems. The availability of high specific power systems will establish whether a permanent human presence at Mars, for example, will become practical or whether it is possible to perform only a very small number of visits for short durations of weeks. Achieving a permanent human presence in space requires very high energy levels, to travel to and from, and to establish settlements in other worlds.

7.1 Limitations of Chemical Rockets

To propel a spacecraft, a propulsion system must operate with both a high-specific impulse I_{sp} and a low-mass powerplant M_w capable of generating large amounts of jet power P_{jet}. Since the thrust-to-engine weight ratio F/M_w of a spacecraft is directly proportional to the engine specific power $(\alpha_p \equiv P_{jet}/M_w)$, large values of α_p are required to provide the acceleration necessary for rapid transportation of a payload (cargo and crew) throughout the Solar System. For the future missions of deep space exploration, high-thrust/high-specific impulse propulsion is imperative.

© The Author(s), under exclusive license to Springer Nature Switzerland AG 2025   283
D. Musielak, *Introduction to Rocket Propulsion for Astronautics*, Synthesis Lectures on
Engineering, Science, and Technology, https://doi.org/10.1007/978-3-031-86141-3_7

The chemical rocket is not suited for crewed flight beyond Mars. It requires too much propellant. Large propellant consumption means that most of the spacecraft mass must be allocated to propellant, thus reducing considerably its payload capability.

The chemical rocket has a high propellant consumption. The thermal energy in a chemical rocket results from the chemical reaction between the fuel and oxidizer (the propellant mass). This thermal energy is converted into kinetic energy of the exhaust gases. Assuming ideal conversion efficiency in a chemical rocket, the exhaust velocity is, from kinetic energy considerations,

$$v_{ex} = \left(\frac{2E}{m}\right)^{1/2} \tag{7.1}$$

where E/m is the energy density—the energy per unit mass released in the chemical reaction.

Another important performance parameter for rocket propulsion is the total impulse I delivered. Long-distance flights or missions that require faster orbital transfer must deliver a high value of total impulse. This quantity is simply the thrust force F multiplied by the thrusting time t_b, that is, $I = F \cdot t_b$. We can rewrite this relationship by substituting the expression for the momentum thrust as $I = \dot{m}v_{ex}t_b$, where $\dot{m}$ is the propellant mass flow rate. And since propellant flow rate multiplied by thrusting time is just the propellant mass m_p at the beginning of the flight, the total impulse delivered by the propulsion system is simply

$$I = m_p v_{ex} \tag{7.2}$$

This expression shows that, for a particular mission, a certain total impulse is required. If the rocket exhaust velocity is low, the propellant mass must be high.

The limit to the exhaust velocity of practical chemical rockets is $v_{ex} < 4.5$ km/s, a velocity too low for ambitious missions beyond Mars. This exhaust velocity limitation can be overcome if electrically charged particles such as ions are exhausted instead of combustion gases. Due to the high kinetic energy and the tiny mass of the ions, it is possible to sustain propulsion for much longer than with chemical propellants. As described in Chap. 6, electrically charged particles can be accelerated to extremely high velocities, and thus an ion-based rocket can accelerate a spacecraft to a higher velocity and for a much longer time. However, electric propulsion requires a large quantity of electric power and the system could become too heavy.

Hence, we must adopt other propulsion technologies to allow for efficient travel across the Solar System. This prompt us to consider other energy sources that could best fulfill the mission requirements such as fusion, fission, and matter-antimatter annihilation. The ideal energy density available for propulsion from those energy sources is obtained from Albert Einstein's energy/matter formula, $E/m = c^2$. The energy-density quantity in the rocket exhaust velocity, Eq. (7.1) can be increased with nuclear energy. In principle, it

Table 7.1 Specific energy release comparison

Energy source	Specific energy (J/kg)
Chemical (hydrogen + oxygen)	1.35×10^7
Fission (uranium $^{235}_{92}U$)	8.30×10^{13}
Fusion (D + helium-3)	3.52×10^{14}
Matter-antimatter annihilation	9.00×10^{16}

would be possible to obtain exhaust velocities orders of magnitude higher by nuclear fission or with fusion reactions. The relative energy yields from various energy sources are given in Table 7.1.

The energy available from fission and fusion reactions is many orders of magnitude greater than that resulting from chemical reactions. As shown, the specific energy release for fission (uranium U-235) and fusion (deuterium and helium-3) is an improvement over the best chemical (hydrogen and oxygen) source by nearly 7 orders of magnitude.

Moreover, using nuclear energy, the theoretical specific impulse is between 960 and 6600 s, depending on the maximum temperature that the exhaust mass could achieve by passing the propellant through a solid core nuclear reactor, for example.

Deep space missions to the outer planets of our Solar System require very high Δv, necessitating vehicles with high energy performance. In addition, jet power levels from 10 MW to 100 GW, produced by 1–10 kW/kg specific power propulsion systems, are necessary. Propulsion systems must also deliver variable specific impulse on the order of 5×10^3 to 10^6 s, with firing durations of months to years, with thrust ranging from 1 N to ~1000 kN. Crewed spacecraft will require minimum 10^+ MW to 100^+ MW to GW of electrical power using highly efficient direct energy converters.

In the following sections we review the different forms of energy that can be used for propulsion and then we discuss proposed propulsion concepts that offer the potential to facilitate travel within our Solar System.

7.2 Non-chemical Energy Sources

We need to identify energy sources for accomplishing missions where v requirements range from 90 km/s to 30,000 km/s. The need to develop space power of this magnitude is a key issue to continue advancing space exploration beyond the confines of the Earth-Moon realm. We must consider both stellar and nuclear energy sources.

7.2.1 Solar Energy

The Sun is a powerful source of energy that is routinely used on nearly all spacecraft today. The International Space Station (ISS), for example, gets its energy from solar arrays that power rechargeable batteries connected to all electrical equipment. The ISS relies on huge solar panel arrays to convert solar energy into electrical power. The solar arrays are large, wing-like structures, each measuring 34 m long and 11 m wide (112 ft × 39 ft). Since each array is extended in opposite directions, the total wingspan is over 73 m (240 ft). Just two arrays can supply nearly 64 kW of power. This is enough to meet the needs of 30 average homes without air conditioning. The 8 eight arrays of the ISS contain 262,400 solar cells, which are electrically connected into the solar panels, and cover an area of about 2500 m^2 (27,000 ft^2)—more than half the area of an American football field! Each solar cell is about 12% efficient.

Total power radiated by the Sun can be estimated with the Stefan-Boltzmann equation, assuming the Sun is a perfect emitter of radius 7.0×10^8 m with a temperature 5500 K:

$$\dot{Q} = \sigma A T^4 = \left(5.67 \times 10^{-8} \ \text{Wm}^{-2} \ \text{K}^{-4}\right)\left(6.16 \times 10^{16} \ \text{m}^2\right)(5500 \ \text{K})^4$$
$$= 3.195 \times 10^{26} \ \text{W}$$

If the Sun's energy is distributed symmetrically over a spherical surface with the Sun at the center, the intensity or power per unit area is

$$I = \frac{\dot{Q}}{A} \tag{7.3}$$

where the area is determined from the mean distance from the Sun, $A = 4\pi R_{sp}^2$.

For example, using the mean distance from the Sun to the Earth, 1.496×10^{11} m, we determine the solar intensity arriving at the Earth is:

$$I = \frac{3.195 \times 10^{26} \ \text{W}}{2.812 \times 10^{23} \ \text{m}^2} = 1.13 \times 10^3 \ \text{W/m}^2$$

Above Earth's atmosphere the intensity is 1.38×10^3 W/m^2 (value known as solar constant). Hence, to gather massive quantities of energy, solar power panels must be large, and the power conversion efficiency must be as high as possible. Using Eq. (7.3) we can easily determine that the intensity of sunlight on Mars (mean distance from the Sun is 2.279×10^{11} m) is less than half of the intensity at Earth. However, since the orbit of Mars is more elliptical than Earth's orbit, at perihelion (closest to the Sun), Mars is 206,600,000 km away, and thus the intensity is a little higher. But at aphelion (farthest from the Sun), Mars is 249,200,000 km away, and the intensity drops beyond the value calculated above. These differences must be included in the design of a solar energy system for an application on Mars.

The power of solar radiation at any point in space, outside any planetary atmosphere in which absorptions occurs, is a fixed quantity given by the formula

$$E_s = \frac{3 \times 10^{25}}{R_{sp}^2} \left(\text{joule/s} \cdot \text{m}^2 \right)$$

where R_{sp} is the distance from the Sun (in meters). At the Earth (1 AU), $E_s \approx$ 1342 joule/s $\cdot$ m^2 = 0.125 kW/ft^2.

> **Did you know?** Jupiter is five times as far from the Sun as our planet, so it receives 1/25 as much solar energy. Solar panels are impractical at Jupiter, even less useful at Saturn where the solar energy is down by a factor of 1/90 compared to Earth.

In principle, a rocket engine can utilize solar energy to heat a propellant, if sufficient solar radiation can be collected and concentrated in such a manner that it transmits to the engine. In the solar thermal rocket concept, a parabolic solar reflector such as a parabolic mirror or Fresnel lens, is used for collecting/concentrating the solar radiation and sent to a receiver heat exchanger. As the liquid propellant (typically hydrogen) passes through the heat exchanger, it becomes a heated gas whose temperature can reach up to 2200 °C. The heated propellant mass is then expanded in a nozzle. The thrust generated is relatively low (1 to 10 N) but the specific impulse is attractive for in-space propulsion applications. One of the technical issues with this concept, is the storage of liquid hydrogen. To date, the solar thermal rocket concept remains an interesting idea.

7.2.2 Nuclear Energy

Nuclear energy is contained within the atomic nucleus. Particles within the nucleus are held together by a strong force. If a large nucleus is split apart (fission), generous amounts of energy can be liberated. Small nuclei can also be combined (fusion) with an accompanying release of energy.

A nuclear power system involves a nuclear reactor to produce heat. This heat is released in the decay of radioisotopes (such as plutonium); in the controlled fission of heavy nuclei (such as uranium-235) in a sustained neutron chain reaction; or in the fusion of light nuclei (such as deuterium and tritium). The heat energy produced can be used directly for propulsion processes or converted into electric power. Current nuclear energy applications in space are based on radioisotope decay and nuclear fission. The following sections provide an overview of these methods.

7.2.2.1 Radioisotope Decay

The thermal energy emitted by the Sun is intense. Solar cells convert the solar energy into electric energy so they are a suitable choice for powering a spacecraft. However, at the great distance from the Sun, solar cells cannot be used for power generation. Energy needed by interplanetary spacecraft to operate its instruments and communicate with Earth is usually provided by radioisotope thermoelectric generators (RTGs). An RTG is an electrical generator that uses an array of thermocouples to convert the heat released by the decay of a suitable radioactive material into electricity. For example, power for the Voyager spacecraft is provided by three RTGs based on plutonium-238 (Pu-238), an isotope of plutonium.

Atoms of the same element can have different numbers of neutrons: the different possible versions of each element are called isotopes. A radioisotope or radioactive isotope, is any of several species of the same chemical element with different masses whose nuclei are unstable and dissipate excess energy by spontaneously emitting radiation in the form of alpha, beta, and gamma rays. In the decay of a radioactive isotope material, heat is produced. This thermal energy can be converted into electrical power and utilized.

Every chemical element has one or more radioactive isotopes. For example, hydrogen, the lightest element, has three isotopes with mass numbers 1, 2, and 3. However, only hydrogen-3 (tritium) is a radioactive isotope, the other two are stable. More than 1000 radioactive isotopes of the various elements are known. Approximately 50 of these are found in nature; the rest are produced artificially as the direct products of nuclear reactions or indirectly as the radioactive descendants of these products.

To refer to a certain isotope, we write it as $_b^a$X, where X is the chemical symbol for the element, b is the atomic number, and a is the number of neutrons and protons combined, called the mass number. For example, $_{92}^{235}$U refers to uranium isotope with 235 particles (neutrons plus protons). The heat produced in the decay of the radioactive isotope is converted into electricity by means of thermoelectric junction circuits or related devices (see Fig. 7.1). Plutonium-238 (Pu-238) is a radioactive isotope of plutonium with a half-life of 87.7 years, a very powerful alpha (α) emitter and—unlike other isotopes of plutonium—it does not emit significant amounts of other, more penetrating, and thus more problematic radiation. Pu-238 decays to uranium-234 by emission of an alpha particle (Helium nucleus) with an energy of 5.5 MeV (8.8×10^{-19} J). This makes the Pu-238 isotope suitable for usage in radioisotope thermoelectric generators (RTGs) and radioisotope heater units—one gram of Pu-238 generates approximately 0.5 watts of thermal power.

Radioisotope decay has a long history of use on space science missions since the Apollo program. The first RTG was used on the Moon in 1969, using heat generated by the decay of radioactive Pu-238 to keep Apollo 11 scientific instruments at a working temperature. On Apollo 12 the heat was converted into electricity to power an instrument package. This was the first use of a miniature nuclear reactor on the Moon—the cylindrical generator measured just 45.7 cm by 40.6 cm (18.2 in by 16.2 in).

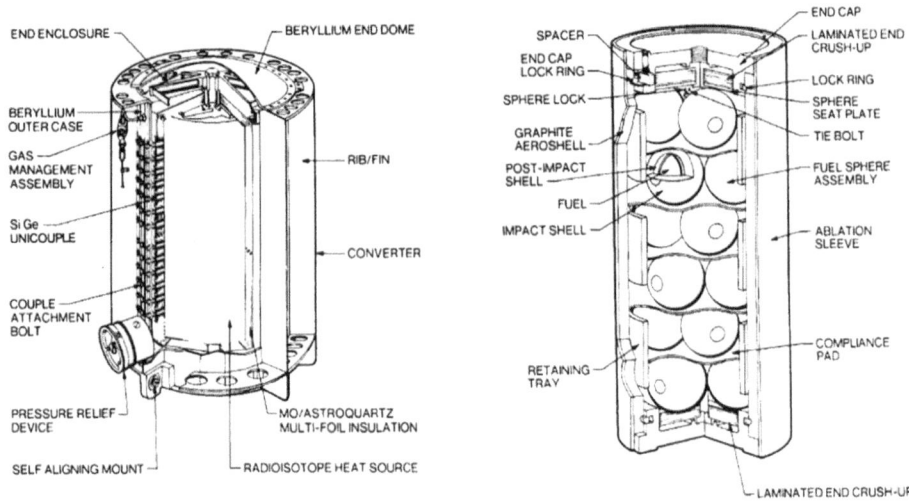

Fig. 7.1 Voyager RTG and heat source. *Credit* Voyager Backgrounder, NASA [33]

Dozens of RTGs have been used for interplanetary missions during the past 4 decades, for both thermal management and electricity production. RTGs are used as power source for deep-space robotic spacecraft needing a few hundred watts (or less) of power for mission durations too long for fuel cells, batteries, or generators to provide power economically, and in places where solar cells are not practical. With a slow radioactive decay, a spacecraft is powered for decades. The twin Voyagers 1 and 2, launched in 1972 and now traveling past the outer edge of the Solar System, are still being powered with RTGs (Fig. 7.1).

Voyager 1 has three large radioisotope thermoelectric generators (RTGs). Each RTG contains 24 pressed plutonium-238 oxide spheres. The heat from the spheres generated about 157 watts of electric power at launch, with the remainder being dissipated as waste heat. The three RTGs provided a total of about 470 watts of electric power. The RTGs of Voyager 1 will continue to support some of its operations through about 2025.

7.2.2.2 Nuclear Fission Process

Nuclear fission is the process of splitting atoms. Fission occurs when a neutron smashes into a larger atom, forcing it to excite and split into two smaller atoms—the fission products. A large amount of energy is released in a fission reaction because the mass of the

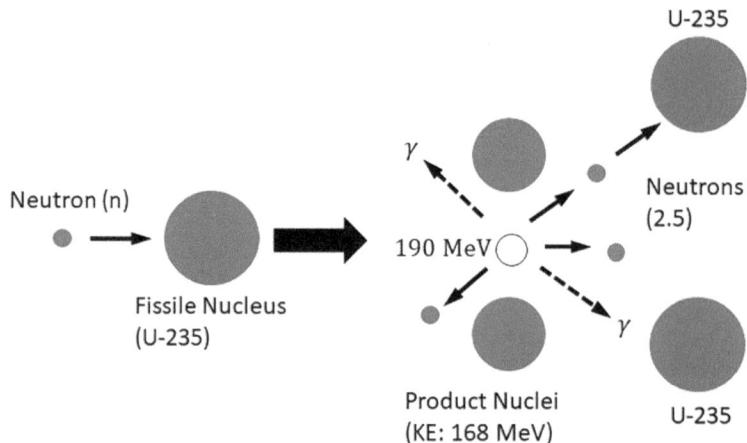

Fig. 7.2 Fission representation of U-235

atomic nucleus is considerably greater than the total mass of the fission fragments plus released neutrons.

Uranium is the main element used in nuclear reactors. Nuclear fission occurs much more readily for $^{235}_{92}U$ than for the more common $^{238}_{92}U$. Fissioning a nucleus after capture of a neutron (n) yields nuclei X_1 and X_2, the fission products, and in the process some neutrons (typically two or three) are also released. A fission reaction is written as

$$n + {}^{235}_{92}U \rightarrow {}^{236}_{92}U \rightarrow X_1 + X_2 + \text{neutrons} \qquad (7.4)$$

At steady-state, one of the two or three neutrons released in the reaction causes a subsequent fission in a "chain reaction," as illustrated in Fig. 7.2.

The fission process occurs very quickly. For example, the nucleus $^{236}_{92}U$ exists for less than 10^{-12} s. The two fission fragments X_1 and X_2 more often split the original uranium mass as about 40%-60%, rather than precisely half and half. Although many reactions are possible, one typical fission reaction is

$$n + {}^{235}_{92}U \rightarrow {}^{141}_{56}Ba + {}^{92}_{36}Kr + 3n \qquad (7.5)$$

A high amount of energy is released in a fission reaction. For example, in the above reaction, the mass of $^{235}_{92}U$ is considerably greater than the total mass of the fission fragments plus released neutrons. Extra mass is a result of the binding energy that holds the protons and neutrons of the nucleus together. Thus, when the uranium atom is split, some of the energy that held it together is released as radiation in the form of heat. Because energy and mass are one and the same, the energy released is also mass released. Therefore, the total mass does decrease a tiny bit during the reaction.

When dealing with energies of electrons, atoms, or molecules the unit electron volt (eV) is used—the joule is a very large unit. One electron volt is the energy acquired by a particle carrying a charge whose magnitude equals that on the electron ($q = e$) by moving through a potential difference of 1 V. Since the change in potential energy is equal to qV, we have, with $e = 1.6 \times 10^{-19}$ C:

$$1 \text{ eV} = \left(1.6 \times 10^{-19} \text{ C}\right)(1.0 \text{ V}) = 1.6 \times 10^{-19} \text{ J}$$

The difference in mass, or energy, between the original uranium nucleus and the fission fragments is about 0.9 MeV per nucleon. Since there are 235 nucleons involved in each fission reaction [see Eq. (7.5)], the total energy released is

$$\left(0.9 \frac{\text{MeV}}{\text{nucleon}}\right)(235 \text{ nucleons}) = 200 \text{ MeV} \tag{7.6}$$

This is a very large amount of energy for one single nuclear reaction. And since the neutrons released can be used to create a chain reaction, a self-sustaining chain reaction system can produce an enormous amount of energy. Creating such fission chain reaction is conceptually very simple. All that is required is the right materials to be placed in the right geometry—no extreme temperatures or pressures are required. This was demonstrated by physicist Enrico Fermi with the construction of the first nuclear reactor in 1942. The idea of using nuclear energy for space propulsion was conceived shortly after Fermi's nuclear reactor was demonstrated.

7.2.2.3 Nuclear Fusion Process

Fusion power is the power generated by nuclear fusion processes. The mass of every stable nucleus is less than the sum of the masses of its constituent protons and neutrons. For example, the mass of the helium isotope ^4_2He is less than the mass of two protons plus the mass of two neutrons. Hence, if two or more light atomic nuclei were to come into contact or fuse to form a helium nucleus, there would be a loss of mass. This mass loss is manifested in the release of a large amount of energy arising from the binding energy due to the strong nuclear force which is manifested as an increase in temperature of the reactants.

The possibility of utilizing the energy released in fusion to make a power reactor for space applications is very attractive. The fusion reactions with higher potential for this application involve the isotopes of hydrogen. Expressing ordinary hydrogen as ^1_1H, deuterium as ^2_1H, and tritium as ^3_1H, it happens that when Deuterium ^2_1H and tritium ^3_1H nuclei fuse, they form a helium nucleus plus a neutron n, and release a certain amount of energy, as indicated in the following reactions:

$$^2_1\text{H} + ^2_1\text{H} \rightarrow ^3_1\text{H} + ^1_1\text{H} \rightarrow 4.00 \text{ MeV} \tag{7.7}$$

$$_{1}^{2}\text{H} + _{1}^{2}\text{H} \rightarrow _{2}^{3}\text{He} + \text{n} \ \rightarrow 3.23 \text{ MeV} \tag{7.8}$$

$$_{1}^{2}\text{H} + _{1}^{3}\text{H} \rightarrow _{2}^{4}\text{He} + \text{n} \rightarrow 7.57 \text{ MeV} \tag{7.9}$$

Theoretically, the energy released in fusion reactions can be greater than the energy obtained with the fission reaction of $_{92}^{235}\text{U}$, for a given amount of fuel mass. However, fusion reactions are difficult. The problem is because all nuclei have a positive charge and repel each other. For the nuclei to get close enough together, they must have large kinetic energy to overcome the electric repulsion. Particle accelerators can do that, but for adequate energy production we must deal with matter in bulk, rather than individual particles, and high kinetic energy means very high temperature. That is why fusion reactors require thermonuclear devices.

Let us assume that the nuclei approach head-on, each with kinetic energy KE, and that the nuclear force dominates when the distance between their centers equals the sum of their nuclear radii. The electrostatic potential energy of the two particles at this distance equals the minimum total kinetic energy of the two particles when far apart. The average kinetic energy is related to temperature as

$$\overline{KE} = \frac{1}{2}\overline{mv^2} = \frac{3}{2}kT \tag{7.10}$$

where the quantity $\frac{1}{2}\overline{mv^2}$ is the average translational kinetic energy of the molecules in the gas, k is the Boltzmann constant, and T is the absolute temperature.

To reach the required high temperatures for fusion, there must be powerful heating, and thermal losses must be minimized by keeping the hot fuel particles away from the walls of the container. This can be achieved by applying strong magnetic fields, in effect creating magnetic containment to prevent the particles from escaping the system. For energy production the plasma must be confined for a sufficiently long period for fusion to occur. The two main approaches for providing the confinement necessary to sustain a fusion reaction are inertial confinement fusion (ICF), and magnetic confinement fusion (MCF). The latter approach uses magnetic fields to confine a plasma. Tokamaks and stellarators are the two leading MCF device candidates as of today.

With Inertial Confinement Fusion (ICF), fusion conditions are achieved by quickly compressing and heating a small quantity of fusion fuel. The result is a fuel at very high pressure, causing the fuel to disassemble. Scientists at Sandia have demonstrated experimentally that an ICF concept called Magnetized Liner Inertial Fusion (MagLIF) can achieve thermonuclear fusion conditions on one of their machines.

Inertial Confinement Fusion (ICF) is considered as an attractive power source for interplanetary crewed spacecraft because ICF offers high power-to-mass ratios and a high specific impulse.

Example 7.1 Estimate the temperature of a deuterium-tritium fusion reaction if the average kinetic energy required is 0.11 MeV.

Solution: From Eq. (7.10), we solve for the temperature T:

$$T = \frac{2\overline{KE}}{3k} = \frac{2(0.11 \text{ MeV})(1.6 \times 10^{-13} \text{ J/MeV})}{3(1.38 \times 10^{-23} \text{ J/K})} = 8.5 \times 10^8 \text{ K}$$

A practical fusion reactor may require temperatures in the range $T \geq 1 - 4 \times 10^8$ K.

7.3   Nuclear Energy Propulsion Concepts

Energy released by a nuclear reaction can be used to generate thrust. Nuclear energy was identified in the 1940s as having great potential for propulsion, precisely because of the huge energy densities. Concepts of nuclear propulsion system using the products of nuclear fission or fusion were considered. In 1963, the Nuclear Engine for Rocket Vehicle Application (NERVA) project was instituted, a jointly funded program by NASA, the Atomic Energy Commission (AEC), and the U.S. Air Force, aiming to develop a nuclear-powered rocket for both long-range missions to Mars and as a possible upper-stage for the Apollo Program. Under the program NERVA/ROVER, a nuclear thermal engine was ground tested in the 1960s. The design requirements of the nuclear rocket included a thrust of 1112 kN (250,000 lbf) and a specific impulse of 850 s, offering a performance that would surpass that of chemical rockets.

Since then, three different types of nuclear energy sources have been investigated for delivering heat to a propellant gas, which subsequently can be expanded in a nozzle and thus accelerated to high ejection velocities. In the fission reactor, the radioactive isotope decay source, and the fusion reactor the heating of the gas is accomplished by energy derived from transformations within the nuclei of atoms. In chemical rockets the energy is obtained from within the propellants by chemical reactions, but in nuclear rockets the power source is separate from the propellant. In the past six decades, a variety of nuclear propulsion concepts have been considered, including nuclear electric propulsion (NEP) and nuclear thermal propulsion (NTP). NEP systems consist of a nuclear reactor combined with a heat-to-electricity power conversion system, and an electrical propulsion system.

7.3.1   Nuclear Thermal Propulsion (NTP)

Conceptually, a nuclear rocket is very similar to a chemical rocket, except that the gas ejected through the nozzle is driven by nuclear energy instead of chemical energy—the source of exhaust gas and the source of energy are independent. The propellant supplies the gas and the reactor supplies the energy to heat the propellant. Nuclear reactions generate far more power than chemical reactions because they convert some of the mass in atomic nuclei into energy in accord with Einstein's formula, $E = mc^2$. In contrast, chemical reactions generate power only by rearranging the energy levels of electrons in atoms or molecules. Fusing a kilogram of hydrogen, for example, generates more than a million times more energy than chemical reactions yield in one kilogram of hydrogen and oxygen.

The most basic and simpler approach in a nuclear rocket is to heat a propellant to very high temperatures by pumping it through a nuclear reactor, then expanding the resulting hot gas through a nozzle to produce the thrust. The heat results from the fission of a nuclear fuel in the reactor. This concept is known as nuclear thermal propulsion (NTP). The propellant need not be chosen based on its energy content (as done in the chemical rocket), but it is selected based on its suitability to provide the highest specific impulse. Since $v_{ex} \propto \sqrt{T_c/\mathcal{M}}$, hydrogen is the best propellant in this respect because it has the lowest molecular mass $\mathcal{M}$, and thus a nuclear rocket using hydrogen heated by the nuclear core to the same temperature found in a conventional rocket engine has the potential to provide the highest exhaust velocity for a given chamber pressure and temperature.

Nuclear thermal rocket engines are defined by the design of its nuclear reactor, which can be a relatively simple solid reactor or a much more complicated but more efficient gas core reactor.

7.3.1.1 Solid Core Nuclear Thermal Rocket

In the nuclear thermal rocket (NTR) concept, hydrogen propellant is heated as it passes through a solid-fuel core reactor. As depicted conceptually in Fig. 7.3, the heated gas is accelerated as it expands in a rocket nozzle to produce thrust. In principle, the hydrogen is first circulated around the nozzle for regenerative cooling, and then it is injected into the reactor core. The reactor is a cylindrical chamber containing the nuclear fuel rods, which occupy about a third of the chamber cross section; the core is surrounded by reflector material on the sides and possibly on one or both ends. The propellant moves through the channels, and absorbs heat. Just as in the electrothermal rocket, in the NTR the propellant is heated by contact with a hot solid—the hot reactor core—so the maximum propellant temperature depends on the nuclear fuel rods.

The rate of fission and thus the heat production is controlled by the reflector. The fissionable material in the graphite fuel element can be particles of uranium carbide coated with pyrolytic carbon. The maximum operating temperature of the propellant must be less than the melting point of the reactor core, the nozzle, and other structural materials. This limits the theoretical specific impulse for a practical NTR.

Fuel elements or fuel rods are designed to withstand very high temperatures (up to 3500 K) and high pressures (up to 200 atm). They are made of very strong materials, either carbon composites or carbides, and normally coated with zirconium hydride. Most

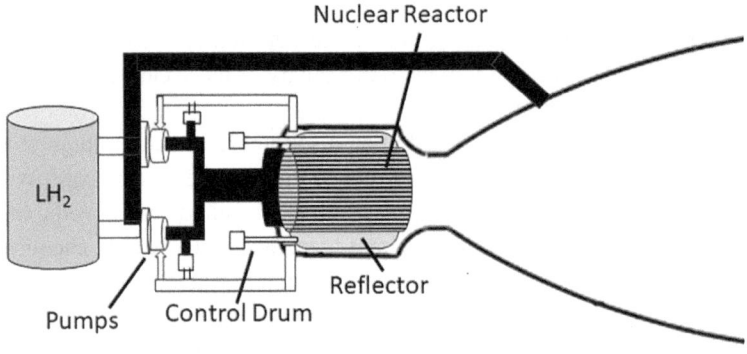

Fig. 7.3 Solid-core nuclear thermal rocket concept

nuclear fuels contain heavy fissile elements that are capable of nuclear fission. When these fuels are struck by neutrons, they are in turn capable of emitting neutrons when they break apart. This makes possible a self-sustaining chain reaction that releases energy with a controlled rate in a nuclear reactor. The solid core of the reactor is comprised of bundles of fuel rods composed of the fuel material, mixed with structural, neutron moderating, or neutron reflecting materials. Table 7.2 provides the melting temperature of some materials used in nuclear reactors.

Table 7.2 Melting temperature of common materials used for solid core fuel elements

Type of material	Name	Temperature (K)
Metal	Uranium	1400
Metal compounds	Uranium nitride (UN)	3160
	Uranium dioxide (UO_2)	3075
	Uranium carbide (UC_2)	2670
Refractory metals	Tungsten (W)	3650
	Rhenium (Re)	3440
	Tantalum (Ta)	3270
	Molybdenum (Mo)	2870
Refractory non-metals	Carbon (C)	3990[a]
	Hafnium carbide (HfC)	4160
	Tantalum carbide (TaC)	4150
	Niobium carbide (NbC)	3770
	Zirconium carbide (ZrC)	3450

[a] Sublimation temperature (direct transition from solid to gas)

The solid core fuel elements of an NTR have a small percentage of U-235 buried well inside an extremely strong carbon or carbide mixture. Unless the physically small reactors have been run for an extended period, the radioactivity of these elements is quite low and would pose a minimal hazard.

The NERVA rocket engine was based upon a solid-core running at high temperatures to heat the liquid hydrogen propellant that moved through the reactor core. With current material technology, a solid-core nuclear thermal rocket could deliver specific impulses up to 1000 s with LH_2 propellant, which is about twice that of LOX/LH_2 chemical rockets. Other propellants have been proposed, such as ammonia, water, or LOX. A reduction in exhaust velocity due to these propellants is compensated by their greater availability, thus reducing payload costs in missions where the velocity change Δv is not too high, such as within cislunar space or for applications between Earth orbit and Martian orbit.

During the 1960s and early 1970s, about 23 nuclear reactors and engines were built and tested using hydrogen cooled reactor technology. The power ranged from 350–4500 MW and delivered thrust from 110–1100 kN (25,000–250,000 lbf). The nuclear rocket was designed to operate at 1500 MW, provide 333 kN of thrust at a specific impulse of 850 s; the engine weight was 10.4 metric tons. It was intended for a 10-h life and 60 operating cycles. The NERVA Rover program was terminated in the early 1970s, before a nuclear rocket ever flew in space.

Studies conducted at NASA suggest that fission electric propulsion systems with a specific mass at or below 50 kg/kW could enhance or enable robotic outer Solar System missions. Nuclear fission propulsion operates independently of solar proximity or orientation, making NTP better suited for deep space missions.

It should be clear that since heat is transferred from a solid reactor to the propellant, there is a limit on the achieved heating of the propellant. The structural components within a nuclear rocket must be hotter than the propellant, and the propellant temperature cannot exceed the limiting temperature of the structure or reactor material. With advances in high temperature materials, it may be possible to achieve high temperatures, at least for short duration operation.

7.3.1.2 Specific Impulse and Power

Conceptually, a nuclear thermal rocket (NTR) resembles the electrothermal thruster we studied in Chap. 6, since the propellant is heated by contact with a hot solid; in the NTR case, the propellant passes through the hot fuel elements or the solid core of the reactor. Therefore, the mechanical integrity of the solid element imposes a limit on the temperature that the core can have and the temperature of the propellant. The thrust depends on the exhaust velocity, which is governed by the temperature and molecular mass of the gas expanded in the rocket nozzle.

For ideal operation, the thrust is simply $F = \dot{m}v_{ex}$, where the nozzle exhaust velocity is $v_{ex} = C_F C^*$. As we found in Chap. 2, the thrust coefficient C_F depends on the nozzle design itself, and C^* is the characteristic velocity from Eq. (2.52):

$$C^* = \left\{ \gamma \left(\frac{2}{\gamma + 1} \right)^{(\gamma+1)/(\gamma-1)} \frac{\mathcal{M}}{\Re T_c} \right\}^{-1/2} \qquad (7.11)$$

The melting temperature of the reactor's fuel elements limits the attainable chamber temperature T_c. Uranium metal, for example, has a melting temperature of just 1400 K, thus it would not be a good fuel element. However, its compounds are a better choice: uranium nitride (UN) and uranium dioxide (UO_2) have melting temperatures 3160 K and 3075 K, respectively. Refractory materials are also good choices as fuel elements (see Table 7.2). Therefore, the attainable exhaust velocity in a nuclear thermal rocket depends on the temperature of the fuel elements of the reactor, which in turn will determine the propellant temperature before entering the nozzle.

Nuclear thermal rockets will have a low thrust to weight ratio. Hence, NTRs could be upper stages where vehicle velocity is near orbital. NTRs could also power space tugs, or launch vehicles lifting from a lower gravity planet, moon, or minor planet where the required thrust is lower. Nuclear thermal rockets are proposed for interplanetary missions to reduce transit time. With a specific impulse potential two to three times that of chemical rockets, NTP is a very attractive option for human exploration missions to the near planets. Theoretical studies suggest that trip time to Mars could be reduced at least 100 days from that of advanced chemical systems. Based solely on the temperature of the propellant, estimates of specific impulse suggest that it is possible to achieve an order of magnitude increase.

For a propulsion system, the amount of power generated is defined by Eq. (6.7),

$$P = F \frac{v_{ex}}{2} = \frac{F g_0 I_{sp}}{2}$$

For a LOX/LH$_2$ chemical rocket delivering 1000 kN thrust and 414 s specific impulse, the power generated is approximately 2000 MW. If a nuclear rocket with a solid-core design can deliver the same thrust and $I_{sp} \sim 850$ s, then the power it needs is about 4000 MW.

The fuel flow rate from the thrust equation for ideal expansion is

$$\dot{m} = \frac{F}{v_{ex}}$$

In this example, the propellant flow rates for the chemical and nuclear rockets would be $\dot{m}_c \cong 250$ kg/s and $\dot{m}_n \cong 118$ kg/s, respectively.

For chemical propulsion, $\dot{m}_c$ includes both fuel and oxidizer propellant, considering the mixture ratio appropriate for optimum combustion. For the nuclear rocket $\dot{m}_n$ represents the amount of only one propellant. This suggests that the propellant tanks and plumbing systems would be less complex for the nuclear rocket.

To achieve high specific impulse, the NTP rocket heats the propellant to extremely high temperature, and this process affects component life and reliability. To cope with these high temperatures, the NTP system requires advanced materials and thermal insulation. The reactor must also be shielded to protect the crew and spacecraft components.

7.3.1.3 LOX-Augmented Nuclear Thermal Rocket (LANTR)

Developed by NASA Glenn Research Center, the LOX-Augmented Nuclear Thermal Rocket or LANTR (Fig. 7.4) is an improved variant of the solid core thermal nuclear rocket. As the name implies, this nuclear rocket concept enhances the performance of a hydrogen-fueled NTP by injecting liquid oxygen (LOX) into the nozzle. The injected LOX acts like an afterburner and operates in a reverse-scramjet mode. i.e. supersonic combustion of oxygen and hot hydrogen takes place in the rocket nozzle. This makes it possible to augment (and vary) the thrust. Of course, the added LOX results in a reduced I_{sp}, but it is compensated by a higher thrust. The performance of the NTR and LANTR is summarized in Table 7.3.

Data from Frisbee [12]

Example 7.2 Determine the minimum amount of $^{235}_{92}U$ fuel that needs to undergo fission to run a 1000 MW reactor for a month of continuous operation. Assume a 30% reactor efficiency.

Solution: For 1000 MW output, the total power generation input must be $\dot{Q}_{in} = 3 \times 1000\,MW = 3000\,MW = 3 \times 10^9$ J/s. This means that 2000 MW must be dumped as waste heat.

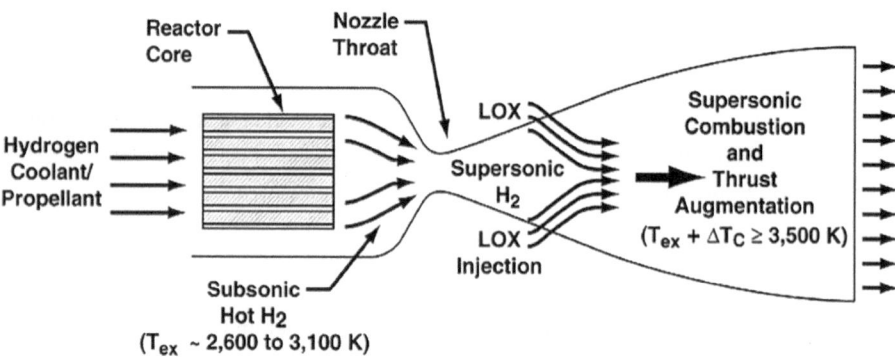

Fig. 7.4 LANTR concept. *Credit* Borowski et al. [5]

Table 7.3 NTR and LANTR performance comparison

Concept	Isp	Thrust
H_2 NTR	940 lbf-s/lbm (9.21 km/s)	67 kN (15,000 lbf)
H_2 LANTR	647 lbf-s/lbm (6.34 km/s)	184 kN (41,300 lbf) with O/F = 3

The total energy release in one month ($t = 2.59 \times 10^6$ s) from fission needs to be

$$E = \dot{Q}_{in} \cdot t = (3 \times 10^9 \text{ J/s})(2.59 \times 10^6 \text{ s}) = 7.77 \times 10^{15} \text{J}$$

According to Eq. (7.11), each fission releases about 200 MeV, that is $e = 2 \times 10^8$ eV of energy. Thus, the number of fissions required for a month is

$$N = \frac{E}{e} = \frac{7.77 \times 10^{15} \text{ J}}{(2 \times 10^8 \frac{\text{eV}}{\text{fission}})(1.6 \times 10^{-19} \text{ J/eV})} = 2.43 \times 10^{26} \text{ fissions}$$

The mass of a single uranium atom is $(235 \text{ u})(1.66 \times 10^{-27} \text{ kg/u}) \approx 4 \times 10^{-25}$ kg. Therefore, the total mass of uranium needed for continuous fission process in one month is

$$m = \left(4 \times 10^{-25} \text{ kg/fission}\right)(2.43 \times 10^{26} \text{fissions}) \approx 97.2 \text{ kg}$$

Example 7.3 Determine the specific impulse of a 940 MW NTP system that uses hydrogen propellant, and the fuel elements in the solid core are made of zirconium carbide (ZrC) which are maintained at a temperature that can heat the propellant to 2400 K.

Solution: The characteristic velocity is given by Eq. (2.52). For hydrogen propellant $\mathcal{M} = 2$, and assuming $\gamma = 1.2$, obtain $C^* = 4810$ m/s. Now assume a thrust coefficient $C_F = 1.85$ for a well-designed modern nozzle, and the effective exhaust velocity is $v_{ex} = C_F C^* = 8898$ m/s. Thus, the specific impulse is

$$I_{sp} = \frac{v_{ex}}{g_0} = \frac{8898 \text{ m/s}}{9.80665 \text{ m/s}^2} = 908 \text{ s}$$

Example 7.4 Calculate the Δv possible with a chemical rocket stage delivering a specific impulse of 414 s; the stage initial and final mass are $m_{0_c} = 119{,}900$ kg, and $m_c = 13{,}311$ kg, respectively. Compare the result with that obtained with a nuclear rocket delivering a specific impulse of 850 s, if $m_{0_n} = 38{,}600$ kg, and $m_n = 17{,}300$ kg.

Solution: From the rocket equation, for the chemical rocket

$$\Delta v_c = v_{ex} \ln \frac{m_{0_c}}{m_c} = 8900 \text{ m/s}$$

and for the nuclear rocket

$$\Delta v_n = v_{ex} \ln \frac{m_{0_n}}{m_n} = 6700 \text{ m/s}$$

The reduced Δv of the nuclear rocket is due to the much higher empty mass of the engine, and to smaller burn time due to the less-dense fuel. However, the amount of propellant mass used by the nuclear rocket is much less, as indicated below:

Mass of propellant for the chemical rocket is, from Eq. (2.19),

$$m_{p_c} = m_{0_c}\left(1 - e^{-\frac{\Delta v}{v_{ex}}}\right) = 106{,}500 \text{ kg}$$

Mass of propellant for the nuclear rocket is,

$$m_{p_n} = m_{0_n}\left(1 - e^{-\frac{\Delta v}{v_{ex}}}\right) = 21{,}317 \text{ kg}$$

Thus, the nuclear rocket would require a smaller propellant tank.

Example 7.5 A round-trip mission to Mars from Earth requires a minimum Δv of approximately 14 km/s. Estimate the amount of propellant required if a 1000 kg spacecraft were propelled by either (a) chemical rockets, (b) nuclear rockets, or (c) ion thrusters.

Solution: Let the initial mass of the rocket be $m_0 = m + m_p$, where m_p is the mass of propellant, and m is the total dry mass (structure plus payload). Then rewrite Eq. (2.19) as

$$\frac{m_0}{m} = \frac{m + m_p}{m} = e^{\Delta v/v_{ex}} \rightarrow m_p = m\left(e^{\Delta v/v_{ex}} - 1\right)$$

so that we determine the mass of propellant required for each propulsion system.

Assume the exhaust velocity v_{ex} is 4.7 km/s for chemical rockets, 9.1 km/s for nuclear rockets, and 20 km/s for ion thruster. The propellant mass for each spacecraft is:

$$m_p = \begin{cases} 1000\left(e^{14/4.7} - 1\right) = 18{,}662.699 \text{ kg, } \textit{chemical} \\ 1000\left(e^{14/9.1} - 1\right) = 3657.42 \text{ kg,} \quad \textit{nuclear} \\ 1000\left(e^{14/20} - 1\right) = 1013.75 \text{ kg,} \quad \textit{ion} \end{cases}$$

Clearly, the ion thruster requires the smallest amount of propellant; it is just 1.37% mass of the total spacecraft to accomplish the Earth-Mars mission.

Example 7.6 Estimate the propellant mass required by a nuclear rocket delivering a specific impulse of 1000 s for a mission with a total velocity change of $\Delta v = 10$ km/s. Compare the result with the propellant mass required with a chemical rocket. Assume the spacecraft final mass is 1500 kg.

Solution: From the rocket equation, the propellant mass for the nuclear rocket is

$$m_p = m\left(e^{\Delta v/v_{ex}} - 1\right) = 2658.61 \text{ kg}$$

where $v_{ex} = g_0 I_{sp} = (9.80665)(1000) = 9806.65$ m/s, and $m = 1500$ kg.

For the same 1500 kg spacecraft propelled by a chemical rocket with $I_{sp} = 500$ s, the propellant mass would be $m_p = 10{,}045.33$ kg. This result implies that, by doubling the specific impulse of propulsion, the propellant mass required for the mission is significantly reduced.

7.3.2   Pulsed Plasma Rocket (PPR)

The Pulsed Plasma Rocket (PPR) is a nuclear propulsion concept that uses a fission-based nuclear power system to quickly promote a phase change in a fuel projectile from solid to plasma during a pulsed cycle. Being developed by Howe Industries and sponsored by NASA, the PPR may generate up to 100 kN of thrust with a specific impulse of 5000 s, and the projected system's high efficiency would allow for crewed missions to Mars to be completed within two months.

Conceptually, pulses of superheated plasma are essential to provide thrust through an electromagnetic nozzle. Howe and his team [15] have designed a nuclear fission system to achieve these cyclic plasma bursts. The system consists of an unmoderated high assay low enriched uranium barrel and moderated uranium projectiles to preferentially heat the projectile instead of the barrel. Other important systems essential for the PPR design include a coil gun injector and a magnetic nozzle. The PPR requires a target projectile to be superheated into a plasma in less than a second to achieve a full-cycle frequency of 1 Hz. Figure 7.5 provides a rendition of the PPR vehicle in which the rear section is comprised of the magnetic nozzle and the uranium barrel, with a propellant tank in the middle, and the payload in front. For additional details, see Howe, et al. [15]. If successful, the PPR promises to advance the propulsion technologies required for fast crewed and robotic interplanetary missions.

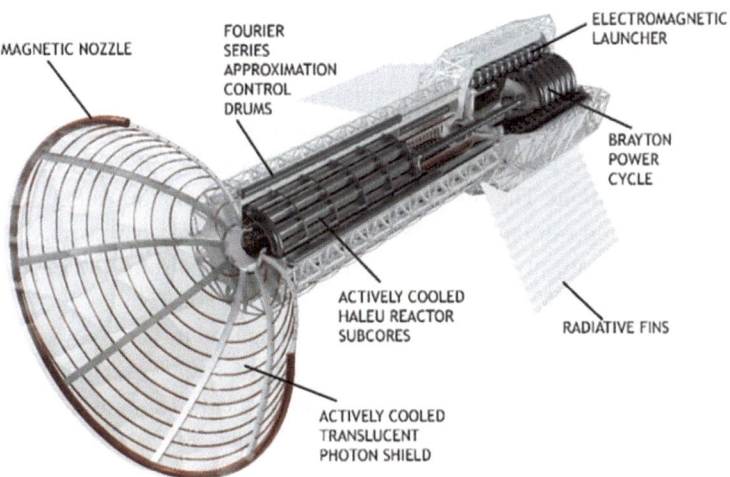

Fig. 7.5 Pulsed Plasma Rocket concept. *Credit* Briana Clements, Howe Industries

7.3.3 Nuclear Fusion Rockets

A nuclear fusion rocket requires a fusion nuclear reactor to heat the propellant to very high temperatures. Fusion power is generated by nuclear fusion reactions in which two light atomic nuclei fuse together to form a heavier nucleus (in contrast with fission power), just as it occurs in the Sun. The result of fusion is the release much larger amount of energy, which arises from the binding energy due to the strong nuclear force, manifested as an increase in temperature of the reactants. Fusing a kilogram of hydrogen, for example, would generate more than a million times more energy than chemical reactions in a kilogram of hydrogen and oxygen.

A fusion rocket design would be driven by fusion power to provide efficient and long-term acceleration in space without the need to carry a large fuel supply. The fusion rocket requires to operate for hours and be capable of being turned off and restarted. However, we do not have the technology to build controlled fusion reactors for space applications, and the required magnetic confinement and other support systems would make fusion rockets much more complex and more massive than any current propulsion system.

7.3.3.1 The VISTA (Vehicle for Interplanetary Space Transport Applications)

In 1987, researchers at the Lawrence Livermore National Laboratory published a study based on a new vehicle concept identified as VISTA. It incorporated inertial confinement fusion (ICF) and was meant to have capability for round-trip missions to Mars in 100 days, including a stay for then days, and carry a 100 MT payload. The Vehicle for Interplanetary Space Transport (VISTA) is sketched in Fig. 7.6.

For VISTA, researchers selected deuterium-tritium (DT) fusion because of its energy release of up to 3.4×10^{11} J/g, providing power-to-mass ratios of ten to several hundred watts per gram (W/g). As conceived, the propulsion system for VISTA would consume 20 metric tons (mt) of tritium. The study assumed a target gain of 1500 and pulse repetition rate of 30 Hz. Carrying 4400 mt of propellant, the engine would have a mass flow rate of 1.5 kg/s and yield a thrust of 2.4×10^5 N and a specific impulse of 17,000 s with a total jet power of 2.0×10^4 MW.

Implied by the size of the spacecraft, VISTA was intended for large power applications. The researchers emphasized the inherent advantages of fusion over other technologies by use of magnetic thrust chambers. Magnetic confinement of the fusion-heated propellants isolates the hot plasmas from first walls and eliminates the thermal constraints that mechanical thrust chambers impose.

VISTA researchers considered using a 6%-efficient excimer-laser driver operating at 1000 K with an output of 5 MJ, and using pellets to allow energy gains from 200 up to 1500 (with maximum pellet repetition rate of 30 Hz). The pellets concept was admittedly highly speculative, as it was based on extrapolations of analytic modeling. Their analysis projected VESTA to yield a specific impulse of about 17,000 s, with a jet efficiency near

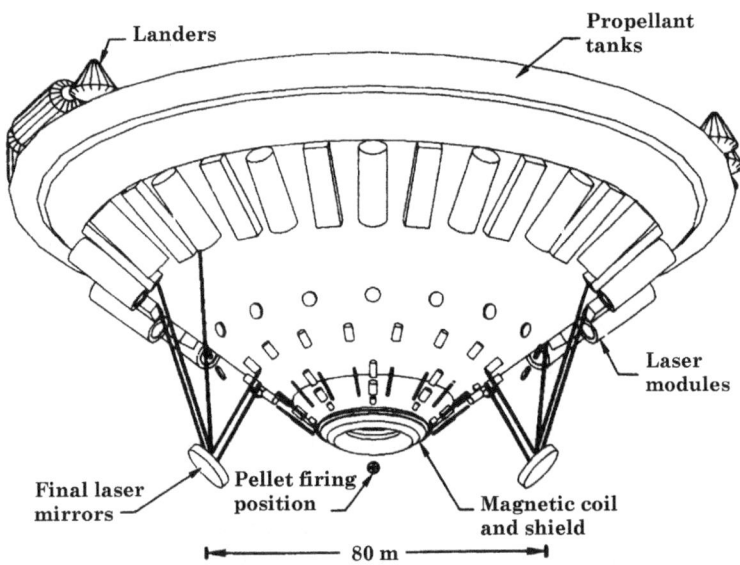

Fig. 7.6 ICF powered crewed VISTA systems layout, showing two final laser focusing mirrors. *Credit* Orth et al. [26]

36%. The power system was calculated to have a power-to-mass ratio near 20 W/g. This projected performance would allow VISTA a round trip to Mars with a 100-metric-ton payload in about 100 days with a launch mass near 6000 metric tons.

7.3.3.2 Orion Project

A popular idea proposed to power a starship is nuclear pulse propulsion (NPP), which would utilize fusion power but without the need for controlled fusion reactors. In principle, NPP requires generating propulsion energy with repeated detonations of relatively small H-bombs. Such rocket would use the highly energetic and efficient energy release from nuclear explosions directly to produce thrust.

In the Project Orion, propulsion would result from explosions taking place a few tens of meters behind the spacecraft. The vaporized debris from each explosion would impact a "pusher plate" on the back of the spacecraft, propelling the spacecraft forward. Conceptually, in an NPP (Fig. 7.7) an individual explosive device, the so-called pulse unit, is ejected from the spacecraft and detonated at a predetermined standoff distance from the rear. The resulting explosion vaporizes the entire pulse unit and causes this "propellant" to expand as a high energy plasma, with some fraction interacting with the vehicle and providing thrust. It requires many pulses, probably at equal intervals or with a predetermined frequency to generate thrust.

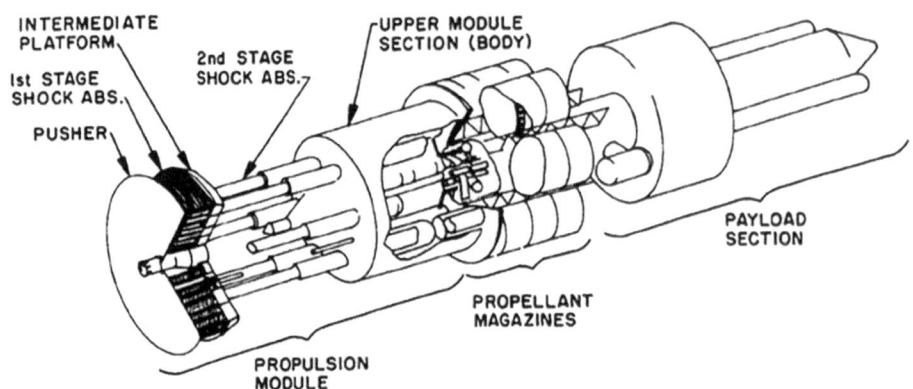

Fig. 7.7 Main components of the Orion propulsion system. *Credit* NASA

Researchers concluded that some materials could survive a nuclear detonation, enough to provide a controllable conversion of blast energy into vehicle kinetic energy. In principle, the NPP could deliver specific impulses between 10,000 s up to 100,000 s with average power densities equal to or greater than chemical rockets. It was expected that an Orion spacecraft could be built with existing technology, although it could be very expensive and would require an exception to the international treaty banning nuclear detonations in space.

Nuclear pulse rockets were originally intended for use on interplanetary missions such as single stage (i.e. directly from Earth's surface) to Mars and back, and a trip to one of the moons of Saturn. The Orion project lasted from 1958 to 1965. British physicist Freeman Dyson conducted an analysis to assess Orion's feasibility to reach Alpha Centauri. In a 1968 paper, Dyson retained the concept of large nuclear explosions but instead of fission bombs he considered the use of one megaton deuterium fusion explosions. Dyson concluded that the debris velocity of fusion explosions was probably in the 3000–30,000 km/s range, i.e. 1% and 10% of the speed of light, and the reflecting geometry of Orion's pusher plate would reduce that range to 750–15,000 km/s. This gave the upper and lower limits for the maximum available exhaust velocity for a nuclear rocket.

Carl Sagan performed other studies and determined that the maximum cruise velocity for the thermonuclear Orion concept would be $0.08c$ to $0.1c$, assuming no propellant was saved for slowing down. Thus, moving at $0.1c$, Orion would require a flight time of at least 44 years to reach the nearest star, not counting time needed to reach that speed, about 36 days at constant acceleration of 1 g. At the same speed, it would take 100 years to travel 10 light years.

7.3.3.3 Daedalus Project, Longshot, and Prometheus

The Project Daedalus considered a fusion-powered rocket for a mission to a nearby star. Conceived in the 1970s by members of the British Interplanetary Society, this propulsion approach relies on generation of a continuous stream of energy from an on-board controlled nuclear fusion reactor. The study is based on a 50-year robotic mission to Barnard's star, only 5.9 ly away. The interstellar ship considered in the Daedalus Project would be rather massive, with an initial mass of 54,000 tons, including 50,000 tons of propellant and 500 tons of scientific payload. For comparison, the International Space Station (ISS) mass is just 400 tons!

Daedalus was designed as a two-stage spaceship. The first stage would operate for two years, moving the huge spaceship to a $0.071c$ velocity. After this the stage would be discarded and the second stage would fire for 1.8 years, accelerating the spaceship to $0.12c$ before shutting down for the 44-year cruise period to reach the star. Figure 7.8 shows an artistic rendition of the 190 m tall Daedalus and its spherical tanks containing the fuel pellets for the nuclear fusion engine. It also shows its relative size, compared with Saturn V, which at 111 m (363 ft) it was about 58 ft taller than the Statue of Liberty from the ground to the torch. Daedalus is beyond our current technological capabilities.

Fig. 7.8 Daedalus compared in size with Saturn V. *Credit* Adrian Mann

Another futuristic interstellar program was Project Longshot, a joint NASA-NAVY study conducted in the late 1980s, intended to reach Alpha Centaury in 100 years. It considered a robotic interstellar mission with a 400 tons spacecraft in which 67% of the initial mass was propellant. Longshot would be as massive as the International Space Station. The primary enabling technology was a pulsed fusion micro-explosion drive delivering a specific impulse of 1,000,000 s, requiring a large, long-life fission reactor with 300 kW power output.

Years later, after cancelling the Longshot project, NASA revitalized interest in nuclear propulsion with its Project Prometheus. Missions planned to involve Prometheus Nuclear Systems and Technology included Jupiter Icy Moons Orbiter, the Jovian moons Europa, Ganymede, and Callisto. Originally planned to be the first mission of Project Prometheus, it was deemed too complex and expensive, and its funding was cut in the 2006 budget.

In principle, fusion-powered spacecraft could probably achieve speeds of about $0.12c$. Such speeds are sufficient for robotic interstellar probes but perhaps not realistic for crewed missions to the near stars. As of today, Daedalus, Longshot and Prometheus are great ideas but beyond our current technological capabilities.

7.3.4   Project Icarus

Icarus is a design study for an uncrewed fusion-powered interstellar probe based on Project Daedalus. Launched in 2009 at the British Interplanetary Society HQ in London, Project Icarus is a five-year volunteer engineering study to design an interstellar spacecraft. Its primary objectives are to obtain the engineering layout, functionality, physics, operation, expected performance and mission profile of an uncrewed interstellar probe. The Icarus project had the goal to design an uncrewed probe capable of delivering useful scientific data about the target star, associated planetary bodies, solar environment and the interstellar medium.

The Icarus spacecraft propulsion "must be mainly fusion based." The main 'combustion' cycle relies upon some mechanism which releases energy through the combined reaction of fusion-based isotopes, usually elements much lighter than iron. This propulsion system is projected to deliver a thrust of 3.6×10^4 N, and a specific impulse of 0.5×10^6 s. This means that the rocket exhaust velocity would be 5000 km/s. About 0.2 years is estimated for the boost phase, assuming the spacecraft accelerates up to the cruise velocity. The Icarus team planned to develop a series of small vehicles to test the new enabling technologies. The minimum preliminary quantitative requirements of the Icarus spacecraft, for a flyby probe to Alpha Centauri only, are summarized in Table 7.4. If the required fusion propulsion technology is available, the Icarus Pathfinder spacecraft can become a precursor to interstellar travel.

In principle, fusion rockets may provide near-optimum performance values for fast interplanetary and deep space missions. However, although fusion reactors have been

Table 7.4 Minimum mission parameters for Icarus spacecraft (Obousy et al. [25])

Cruise duration	Total Δv	Cruise velocity (km/s)	Size (m)	Total spacecraft mass (ton)	Propellant mass (ton)	Payload mass (ton)	Min KE to cruise speed (J)	Min power to cruise (TW)
22 yr	0.04c	13,000	10 × 5	40,000	37,000	2.0	1.7 × 10^{18}	0.30

under development for decades—aiming to demonstrate controlled fusion for terrestrial power plants—an enormous amount of research and development remains before sustained fusion is achieved. Development of space-based systems represent a significant challenge and the realization of a fusion-powered rocket is still considered relatively far term.

7.4   Propulsion Without Propellant Reaction Mass

The ultimate limit of propulsion performance is set by the rocket equation, which is a function of Δv, I_{sp}, and most critically of the overall mass ratio. Hence, we strive to optimize I_{sp} by selecting a propulsion device that can exhaust the propellant mass at higher velocities, minimizing the mission total Δv by incorporating gravity-assist, for example, to the spacecraft mission trajectory. Ultimately, we can also conceive propulsion methods that do not rely in a stored propellant reaction mass, producing thrust without ejecting mass, thus optimizing the spacecraft mass ratio. For example, we can harness solar energy to provide power and the force of solar pressure to propel spacecraft. The solar sail is essentially a big photon reflector surface in which the external power source is the Sun.

7.4.1   Solar Sails

Centuries ago, Johannes Kepler observed the tails of comets blown by the solar breeze and suggested that perhaps vessels might also navigate through space using solar sails. Today we know that solar radiation causes the volatile materials within the comet to vaporize and stream out of the nucleus, carrying dust away with them. The streams of dust and gas released take the form of a huge, extremely tenuous atmosphere around the comet called the coma, and the force exerted by the Sun's radiation pressure and solar wind on the coma cause an enormous tail to form. This phenomenon led to the idea of propelling a spacecraft without propellant.

In principle, spacecraft with large, lightweight sails could be propelled by the solar wind and the radiation pressure of sunlight. A solar sail takes advantage of the fact that, although light has no mass, it has momentum, and therefore can exert pressure. Of course, sunlight pressure is extremely small, but it can be enough for propulsion if the sail is big enough and very long trip times are acceptable for a mission. NASA proposed to develop an interstellar spacecraft propelled by sunlight reflected from an ultrathin sail. Nearly half a kilometer wide, the delicate solar sail would be unfurled in space. Once opened, the spacecraft would slowly revolve around the Sun, gaining more and more momentum as it moves. After several years orbiting the Sun, the sail-driven spacecraft would spiral out of the Solar System and reach the stars. NASA engineers assumed that the continuous pressure from sunlight would ultimately accelerate the craft to speeds about five times higher than possible with conventional rockets—without requiring any fuel. Figure 7.9 shows an artistic rendition of a solar sail.

In its most basic concept, the solar energy would give an optimally designed sailing spacecraft a tiny but continuous acceleration. Therefore, by the time the spacecraft reaches a distance at which the solar wind could no longer provide acceleration, it might already be traveling at very high speeds. To determine the size of the sail we need to calculate the force of the solar radiation pressure.

All electromagnetic (EM) waves carry energy, and they also carry linear momentum. Hence, when electromagnetic waves encounter the surface of an object, a force will be

Fig. 7.9 An artist's concept of a solar sail in Earth orbit. *Credit* NASA

exerted on the surface due to momentum transfer, just as when a moving particle strikes another. The force exerted by the electromagnetic waves is called *radiation pressure*.

Maxwell showed that if a beam of EM radiation is completely absorbed by an object, then the momentum transferred is

$$\Delta p = \frac{\Delta U}{c} \tag{7.12}$$

where ΔU is the energy absorbed by the object in a time interval Δt, and c is the speed of light.

If instead, the radiation is fully reflected, then the momentum transferred is twice as great,

$$\Delta p = \frac{2\Delta U}{c} \tag{7.13}$$

Using Newton's second law, the force and pressure exerted by EM radiation on an object can be easily found. The force F is expressed as

$$F = \frac{dp}{dt} \tag{7.14}$$

The average rate at which energy is delivered to an object is related to the Poynting vector. This represents the directional energy flux density or the rate of energy transfer per unit area:

$$\frac{dU}{dt} = IA \tag{7.15}$$

where I denotes the intensity of radiation, and A is the cross-sectional area of the object intercepting the radiation.

The solar radiation pressure p_s, assuming full absorption, is

$$p_s = \frac{F}{A} = \frac{1}{A}\frac{dp}{dt} = \frac{1}{Ac}\frac{dU}{dt} = \frac{I}{c} \tag{7.16}$$

Now, if the radiation is fully reflected, the rebounding photons would cause the pressure to be doubled, that is,

$$p_s = \frac{2I}{c} \tag{7.17}$$

This expression provides a quick estimate of magnitude of the propulsion force on the sail. Radiation from the Sun that reaches the Earth's surface, after passing through the atmosphere, transports energy at a rate of about 1130 W/m^2. The intensity of solar radiation at the orbital altitude of the ISS as about 1366 W/m^2.

Solar sailing is proposed as an inexpensive way of navigating within the Solar System, and more sophisticated approaches consider solar sails for interstellar probes. As interplanetary propulsion systems solar sails are limited: even close to the Sun the solar pressure is weak and as one gets too far from it, for example beyond the orbit of Mars, the pressure becomes so feeble that propulsion is no longer viable. Moreover, this method of propulsion takes a long time to accelerate a spacecraft to a reasonable velocity necessary to traverse vast interplanetary distances, thus unable to reduce the transit time. Of course, using lasers or microwave transmitters would power the solar sails along at a faster rate than sunlight alone.

A major problem with a solar sail-driven spaceship is the difficulty of stopping or deceleration. Engineers propose to reverse the direction of the sail and use the light pressure of the destination star to slow down. It is also proposed to sail around a star, using the star's gravity to create a slingshot effect for the return voyage.

The development of solar sails has encountered some challenges. In 1993, Russia deployed a sixty-foot Mylar reflector from the Mir space station to demonstrate deployment. A second attempt failed. In 2004, the Japanese successfully launched two solar sail prototypes also to test deployment, not propulsion. In 2005, the Planetary Society Cosmo Studios and the Russian Academy of Sciences teamed up to deploy a solar sail called Cosmo I. However, the Volna rocket misfired and failed to reach orbit. And in 2008 NASA tried to launch a solar sail called NanoSail-D, but it was lost when the SpaceX Falcon 1 launch vehicle failed.

In 2010, the Japan Aerospace Exploration Agency (JAXA) successfully launched the IKAROS, the first spacecraft to use solar sail technology in interplanetary space. It has a squared shaped sail, 20 m (60 ft) on the diagonal, and used solar light for propulsion to travel on its way to Venus. The Japanese plan to send a similar probe to Jupiter.

In 2011, NASA's NanoSail-D spacecraft unfurled a gleaming sheet of space-age fabric 650 km above Earth, becoming the first-ever solar sail to circle our planet (see Fig. 7.9). The NanoSail-D is designed to remain in the atmosphere. Its mission is to circle Earth and investigate the possibility of using solar sails as a tool to de-orbit old satellites and space junk.

7.4.2   Beam Propulsion

Beam-energy propulsion is a concept in which a beam of energy is directed at a spacecraft either to heat up its propellant or to deliver electricity or power to its engine. By removing the energy source from the rocket, itself, beam-energy propulsion has some potential benefits.

The idea of beam propulsion is to dispense with reaction mass entirely. In this concept a large, primary spacecraft sends out a directed beam of rarefied gas or laser light to be intercepted by a second spacecraft. Momentum transfer during interaction then propels a second spacecraft to high velocity, as it carries neither propellant nor a propulsion power source.

As interstellar propulsion systems solar sails are not effective: even close to the Sun the solar pressure is weak, and as one gets too far from it, for example beyond the orbit of Mars, the pressure becomes so feeble that propulsion is no longer viable. However, using lasers or microwave transmitters would power the solar sails along at a faster rate than sunlight alone. The sails would be powered like those designed to use solar photons, with the lasers or microwave transmitters beaming at the sail for a few days or weeks, providing the sail with the light needed to create thrust. Theoretically, microwave transmitters can be used to blast the sail until it heats to 2000 K, which would accelerate the sail at hundreds of g's to get the sail up to $0.1c$. Either system—high-powered lasers or microwave transmitters—could be used but both require advances in technology not foreseeable any time soon.

In principle, beamed energy propulsion could accelerate a spaceship to relativistic speeds. However, the required laser needs to be extremely powerful. It is estimated that to accelerate a spacecraft to $0.5c$ within a few years requires a laser that uses 1000 times more power than all current human power consumption. If such laser were to be developed, it would just solve half the problem. A method to decelerate the spacecraft upon approaching its destination must also be available.

Physicist and science fiction writer Robert L. Forward conceived a means for decelerating an interstellar light sail without requiring a laser array as others suggested. In Forward's concept, a smaller secondary sail would be deployed at the rear of the spacecraft, while the large primary sail detached from the vehicle to keep it moving forward on its own. In other words, light reflected from the large primary sail into the secondary sail would decelerate the spacecraft.

In another application of beamed energy, an external pulsed source of laser energy is used to provide power for producing thrust. In the "lightcraft" concept, the laser is focused on a parabolic reflector on the underside of the vehicle to produce a region of extremely high temperature. Air is heated and expands violently, producing thrust with each pulse of laser light. In space, a lightcraft would need to provide this gas itself from onboard tanks or from an ablative solid. Its proponents believe that this concept makes it ideal for launch applications since, by leaving the vehicle's power source on the ground and by using ambient atmosphere as reaction mass for much of its ascent, a lightcraft would be capable of delivering a very large percentage of its launch mass to orbit. Leik Myrabo, founder of Lightcraft Technologies, demonstrated that one can propel a small "lightcraft" 71 m in the air by using pulses of light that heat the propellant.

Beamed energy has some major issues that are cause for concern and have not been addressed. For example, an energy source powerful enough to propel a rocket could also burn it up. Also, there is potential for atmospheric interference with the beam. And of course, there is the issue of the power required for the laser itself. Beamed energy propulsion is an interesting concept for application to interplanetary travel, but its development remains far in the future.

The Diffractionless Beamed Propulsion for Breakthrough Interstellar Missions (PROC-SIMA) concept proposed by Limbach and Hara [18], increases the distance over which spacecraft is accelerated, while simultaneously reducing the beam size at the transmitter and probe from 10 s of kilometers to less than 10 m.

Other approaches for propulsion using nuclear explosions and pulsed nuclear fusion have been conceived, but are not yet technologically feasible. Currently, the interest is in developing concepts for transmitting radiation energy (by lasers or microwaves) from Earth stations to satellites. Since the focus of this book is rocket propulsion, no further discussion of these concepts is provided.

Example 7.7 Consider a sail made of an aluminized material moving in the vicinity of the Earth's orbit and facing the Sun. Calculate the sail's acceleration, neglecting the gravitational effects. The thin sail has an areal density (mass per unit area) equal to 1.0×10^{-4} kg/m^2.

Solution: The radiation pressure on a sail at 1 AU from the Sun is, from Eq. (7.17),

$$p_s = \frac{2I}{c} = 0.91 \times 10^{-3} \text{ N/m}^2$$

The sail's acceleration is equal to the pressure divided by the areal density ρ of the sail,

$$a = \frac{p_s}{\rho} = 9.1 \text{ m/s}^2 \rightarrow a = 0.92 \, g_0$$

Example 7.8 Consider a solar sail to propel a spacecraft across the Solar System. (a) Determine the magnitude of the force that would be applied on a solar sail made of highly reflective material and that measures 1000 m $\times$ 1000 m. (b) Estimate the change in velocity Δv for a 5000 kg spacecraft in a one-year mission.

Solution: (a) The force applied on the sail is $F = p_s \cdot A$, where the solar pressure p_s is, from Eq. (7.17),

$$p_s = \frac{2I}{c} = 6 \times 10^{-6} \text{ Nm}^{-2}$$

Hence, the force exerted by solar radiation is $F = p_s \cdot A = 6$ N.

(b) From Newton's second law, the acceleration of the spacecraft due solely to this force is

$$a = \frac{F}{m} = 1.2 \times 10^{-3} \text{ms}^{-2}$$

The velocity increase due to this acceleration in one year is simply,

$$\Delta v = v - v_0 = at = 4 \times 10^4 \text{ m/s}$$

Starting from rest, this acceleration would result in a travel distance in one year of about 5.95×10^{11} m, which is approximately 4 AU from Earth, beyond the orbit of Mars.

7.5   Interstellar Ramjet

Most propulsion concepts have limited capabilities, i.e. rockets can carry only a limited amount of propellant no matter what their source is, and beamed energy becomes too weak to accelerate a vehicle when it moves too far from the source. One approach to overcome those limitations is a propulsion system that collects the propellant as it moves through space.

In 1960, Robert W. Bussard proposed such propulsion system similar in principle to an air-breathing ramjet engine but intended for interstellar travel. The ramjet engine takes in the air in the atmosphere and uses it as the oxidizer to combust a fuel. The chemical reaction of fuel and atmospheric oxygen results in a hot gas that is expanded through a nozzle and creates thrust. Bussard's idea used the same approach. However, instead of air, the interstellar ramjet would scoop hydrogen from interstellar medium, and use it as propellant for a fusion nuclear reactor. Once inside the reactor, the hydrogen atoms would be compressed and heated by electromagnetic fields until the hydrogen fused into helium, releasing enormous amounts of energy in the process. The interstellar ramjet could accelerate continuously if it collected enough propellant for fusion reactions.

In the Sun (and other main stars), the nuclear reaction fuses two hydrogen atoms together to produce helium: $H + H \rightarrow H_e + E$, where the amount of released energy E is calculated with the mass-energy conversion formula, $E = mc^2$. It is estimated that the Sun fuses about 600 million tons of hydrogen every second, yielding 596 million tons of helium. The remaining four million tons of hydrogen are converted to energy, which makes the Sun shine. Most of this energy is in the form of gamma-rays and X-rays.

Bussard estimated that if a 1000-ton ramjet engine could maintain the acceleration of 1 g, then it would approach relativistic speed in just one year. Since the hydrogen atoms are limitless the ramjet could run forever, and the spaceship could transverse the Galaxy. In principle, such a spaceship could accelerate at 1 g for half the voyage to its destination and then turn around and decelerate at 1 g until arrival. This approach would provide an Earth-like gravity field for the full trip, making it comfortable ambient for the crew. If achievable, during most of the voyage, the astronauts onboard would be traveling relative to Earth (and the destination) at a speed very close to the speed of light, so time on the ship would pass very slowly compared to time on Earth. Figure 7.10 depicts a relativistic vehicle propelled by an interstellar ramjet.

The technological challenges for the interstellar ramjet include the fusion process itself and the design of the vehicle. First, although a fusion reactor on Earth could produce a large amount of energy by fusing deuterium and tritium, interstellar space contains mainly

Fig. 7.10 Bussard interstellar ramjet concept. A huge magnetic funnel collects hydrogen from space to use as fuel for the interstellar ramjet. *Credit* Adrian Mann

protons of hydrogen, and fusing protons with protons would not yield sufficient energy. Bussard considered adding some carbon to the mixture to serve as a catalyst to increase the energy produced, sufficient to drive the starship. But this would require carrying extra mass and would complicate the design.

Furthermore, because the density of interstellar space is so low, the scoop on the forward side of interstellar ramjet must be huge to collect enough hydrogen atoms for fusion; a concave disk of 160 km was estimated. Such large scoop would incur a huge drag, preventing the spaceship from accelerating to relativistic speeds. In interstellar space the drag results from the resistance that a huge body would encounter as it moves in a field of hydrogen atoms.

Interstellar ramjets therefore seem as far in the future as any of other fusion rockets. However, in principle the interstellar ramjet has one major advantage over other concepts: the ability to accelerate continuously, provided it collects enough propellant for fusion reactions. Continuous acceleration to high relativistic speed is theoretically possible. However, it is estimated that the thrust generated would be much less than the drag caused by the scoop.

7.6   The TAU Mission

In 1987, K.T. Nock from NASA/JPL proposed to send a space probe to a region in space 1000 AU from the Earth, for a mission known as TAU (Thousand Astronomical Units). This distance is beyond the Kuiper Belt, a region of the Solar System full of icy objects, extending from the orbit of Neptune to approximately 50 AU from the Sun.

Nock proposed to use nuclear electric rockets with a 1 MW fission reactor and an ion drive with a propulsion time of 10 years to reach the 1000 AU distance in 50 years. The TAU vehicle would be accelerated by an ion NEP system for about ten years before escape velocity could be attained because of the extremely low thrust of the xenon-fueled ion engines. At the end of the thrusting phase, the NEP system would be jettisoned to allow the TAU spacecraft and science experiments to coast to 1000 AU. Before the TAU proposal, another study proposed a spacecraft to 550 AU from the Sun to exploit the magnification provided by the gravitational lens of the Sun. These and many other exciting futuristic ideas for space exploration require funding to develop advanced propulsion and supporting technologies.

7.7   Technology Challenges for Advanced Propulsion

Solar energy is and will continue to be used in space applications. However, we must remember that to gather massive quantities of energy, solar power panels must be large, and power conversion efficiency must be high. On the other hand, because a nuclear reactor has a very large amount of energy per unit mass, it is an ideal power source in space. That is why nuclear electric powered propulsion is considered for reducing human transit times between planets and to propel robotic cargo missions with a very large payload mass fraction.

Let us now concentrate on the nuclear energy source and not the nuclear rocket. To date, the power conversion efficiency of nuclear energy systems is rather low, ranging from 5% for a static thermal to electric conversion system to 25% for a dynamic system such as a Brayton power conversion device. Hence, a significant amount of energy will be wasted, requiring extra systems to radiate the waste heat to space, and optimize available energy for thrust.

7.7.1   Power Generation System

There are several sources of electrical energy in space to power spacecraft. Power generation systems include batteries (chargeable and non-chargeable), solar cells, photovoltaic devices, and nuclear power systems. Solar cells are the most reliable energy generation systems for aerospace applications. Since the mass of the power supply contributes directly to the vehicle mass ratio for any maneuver, power generation systems must have a high power-to-mass ratio.

Due to their high-power conversion efficiency and certified reliability while operating in orbit, Silicon-based multi-junction solar cells are now the standard technology for powering spacecraft. However, new materials for solar cells are now considered. For example, perovskite solar cells (PSCs) are promising candidates. A PSC is a type of solar cell which includes a perovskite structured compound, most commonly a hybrid organic–inorganic lead or tin halide-based material, as the light-harvesting active layer. Perovskite materials possess intrinsic properties like broad absorption spectrum, fast charge separation, long transport distance of electrons and holes, long carrier separation lifetime, that make them very promising materials for solid-state solar cells. For aerospace, PSCs are appealing being light-weight with cost-effective manufacturing, and exceptional radiation resistance (see Verduci et al. [30]).

7.7.2 Radioisotope Energy for Interplanetary Spacecraft

The half-life of Pu-238 is about 88 years. Half-life is the time it takes for half of the atoms of an element to decay. That means the isotope's heat output will not be reduced to half for 88 years. When its mission began, the RTGs of the Voyager delivered 470 W of 30 V (direct current). Early in 2008, the power generated by Voyager 1 had dropped to ~285 W and to ~287 W for Voyager 2. These power levels represent better performance than the pre-launch predictions, which included a conservative degradation model for the bi-metallic thermocouples used to convert thermal energy into electrical energy. And although the Voyager will continue to be powered for many more years, as the electrical power is reduced, power loads must be turned off to avoid having demand exceed supply.

In addition, Pu-238 is stable at high temperatures, can generate substantial heat in small amounts, and it emits relatively low levels of radiation that is easily shielded, so spacecraft critical instruments and equipment are not affected.

On the other hand, the heat generated by the RTGs can cause a spacecraft to decelerate. The so-called Pioneer anomaly is now attributed to the heat from the RTGs reflected off the backside of the large antenna that affected the two Pioneer spacecraft as they left the Solar system. Researchers observed that the twin Pioneer 10 and 11 spacecraft were not at a location in space where they were expected to be. Launched in 1972 and 1973, the Pioneer have covered hundreds of millions of kilometers, heading toward the edge of our Solar System. But heat from RTGs holds them back. Each year, they fall behind in their projected travel by about 5000 km (3000 miles). After a study of thermal analysis combined with Doppler data, researchers concluded that this heat flux is "pushing" the spacecraft back toward the Sun (or decelerating it with respect to Earth). Some of the heat from the plutonium inside the generators is converted into electricity while the rest of it radiated into space. If this is done unevenly, radiating more heat in one direction than in another—only a 5% difference is required—that apparently is sufficient to give rise to the Pioneer anomaly.

NASA is now developing an advanced Stirling radioisotope generator (ASRG). This is a radioisotope power system that uses a Stirling power conversion approach to convert radioactive-decay heat into electricity. Researchers estimated that an ASRG can be four times more efficient than RTGs. The ASRG's energy conversion process needs about one quarter of the plutonium-238 used in previous radioisotope systems to produce a similar amount of power.

7.7.3 Other Challenges for Advancing Propulsion

Interplanetary distances are very large. Earth is about 149 million km away from the Sun (1 AU). Our neighbor planet Mars is about 56 million km away when it is at favorable opposition (the closest). Pluto is at an average distance of 5,763,920,000 km from the Earth, so far away from us that it is not visible in the sky. Having a sense of the huge distance that separates us from other planets in the Solar System, we begin to understand how gigantic our galactic neighborhood truly is.

If a spacecraft can reach velocities of 50–100 km/s in the future, voyages to the nearest planets in the Solar System will require a few years. For interplanetary robotic missions these speeds are therefore quite adequate. Nuclear electric propulsion (NEP) for example, in which a nuclear reactor supplies power for an electric thruster, can provide high specific impulse values of about 1000 s. Although this projected performance represents a promising option for future spaceflight, it is not optimum for crewed deep space exploration.

To leave our Solar System, velocities of 100 km/s are insufficient. To traverse interstellar distances, we require speeds which are a substantial fraction of the speed of light. Nuclear fusion can provide the energy for interplanetary and interstellar travel. Fusion rockets may provide near-optimum performance values for the most demanding space missions. In principle, fusion-powered spacecraft could probably achieve speeds of about $0.12c$. Such speeds are sufficient for robotic interstellar probes but perhaps not realistic for crewed missions to the near stars.

However, although fusion reactors have been under development for several decades to demonstrate controlled fusion for terrestrial power plants, an enormous amount of research and development remains before sustained fusion is achieved. We still do not know how to contain the radioactive material if it were to escape from the rocket's core. Development of space-based systems therefore represent a significant challenge. In fact, most if not all space-based fusion propulsion systems considered to date are highly speculative, and the realization of a fusion-powered rocket is still considered relatively far term.

If a crewed spacecraft could achieve relativistic speeds, an interstellar trip would be one-way as such voyage time would take the span of a human's lifetime. Time dilation allows such a trip, but the enormous practical problems of achieving relativistic speeds are too difficult to overcome with current technologies. If a rocket accelerates at 1 g (~ 9.81 m/s^2), astronauts would experience a gravitational field with the same strength as that we feel on Earth. If this could be maintained for long enough time they would benefit from the relativistic effects and improve the effective rate of travel to the stars.

Glossary

Advanced Stirling Radioisotope Generator (ASRG) An energy producing device or radioisotope generator based on a Stirling cycle powered by a large radioisotope heater unit. The ASRG can reduce the amount of Pu-238 required for a given electric power output.

Nuclear Fission Fission is the division into approximately two parts of unstable nuclei (such as uranium 235), usually caused by entrance of slow neutrons. Large amounts of kinetic and radiant energy are released as well as neutrons, which may be moderated in speed to induce similar reactions in nearby fissionable nuclei. In a properly assembled distribution of fissionable mass and moderator, the process can chain-react, or be self-sustaining. Critical fission mass is the minimum mass of fissionable material required to sustain a reaction.

Nuclear Fusion Fusion is the process of joining two rapidly moving nuclei (e.g. deuterium) to form a new nucleus resulting in release of large amounts of energy (greater per gram of reacting material than that of nuclear fission). The kinetic energy required to begin the reaction may be produced by thermal random motion at very high temperature. If high temperature is maintained by energy released in fusion reaction, the process may be self-sustaining and thermonuclear reactions will take place over an extended volume. There is no critical mass, nor are there toxic reaction products.

Nuclear Thermal Propulsion (NTP) A rocket concept wherein the thermal energy required to heat a propellant is taken from the fission of a nuclear fuel in a nuclear reactor.

Radioisotope Decay The spontaneous transformation of one radioisotope into one or more different isotopes known as "decay products." All radioisotopes decay or undergo a nuclear transformation at a unique rate. The number of atoms transformed per unit time decreases exponentially.

Radioisotope Thermoelectric Generators (RTG) an energy producing device that uses an array of thermocouples to convert the heat released by the decay of a suitable radioactive material into electricity. RTGs are used to supply electrical energy to space probes.

Uranium 235 an isotope of uranium, comprising about 0.72% of natural uranium. It has a nucleus containing 92 protons and 143 neutrons. Uranium-235 is the only naturally occurring fissile isotope of uranium—it can sustain a nuclear chain reaction—and it is of interest for nuclear propulsion.

Recommended Reading and References

1. Adams, R, et al., (2003). "Conceptual design of in-space vehicles for human exploration of the outer planets." NASA/TP-2003-212691. November 2003.
2. Beals, K.A et al., "Project Longshot, an unmanned probe to Alpha Centauri", N89-16904, 1988.
3. Belair, M. L., Sarmiento, C. J., and Lavelle, T. M. (2013). "Nuclear Thermal Rocket Simulation in NPSS." 49th AIAA/ASME/SAE/ASEE Joint Propulsion Conference, July 14–17, 2013, San Jose, CA, https://doi.org/10.2514/6.2013-4001.
4. Borowski, S.K., et al. (1993). "Nuclear Thermal Rocket/Vehicle Design Options for Future NASA Missions to the Moon and Mars," National Aeronautics and Space Administration, Technical Memorandum 107071, AIAA-93-4170, September 1993.
5. Borowski, S.K., et al. (1994). "A Revolutionary Lunar Space Transportation System Architecture Using Extraterrestrial LOX-Augmented NTR Propulsion," AIAA-94-3343, American Institute of Aeronautics and Astronautics (June 1994).
6. Borowski, S.K., Ryan, S.W., Burke, L. M., et al. (2018). "Robust Exploration and Commercial Missions to the Moon Using Nuclear Thermal Rocket Propulsion and Lunar Liquid Oxygen Derived From FeO-Rich Pyroclastic Deposits," NASA/TM—2018-219725.
7. Castillo, A., Kumar, P., Limbach, C.M. and Hara, K. (2022). "Mutually Guided Light and Particle Beam Propagation," Scientific Reports 12, 4819 (2022). https://doi.org/10.1038/s41598-022-08802-z.
8. Dyson, George (2002). *Project Orion: The True Story of the Atomic Spaceship*. New York, N.Y.: Henry Holt and Co.
9. Dyson, F.J., *Interstellar Transport*, Physics Today, October 1968, p. 41–45.
10. Fearn, H., Hudson, G.C., et al. (2018). "Mach Effects for in Space Propulsion: Interstellar Mission." Contractor Report HQ-E-DAA-TN58808. NIAC Case Study: Award NX17AJ78G. https://ntrs.nasa.gov/api/citations/20190000915/downloads/20190000915.pdf.
11. Freeh, J. E., Burke, L. M., et al. (2015). "Comparison of Solar Electric and Chemical Propulsion Missions." IAC-15-D1.4.3.
12. Frisbee, R.H. (2003). *How to Build an Antimatter Rocket for Interstellar Missions; Systems Level Considerations in Designing Advanced Propulsion Technology Vehicles.* Presented at the 39th AIAA/ASME/SAE/ASEE J. Propulsion Conference, Huntsville, AL, July 20–23, 2003, AIAA 2003-4696.
13. Gruntman et al. (2005). "Innovative Explorer Mission to Interstellar Space."
14. Houts, M.G. and Borowski, S.K. (2010). Advanced Space Fission Propulsion Systems, NASA Report M11-0080, 2010.
15. Howe, S. D., Howe, T., Bennett, F. G., et al. (2022). "Pulsed Plasma Rocket—Developing a Dynamic Fission Process for High Specific Impulse and High Thrust Propulsion." Version of Record: https://www.sciencedirect.com/science/article/pii/S0094576522001187.
16. Lawrence, T. J. (2008) *Nuclear-Thermal-Rocket Propulsion Systems*, in Nuclear Space Power and Propulsion Systems, ed. by C. Bruno (Am. Inst. Aeronautics and Astronautics, 2008).
17. Liewer, P.C., Mewaldt, R.A., Ayon, J.A., and R.A. Wallace, R.A. (2000). "NASA's interstellar probe mission," STAIF-2000 Proc., 2000.

18. Limbach, C. and Hara, K. (2019). "PROCSIMA: Diffractionless Beamed Propulsion for Break-through Interstellar Missions." NIAC Phase I, NASA Technical Report, 2019.

19. Limbach, C.M. and Morgan, H.P. (2022). "Phenomenology and Capabilities of Mutually Guided Laser and Neutral Particle Beams for Deep Space Propulsion," Acta Astronautica 197, 298–309 (2022) https://doi.org/10.1016/j.actaastro.2022.04.006.

20. Long, K. F. (2016). "Project Icarus: Specific Power for Interstellar Missions using Inertial confinement Fusion Propulsion," JBIS, 69(5), 190–194, May 2016.

21. Mann, A. https://i4is.org/who-we-are/interstellar-artists/adrian-mann/#gsc.tab=0.

22. Mitchell, E.D. and Staretz, R. (2010). *Energy and Interstellar Travel*, Journal of Cosmology, 2010, Vol. 12, 3537–3548. http://journalofcosmology.com/Mars132.html.

23. Morgan, D. L., Jr. (1982). *Concepts for the design of an antimatter annihilation rocket*. NASA-CR-168660, NAS 1.26:168660, 1982.

24. Nock, K.T. (1987). "TAU—A mission to a thousand astronomical units." Paper presented at 19th AIAA/DGLR/JSASS International Electric Propulsion Conference, Colorado Springs, CO, AIAA-87-1049, May 11–13, 1987.

25. Obousy, R.K., Tziolas, A.C., Long, K.F., Galea, P., Crowl, A., Crawford, I.A., Swinney, R, A. Hein, A, Osborne, R and Reiss, P. (2011). "Project Icarus: Progress Report on Technical Developments and Design Considerations," JBIS, Vol. 64, 2011.

26. Orth, C., et al, (1987). "The VISTA Spacecraft-Advantages of ICF for Interplanetary Fusion Propulsion Applications." Lawrence Livermore National Laboratory, UCRL-96676, October 2, 1987.

27. Project Longshot: An Unmanned Probe to Alpha Centauri. NASA-CR-184718.

28. Schulze, N. R. (1991). "Fusion Energy for Space Missions in the 21st Century." NASA TM 4298, Agust 1991. https://ntrs.nasa.gov/api/citations/19920002565/downloads/19920002565.pdf.

29. Taylor, B., Cassibry, J. and Adams, R. (2020). "Ignition and Burn in a Hybrid Nuclear Fuel for a Pulsed Rocket Engine." Acta Astronautica, Vol. 175, October 2020, pp. 465–475.

30. Verduci, R., Romano, V., Brunetti, G., et al. (2022). "Solar Energy in Space Applications: Review and Technology Perspectives." Advanced Energy Materials, Vol. 12, Issue 29, Aug. 2022. https://doi.org/10.1002/aenm.202200125.

31. Walter, U. (2006). "Relativistic Rocket and Space Flight," Acta Astronautica 56 (2006), pp. 453–461.

32. Westmoreland, S., *A Note on Relativistic Rocketry*, Acta Astronautica, Vol. 67, Issues 9–10, November-December, 2010, pp. 1248–1251.

33. Voyager Backgrounder (1980). NASA News Release 80-160, 1 October 1980, Report P80-10172. https://ntrs.nasa.gov/api/citations/19810001583/downloads/19810001583.pdf.

34. Voyager Status: https://voyager.jpl.nasa.gov/mission/status/.

Our Future Is in the Sky

What will the year 2125 be like for astronautics? I hope by then humanity has a permanent presence in the Moon, has built a base on Mars, explored further the outer planets, found signs of alien life on the moons of Jupiter, Uranus, and elsewhere, and has begun interstellar missions, sending fast probes to explore other worlds among the stars. Will this scenario become reality for the habitants of Earth in the twenty-second century?

The phenomenal revolution sparked by people's reach of space incites us to raise our eyes to the stars with awe and reverence, and with unexplained longing. The first launch of an artificial satellite to Earth orbit in the twentieth century gave birth to the Space Age. Since then, new astronomical discoveries are announced almost every day, as space telescopes peer deeper into the farthest galaxies, amazing us with new facts that challenge our understanding and spark our curiosity. For the last thirty years the missions of exploration and space probes have given us spectacular and breathtaking views of the deep sky, revealing the ethereal and exquisite beauty of our Universe. What new epoch awaits us now that we begin new programs of human exploration aided with more powerful rockets to reach farther cosmic frontiers?

We live amid a most extraordinary era of space discoveries of unprecedented magnitude, revealing every day a bit more of the marvelous creations of God. Engineers, scientists, mathematicians, and space technologists all over the world are coming together to develop improved propulsion systems, more capable launch vehicles, interplanetary spacecraft, and space robots to take us farther and unravel many mysteries of the cosmos. We continue pushing the limits of propulsion technologies to power the vehicles that soon will take human beings to the Moon, to Mars and beyond.

We have learned so much stimulated by our desire to reach the stars. Adding space observatories and automated spacecraft to the network of telescopes on the ground has opened more windows to the heavens. Just eighty years ago we did not know how big the Universe really is. Now we have mapped its evolution and have seen the birth of stars. Just thirty years ago astronomers began the search for planets circling around distant stars.

D. Musielak, *Introduction to Rocket Propulsion for Astronautics*, Synthesis Lectures on Engineering, Science, and Technology, https://doi.org/10.1007/978-3-031-86141-3

By 2024, more than 5780 exoplanets have been detected in 4314 planetary systems, and many more alien worlds are waiting to be exposed, hiding behind the bright glow of their parent stars. All that was possible thanks to having the rocket propulsion systems to place telescopes in orbit.

We have vastly expanded our sense of space and determined clearly our place in the Universe. We have learned that we are not lost in the infinite cosmic sea. That with our star the Sun we are members of a beautiful and majestic galaxy that the ancient astronomers named the Milky Way. With every new discovery, space beckons us. The stars blink brightly and call us, challenging us all to discover their secrets.

Can we make an interstellar voyage this century? Alas, human beings cannot navigate among the stars. Not now. Maybe one day in the distant future. We realize the distance between stars is enormous, too large to make it across within the frame of reference of our human life span, even moving at light speed. Today, interstellar travel would take hundreds of years using the fastest spaceship that could be built with current technologies. It took Voyager thirty years to reach the outer regions of our Solar System. It would have to travel for another 73,000 years to arrive near the closest stars.

By comparing the highest departure velocity from Earth by any spacecraft built to date—about 17.57 km/s for the Voyager 1—with the speed of light ($c \sim 30,000$ km/s), we quickly determine that to make it to the nearest stars we must have a relativistic rocket that could move at a velocity that is 4 orders of magnitude higher than the fastest spacecraft.

Nevertheless, we can travel to the Moon and make interplanetary trips with rockets we already have. To go to the closest planets in our Solar System we must accelerate scientific research and we must continue vigorously developing new technologies to perfect the systems of propulsion. I believe that we are skillful and clever enough to extend our presence in the sky, and we must attempt in this century to go at least to Mars and build a settlement there.

In the past, many doubted that human beings could travel to the Moon. But humans did, as engineers built powerful rockets that enabled space vehicles to break free of the grip of Earth's gravity, taking the first men to land on the Moon. Since then, many women and men have been privileged to travel and live in space stations that orbit the Earth in a synchronous waltz that represents one of the greatest achievements of human ingenuity.

A human trip to another planet involves many challenges too difficult to quantify here. However, I believe human voyages to Mars are attainable within a few years, hopefully within my life time. First, however, people must return to the Moon. There is so much to do there to prepare humanity for reaching the next space frontier.

We aim for a sustained human presence in space. Rather than visit the Moon for hours or a few days, the future space explorers must embark on long term missions. They need new tools and technologies for living on the new worlds. At the same time, engineers must build the potent vehicles than can transport the infrastructure for the settlement and life-supporting supplies.

The voyage to Mars will be a much more complex endeavor, both in scale and in distance than any robotic or human space trip achieved so far. Yet, it is doable. Our challenge now is to mature and advance required technologies. Aside from sample-return automated spacecraft, most space robots go away holding a one-way ticket—they are not coming back! But for any human mission, we need a round-trip ticket.

Chemical rockets will continue to power the launch of spacecraft from Earth because chemical rockets provide the most lift thrust per weight capability. Then again, for crossing the vast distances between the planets in the Solar System, we need to conceive a different type of propulsion that can achieve higher vehicle velocities and use less propellant.

To reach any interplanetary destination beyond the Moon we must considered other energy sources to significantly reduce trip time and the initial mass of the vehicle. Plasma rockets and thermo-nuclear rockets are two potential candidates that could be realized in the next few decades.

Of course, we also need to mature many other technologies, including life support systems, to build a lunar base so that we learn how to live in another world and eventually colonize the Red Planet. It will not be easy and it will not happen overnight. Maybe it will take hundreds of years for humanity to disperse and have our descendants born in other planets. However, the germination of ideas to realize space flight started with the bold imagination of a few bright, intelligent individuals. Now we need a new crop of smart, well-educated engineers, technologists, and scientists to realize our wildest dreams of deep space exploration, because to develop new ideas for crewed interplanetary travel, we must think beyond what we know today.

There was a time, during the first years of the space race, when the Moon must have seen as the farthest destination in the sky, the most difficult goal that humanity could aspire to reach. Today, we think the Moon is not such a great challenge and we believe instead that human beings must intent to dominate the cosmic space beyond.

To land and walk on the surface of the Moon was the most splendorous technical achievement for the people of the twentieth century, but now we wish to conquer other challenges even more rigorous to realize greater goals for the future of space exploration. Now let us voyage, land and walk on the red surface of Mars. Astronauts could reach a huge asteroid and change its trajectory if it is determined that it moves dangerously towards the Earth. In the future, humans could also build nuclear reactors in space to produce the necessary energy for rockets that could go beyond the asteroids.

In this century I hope that our citizens will travel to Earth orbit, go on vacations to the Moon or even live in another world in the sky. We are now ready to establish research settlements on the Moon and Mars, and I hope humans attempt many other activities of exploration in the sky to put in practice may ideas that we have already imagined. However, before space can be habitable, we must resolve many challenges. Engineering will develop new and better rocket propulsion systems. Space medicine will establish the treatments or protocols to overcome the physiological effects in the human body, and space psychology will help astronauts cope with the isolation and other mental health issues while traveling in long interplanetary routes.

Human beings, whose sense of distance may be defined by the unbounded view of the sky, may very well wonder how far we can go, how far we can really reach out if we build a spaceship to the stars. Some may feel bewildered and think that we cannot build a vehicle fast enough to go far enough within the Galaxy. Others may say that the light speed limit may stop our quest.

In my research, I study an interstellar vehicle propelled by a rocket driven by matter-antimatter annihilation reactions. This is an attractive solution to achieving relativistic motion because of the large amount of energy released from a small amount of reacting mass in this rocket system. To develop such antimatter rocket, we are confronted with several challenges: producing the required antimatter, storing it in a compact device onboard the spacecraft, and manipulating the exhaust mass to achieve the high velocity for the interstellar mission. There are opportunities for interdisciplinary research to advance the required technologies. I strongly believe that an antimatter rocket will help us take a huge step forward in space exploration, as we could reach the nearest star in less than forty years.

The exploration of space inevitably will lead us to conceive new ideas. Just trying something new or different will stretch our imagination. The limit will not be known until we try new ways and invent new technologies. In less than a hundred years humans applied the laws of physics that govern space travel, developed spacecraft to overcome the grip of Earth's gravity, sojourned in space and even walked on the Moon! So, I am sure than in a thousand years our descendants will discover new ways to tour the cosmos by developing the technologies to move from star to star, not just from planet to planet.

The future of space exploration offers the visionaries of the world many new possibilities. This is, after all, the essence of the quest, the calling of our souls to reach the heavens. As we gain more insight into the principles that govern the Universe, after we uncover more of its secrets, new ideas will germinate in our minds. As we learn more and are receptive to different possibilities, new phenomena we discover may open new doors to the unknown of today.

The exploration of space has, and will continue to require, the best efforts and intellectual talents the world has available to make it reality. Remember, the Universe is indeed full of riddles and mysteries to be unraveled. Every day new discoveries uncover more secrets, and maybe you will be part of the new discoveries. I hope you are among those who resolve the technical challenges of rocket propulsion.

My wish is that the concepts and ideas in this book will prompt you to take the first step (or accelerate it if you have already started) on the road of discovery. I hope you do because, in the end, we can no longer ignore our dreams and yearnings to explore the cosmos. The stars of our ancestors watch over us and beckon us. You and I must continue this quest to seek the truth for our existence in this world, so that we may understand our role in the scheme of our vast and magnificent Universe. *Ad Astra*!

Dora Musielak.

April 2025.

Appendix

Unit System for Rocket Propulsion and Unit Conversion Guide

In rocket propulsion analysis, we must remember the distinction on unit of force in the English Engineering (EE) system and the International Standard or SI system of units. Newton's second law of motion relates force to mass, length, and time. The sum of forces in a system is proportional to rate of change of momentum (mv):

$$\sum F = \frac{d(mv)}{dt}$$

In the SI system, units of mass (kg), length (m), and time (s) yield unit of force as newton (N), that is:

$$N = \frac{kg \cdot m/s}{s} = \frac{kg \cdot m}{s^2}$$

When calculating rocket engine Thrust using the SI System, units should be newton (N) using kg/s for the mass flow rate, m/s for velocity, meter square for area, and newton per meter square (or Pascal) for pressure:

$$F = \dot{m} v_{ex} + (p_e - p_a)A_e$$

$$N = \frac{kg}{s} \cdot \frac{m}{s} + \frac{N}{m^2} m^2 = \frac{kg \cdot m}{s^2} + N$$

In the EE system of units, units of mass (lbm), length (ft), and time (s) yield the unit of force as the lbf. Hence, we must incorporate the constant of proportionality

$$g_c = 32.174 \frac{ft \cdot lbm}{lbf \cdot s^2}$$

so that when calculating the thrust in EE units, it gives units of lbf:

© The Editor(s) (if applicable) and The Author(s), under exclusive license to Springer Nature Switzerland AG 2025
D. Musielak, *Introduction to Rocket Propulsion for Astronautics*, Synthesis Lectures on Engineering, Science, and Technology, https://doi.org/10.1007/978-3-031-86141-3

$$\text{lbf} = \frac{\text{lbm ft}}{\text{s}}\cdot\frac{1}{32.174\ \text{ft}\cdot\text{lbm/lbf}\cdot\text{s}^2} + \frac{\text{lbf}}{\text{ft}^2}\text{ft}^2 = \text{lbf} + \text{lbf}$$

The unit of mass in the EE system is derived from the law of motion, and has the dimensions $\text{lbf}\cdot\text{s}^2/\text{ft}$:

$$1\frac{\text{lbf}\cdot\text{s}^2}{\text{ft}}\text{(mass unit)} = 1\ \text{lbf (force unit)} \div 1\frac{\text{ft}}{\text{s}^2}\text{(acceleration unit)}$$

The pound mass (lbm) is very commonly used for mass measurements in rocket propulsion. Thus, we establish the unit conversion factor between the mass units of lbm and $\text{lbf}\cdot\text{s}^2/\text{ft}$ when using the EE system. Note that 1 lbf is defined as the force of gravity (weight) acting on 1 lbm at standard sea level, so it follows that 1 lbf accelerates a 1 lbm at 32.174 ft/s^2, and therefore that

$$1\frac{\text{lbf}\cdot\text{s}^2}{\text{ft}}\text{(mass unit)} = 32.174\ \text{lbm (mass unit)}$$

Keep in mind the fact that 1 lbm mass at (or near) the surface of the Earth weighs 1 lbf.

To facilitate conversion, you could use this table

Dimension	Multiply	By	To Convert to
Area	ft^2	0.09290304	m^2
Density	lbm/in^3	27679.905	kg/m^3
Distance	miles	1.60934	km
Energy	BTU	1055.06	J
Force	lbf	4.448221615	N
Impulse	lbf·s	4.448221615	N·s
Specific impulse	lbf·s/lbm	9.8066516	N·s/kg
Length	ft	0.3048	m
Mass	lbm	0.45359237	kg
Momentum	lbf·ft/s	1.355818	N·m/s
Pressure	psi	6894.7572	N/m^2
Speed	ft/s	0.0003048	km/s
Thrust	lbf	0.004448221615	kN
Volume	ft^3	0.0283168	m^3
Velocity	miles/hr	1.60934	km/hr

Derived units can always be represented as products of powers of the base units. For example, the standard (SI) unit of measure for energy is the Joule (J). This unit derives from the definition of energy, which is the ability to do work. Work is measured also in

Joules, which results from multiplying Force (in newtons) by distance (in meters). For pressure (and stress) the SI unit is pascal (denoted as Pa), which is equal to one newton per square meter.

Joule	J	Energy, work, amount of heat	$\frac{kg \cdot m^2}{s^2}$	$N \cdot m = Pa \cdot m^3$
Pascal	Pa	Pressure, stress	$\frac{kg}{m \cdot s^2}$	$\frac{N}{m^2} = \frac{J}{m^3}$
Newton	N	Force, thrust	$\frac{kg \cdot m}{s^2}$	

Index